AF389480

Challenges and Prospects of Steelmaking Towards the Year 2050

Challenges and Prospects of Steelmaking Towards the Year 2050

Editor

Lauri Holappa

MDPI • Basel • Beijing • Wuhan • Barcelona • Belgrade • Manchester • Tokyo • Cluj • Tianjin

Editor
Lauri Holappa
Chemical and Metallurgical
Engineering
Aalto University School of
Chemical Engineering
Espoo
Finland

Editorial Office
MDPI
St. Alban-Anlage 66
4052 Basel, Switzerland

This is a reprint of articles from the Special Issue published online in the open access journal *Metals* (ISSN 2075-4701) (available at: www.mdpi.com/journal/metals/special_issues/metal_steelmaking).

For citation purposes, cite each article independently as indicated on the article page online and as indicated below:

LastName, A.A.; LastName, B.B.; LastName, C.C. Article Title. *Journal Name* **Year**, *Volume Number*, Page Range.

ISBN 978-3-0365-2777-2 (Hbk)
ISBN 978-3-0365-2776-5 (PDF)

Contents

About the Editor

Lauri Holappa

Lauri Holappa is professor emeritus at Aalto University since 2009. Before that he was professor in metallurgy at Helsinki University of Technology, Department of Materials Science and Engineering from 1979 until retirement in 2009. He obtained M.Sc. in 1964 and D.Sc. (Tech.) in 1970 at Helsinki University of Technology. He worked at OVAKO OY AB Imatra Steelworks as research manager and chief metallurgist until 1979, when he was appointed professor. Current research activities are directed in fundamentals of metallurgical processes, sustainability in metals production and clean steel production. He was Editor of Vol.3: Industrial Processes, Part A Iron and Steel Technology in Treatise on Process Metallurgy, Seetharaman S. (Ed.); Elsevier 2014 — and he is continuing for the coming revised 2nd Edition in 2022 as well. He has obtained several Special Awards: Professor chair at Dongbei University (NEU), Shenyang, China, 1991; Metal Separation Technologies III Symposium in Honor of Lauri E. Holappa, Copper Mountain, Colorado, June 20–24, 2004 and John F. Elliott Lectureship Award by AISTech, USA, 2012. He was nominated for honorary member of the Finnish Association of Mining and Metallurgical Engineers and honorary member of the Academy of Engineering Sciences, Finland, 2013.

Editorial

Challenges and Prospects of Steelmaking towards the Year 2050

Lauri Holappa 

Department of Chemical and Metallurgical Engineering, Aalto University School of Chemical Engineering, 02150 Espoo, Finland; lauri.holappa@aalto.fi; Tel.: +358-50-5608377

check for updates

Citation: Holappa, L. Challenges and Prospects of Steelmaking towards the Year 2050. *Metals* **2021**, *11*, 1978. https://doi.org/10.3390/met11121978

Received: 4 November 2021
Accepted: 29 November 2021
Published: 8 December 2021

1. Introduction

The world is experiencing a period of imminent threat owing to climate change. The 2018 IPCC report defined the jointly approved target of limiting global warming to 1.5 °C by 2050, which means deep cutting of CO_2 emissions comprehensively. "Rapid and far-reaching transitions in land, energy, industry, buildings, transport, and cities are required". These challenges concern all human activities, including steel production.

Steel is a central material to modern society. It is necessary for infrastructure, buildings, transportation vehicles, and energy production. The annual consumption of steel has exceeded 1.8 billion tons and is further growing due to global development. Steel production is an energy-intensive branch of industry. Due to the central role of coal/coke in ironmaking, carbon dioxide emissions are large, corresponding to approximately 7% of the total anthropogenic CO_2 emissions. On the other hand, steel is a necessary material and enabler in solving the global dilemma of radically decreasing the use of fossil energy and increasing the share of renewable energy. Thus, the challenge is dual. On one hand, the industry must strongly cut its own CO_2 emissions by improving and developing the process route towards carbon neutrality. On the other hand, it has the role of a problem solver as a material supplier and by developing new steel grades with improved product properties which result in material and energy savings, longer life cycle and better recyclability.

This Special Issue was initiated to review the present situation of steel production, energy consumption, and CO_2 emissions. The potential methods to decrease CO_2 emissions in current processes via improved energy and materials efficiency, increasing recycling, and utilizing alternative energy sources including hydrogen were considered. Achievements in current processes and potentials of alternative energy sources as well as novel innovative processes were surveyed. In addition, the role of steel as an integral part of the global circular economy was discussed. En bloc the target of the Issue is to give a holistic overview of the current situation and challenges, and a comprehensive cross-section of the potential technologies and solutions for the global CO_2 emissions problem.

2. Contributions

The review article by Jean-Pierre Birat [1] is an extensive overview on the multifold influences of steel production. It does not confine itself only to CO_2 emissions but includes all noteworthy emissions, environmental effects as well as materials and energy issues. Different possible ways to cut CO_2 emissions from iron and steel production are reviewed with main emphasis on the experiences and results of European research projects (ULCOS). At the same time the article reflects on the efforts and prospects of the European steel industry, which accounts for 8.4% of the world production.

The next two papers highlight the situation in the two biggest steel-producing countries [2,3]. China is the overwhelming producer nearing one billion tons per year. After very rapid growth the facilities are mostly rather new, but production is based on blast furnaces with relatively low energy efficiency and high CO_2 emissions. Coal plays a major role also in the power sector and electricity generation. For structural reasons availability of recycled steel scrap is minimal, approximately 10% of the iron raw material. Substantial improvements in energy intensity and specific emissions have been achieved. Notwithstanding, the Chinese steel industry is facing big problems in trying to peak the emissions, cutting them down in accordance with the national policy and carbon neutrality targets [2]. Similar problems affect India which has risen to second place among steel-producing countries [3]. Its production structure is, however, totally different. A "conventional" BF-BOF production route accounts for 45% of the production, whereas a bigger share comes from electric furnaces, mini EAFs and a huge number of "micro mills" with induction furnaces. Their main raw material is direct reduced iron from rotary kilns. Altogether, the industry is strongly reliant on coal. India is an exception also in that regard; it has firm plans to strongly enhance steel production from the current level 111 Mt in 2019 to 250 Mt/year in 2030. This makes the target to peak the CO_2 emissions, suppressing them down to acceptable levels, extremely challenging.

Biofuels are regarded as low CO_2 emitting energy sources due to the carbon cycle. Historically, charcoal was used in ironmaking until the late 19th century when coke fully replaced it. Nowadays, the endeavors to get away from fossil energy have raised interest in charcoal again as a reductant and fuel in ironmaking. Brazil, with many eucalyptus plantations is producing and using large amounts of charcoal in metallurgical industries. Mini blast furnaces can operate with charcoal alone, and in big furnaces injection of powdered charcoal can substitute for injected fossil coal [4]. The paper by Adilson de Castro et al. describes recent research on mini blast furnace operation with possible applications of charcoal. Substitution of metallurgical coke and coal injection with bioenergy in conventional blast furnaces has been discussed in several papers in the issue.

Suspension smelting is an established technology for processing of non-ferrous sulfidic concentrates. The same principle has been applied in flash iron making by H-Y. Sohn at the University of Utah [5]. This technology might surpass fine ore agglomeration and coke making by using fine concentrates and natural gas as the main inputs resulting in remarkable energy saving. It has been tested on a pilot-scale and even with pure hydrogen but has not yet been commercialized.

Full decarbonization of steel making is possible only by discarding carbon containing fossil energy in iron and steel processes. An axiomatic solution is hydrogen. The principle of hydrogen reduction is well-known and even industrially proven in shaft furnace processes by using hydrogen instead of natural gas. The core problem is that hydrogen is not available as a natural resource like coal but has to be produced. Natural gas is a possible raw material, but it is a fossil fuel and not carbon free. Water is thus the obvious resource and water splitting by electrolysis the method to produce hydrogen gas on a mass scale. Patisson et al. give an overview on bases of hydrogen reduction [6] and Pei et al. [7] describe an ambitious project "Toward a Fossil Free Future with HYBRIT-Development of Iron and Steelmaking Technology in Sweden and Finland". Accordingly, iron and steelmaking will make a stepwise transition from the BF–BOF route to direct hydrogen reduction–electric melting route until the 2040s. The plan even includes massive hydrogen production and fossil-free electricity for water electrolysis, steel melting and other processes.

As the transition from coal to hydrogen in the global steel industry entails huge investment and cost with limited time, all rapid actions to mitigate CO_2 emissions are precious. Capture, utilization and storage of CO_2 are important issues. Mineralization of CO_2 by reacting it with steelmaking slag to form precipitated calcium carbonate product is an example of "from waste to valuable by-product" [8]. Further examples of reuse, recycling and productization of wastes via industrial symbiosis are described by Branca et al. [9]. In another article, recent European research activities in iron and steel making from the viewpoint of digitalization are reviewed with a strong focus on sustainability and low-carbon technologies [10].

The final chapter is a kind of summarizing review of the whole issue area [11]. First, the current state of the global steel industry and potentials for energy saving and emissions mitigation by retrofitting existing plants with the best available technologies and by utilizing energy sources with the smallest emissions are discussed. New and emerging means, such as CO_2 capture and storage, increasing recycling ratios and anticipated breakthroughs in hydrogen technology are reviewed as well. The review ends with a hybrid scenario to the year 2050 in which CO_2 emissions would be cut by 70% from the current level although steel production is predicted to grow by up to 2.5 billion tons/year. It is a challenging goal, but only an intermediate point toward carbon-neutral steel.

Funding: This research received no external funding.

Acknowledgments: As Guest Editor I would like to express my gratitude to all the contributing authors and reviewers. You did a great job and enabled the publication of this special issue. I am also indebted to Sunny He and the staff at MDPI for their valuable and friendly support and active role in the publication. Now that the book is coming out, our joint journey is ending but we all continue our efforts toward sustainable steel.

Conflicts of Interest: The author declares no conflict of interest.

References

1. Birat, J.-P. Society, materials, and the environment: The case of steel. *Metals* **2020**, *10*, 331. [CrossRef]
2. He, K.; Wang, L.; Li, X. Review of the energy consumption and production structure of China's Steel Industry: Current situation and future development. *Metals* **2020**, *10*, 302. [CrossRef]
3. Shanmugam, S.P.; Nurni, V.N.; Manjini, S.; Chandra, S.; Holappa, L.E.K. Challenges and Outlines of Steelmaking toward the Year 2030 and beyond—Indian Perspective. *Metals* **2021**, *11*, 1654. [CrossRef]
4. Adilson de Castro, J.; Medeiros, G.A.d.; Oliveira, E.M.d.; de Campos, M.F.; Nogami, H. The mini blast furnace process: An efficient reactor for green pig iron production using charcoal and hydrogen-rich gas: A study of cases. *Metals* **2020**, *10*, 1501. [CrossRef]
5. Sohn, H.Y. Energy Consumption and CO_2 Emissions in ironmaking and development of a novel flash technology. *Metals* **2020**, *10*, 54. [CrossRef]
6. Patisson, F.; Mirgaux, O. Hydrogen ironmaking: How it works. *Metals* **2020**, *10*, 922. [CrossRef]
7. Pei, M.; Petäjäniemi, M.; Regnell, A.; Wijk, O. Toward a fossil free future with HYBRIT: Development of iron and steelmaking technology in Sweden and Finland. *Metals* **2020**, *10*, 972. [CrossRef]
8. Zevenhoven, R. Metals production, CO_2 mineralization and LCA. *Metals* **2020**, *10*, 342. [CrossRef]
9. Branca, T.A.; Colla, V.; Algermissen, D.; Granbom, H.; Martini, U.; Morillon, A.; Pietruck, R.; Rosendahl, S. Reuse and recycling of by-products in the steel sector: Recent achievements paving the way to circular economy and industrial symbiosis in Europe. *Metals* **2020**, *10*, 345. [CrossRef]
10. Branca, T.A.; Fornai, B.; Colla, V.; Murri, M.M.; Streppa, E.; Schröder, A.J. The challenge of digitalization in the steel sector. *Metals* **2020**, *10*, 288. [CrossRef]
11. Holappa, L. A General vision for reduction of energy consumption and CO_2 Emissions from the steel industry. *Metals* **2020**, *10*, 1117. [CrossRef]

 *metals*

Review

Society, Materials, and the Environment: The Case of Steel

Jean-Pierre Birat

IF Steelman, Moselle, 57280 Semécourt, France; jean-pierre.birat@ifsteelman.eu; Tel.: +333-8751-1117

Received: 1 February 2020; Accepted: 25 February 2020; Published: 2 March 2020

Abstract: This paper reviews the relationship between the production of steel and the environment as it stands today. It deals with raw material issues (availability, scarcity), energy resources, and generation of by-products, i.e., the circular economy, the anthropogenic iron mine, and the energy transition. The paper also deals with emissions to air (dust, Particulate Matter, heavy metals, Persistant Organics Pollutants), water, and soil, i.e., with toxicity, ecotoxicity, epidemiology, and health issues, but also greenhouse gas emissions, i.e., climate change. The loss of biodiversity is also mentioned. All these topics are analyzed with historical hindsight and the present understanding of their physics and chemistry is discussed, stressing areas where knowledge is still lacking. In the face of all these issues, technological solutions were sought to alleviate their effects: many areas are presently satisfactorily handled (the circular economy—a historical' practice in the case of steel, energy conservation, air/water/soil emissions) and in line with present environmental regulations; on the other hand, there are important hanging issues, such as the generation of mine tailings (and tailings dam failures), the emissions of greenhouse gases (the steel industry plans to become carbon-neutral by 2050, at least in the EU), and the emission of fine PM, which WHO correlates with premature deaths. Moreover, present regulatory levels of emissions will necessarily become much stricter.

Keywords: steel; environment; mining; production; circular economy; lean and frugal design; ecology transition; climate change; pollution; toxicology

1. Introduction

The present article discusses the connection between the environment and steel, both production and use. It argues about the sustainability of steel as a material, and more precisely discusses the triptych: Society, the environment, and materials [1].

To define the topics that ought to be covered in this review, we refer to a simple ecological model of the planet, the *spheres model*, which distinguishes between the geosphere, the biosphere, and the anthroposphere [2]. The environment consists of the concatenation of the biosphere and geosphere.

An activity like steel, firmly anchored in the anthroposphere, interacts with:

- The geosphere because raw materials and energy resources stem from there and much of the waste generated by mining, industry, and end-of-life returns there;
- The biosphere because emissions to air, water, and soil leak into the atmosphere, the hydrosphere, and the top of the geosphere. This should also include Greenhouse gas (GHG) emissions. Steelmaking activities pollute and affect ecosystems and biodiversity;
- The anthroposphere itself, because steel-related elements and compounds influence the health of people (toxicological effects) while steel-related activities create economic and social benefits, directly through the use of steel in the anthroposphere, or indirectly, through its eco-socio-systemic services [3]—two sides of the balance sheet with the pros and cons of steel in the economy and society.

Steel ought to be appraised not simply through its direct production, i.e., through the activities of the steel industry, but through its whole value chain, from raw materials to consumer and investment goods, as well as through its lifecycle, thus tracking the material in goods until and including end of life and its involvement in the circular economy. This is similar to a Life cycle assessment (LCA) approach.

Finally, the theme is time dependent, with a strong historical dimension, both in the long term and short term.

This review is original in the way it brings together the complex set of environmental issues as exhaustively as possible, by going beyond the steel mill itself and into the whole value chain of steel, thus mining, use of steel, pollution, reuse and recycling and health matters, etc. It also addresses the most pressing contemporary environmental topics, like climate change and air pollution, while the discussion stresses the role of society in framing issues and looking for mitigation solutions.

2. Steel and Raw Materials

The oldest environmental issue related to steel is the matter of its resources.

Historically, iron ore was ubiquitous in all the parts of the world and, therefore, steel production was located near energy resources rather than ore deposits, thus near water streams for hydraulic power, then near forests for access to large quantities of charcoal, and, since the use of earth coal, near coal mines. The 20th century saw the "discovery" of larger deposits, capable of mines that would be exploited for a long time. In the second half of the century, the scales were turned around with the discovery of high-grade ore deposits (almost pure hematite, more rarely magnetite) and the buildup of a logistical chain based on gigantic ore carriers and new high-capacity export and import harbors. The steel business moved to large sea harbors or to very large waterways inland, like the Great Lakes, the Mississippi, the Rhine, the Danube, or the Yangtze rivers. This was organized with economic targets in mind, based on the rationale of economies of scale, as, indeed, the integrated steel mills grew in size accordingly while the steel market exploded: this was the first wave of globalization, before a second wave of globalization moved goods around the world, beyond raw materials.

Assumed resources of iron ore, worldwide, ought to be able to meet demand for years to come, a cautious expression that should be taken to mean indefinitely—indeed, according to the USGS, world reserves of iron ore were estimated for 2018 at 170×10^9 tons, containing 84×10^9 tons of iron and resources at more than 800×10^9 tons of crude ore containing over 230×10^9 tons of iron [4]. Taking reserves into account strengthens the conclusion that it not a critical raw material, based on a definition of criticality close to that of scarcity. However, one can distinguish between physical *long-term scarcity* and economic *short-term scarcity* [5]. While there is no risk of long-term scarcity, the discrepancy between the mining and steel sectors' characteristic business times (typically, 5 and 20 years, respectively) may create a short-term scarcity when steel demand is high while the ore offer/supply from exploited mines has not caught up yet, such as in 2007 and 2008 prior to the economic crisis, and, therefore, price excursions may occur [6].

Note also that the circular economy is already changing the deal, as large quantities of recycled material (36% worldwide) feed the steel business as a secondary raw material in parallel with the primary one [7].

Among the other raw materials that the steel sector uses, coal is not critical either, although coking coal was put on the European list of critical raw materials [8], and neither is lime. Some alloying elements may be considered as critical, such as vanadium for example [9].

Producing a material like steel involves engaging much larger quantities of raw materials than the crude steel produced: the "useful" element of economic value, iron, is concentrated down the process route while the unneeded material is sent back to the "environment", usually to landfills, and is labelled as waste.

Figure 1 shows a typical plot of the mass engaged for the production of steel in an integrated steel mill, from the mine to the exit of the steelmaking shop, where crude steel is generated [10]: In this example, the ratio of engaged material vs. useful material is 16.5 to 1. It is a fairly typical set of data, although the upstream figures (here a burden/ore ratio of 3, slightly on the high side) vary with the iron mine.

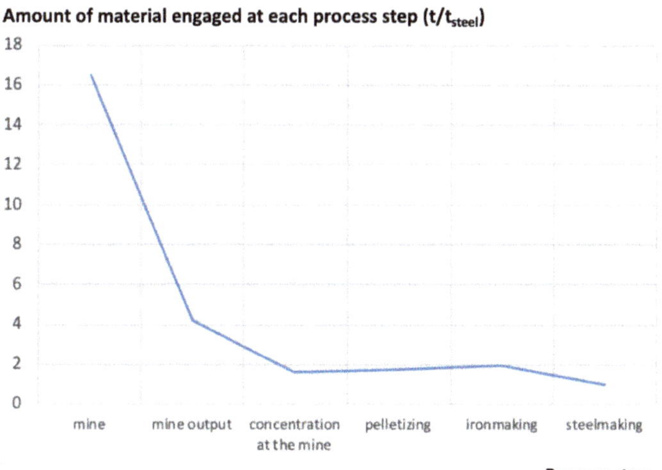

Figure 1. Amount of material engaged at each process step (in t). The final output of the process line is 1 t of crude steel (source: author).

Note that waste is generated mainly at three steps, cf. Figure 2. At the mine, the overburden (cf. Figure 3) represents the largest amount of waste (12.3 t); it corresponds to what is left once the ore proper has been separated. It is thus physically identical to the rock present in the mine, minus the ore. Further, in the mining facilities, during the process of beneficiation, a second type of waste is generated after the crude ore is crushed and only the larger-sized ore is retained while the rest, of smaller size, is washed away with water. The output is a slurry called tailings or tailings fines (2.6 t), sometimes contaminated with mineral or chemical additives: It is useless as an iron source with present beneficiation and steelmaking technologies, although it contains iron at the level of a low-grade ore. The third kind of waste is generated in the steel mill itself. It is composed of slag, dust, and millscale, simply called slag here (0.385 t).

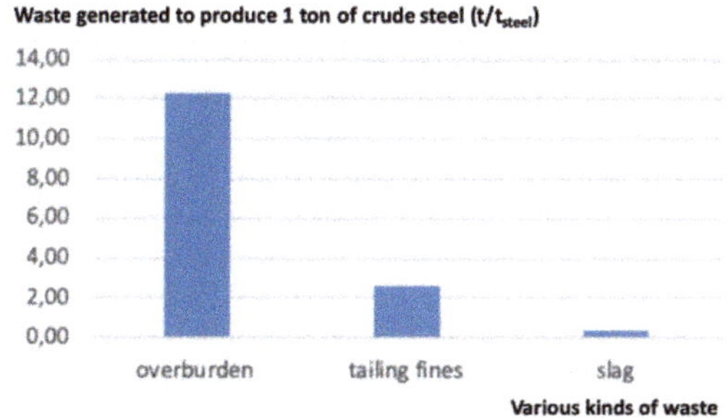

Figure 2. Waste generated in the production of 1 t of crude steel (source: author).

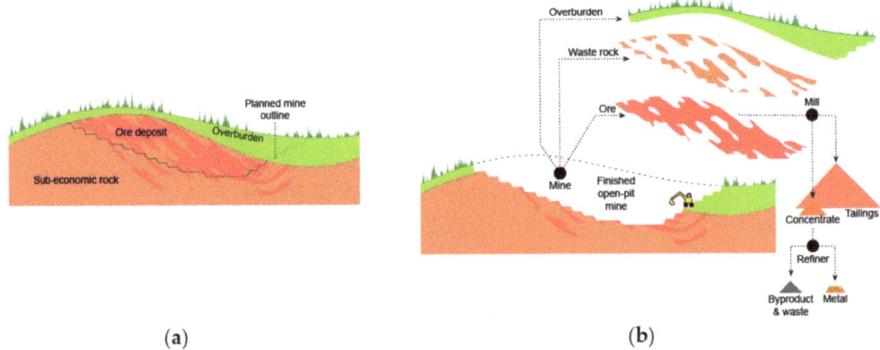

(a)   (b)

Figure 3. Basic mining and metal refining terminology (adapted from groundtruthtrekking.org). Before (**a**) and after (**b**) starting mining operation.

Tailings raise environmental issues because they are often contaminated and are usually stored as slurries in tailings ponds confined behind a dam or an impoundment—in 2000, there were about 3500 active tailings impoundments in the world.

Tailings dam failures are a major risk and have led to numerous mining disasters [11–13]. There were 31 recorded major failures between 2008 and 2018, and 5 already took place in 2019, including the Brumadinho disaster in Brazil shown in Figure 4. Dam disasters have occurred in connection with the mining of all metals beyond iron. In addition to claiming lives, they pollute waterways, land, and ecosystems. The cost of dam failures to the mining business is also high.

Figure 4. Brumadinho dam disaster, which occurred on 25 January 2019 when Dam I, a tailings dam at the Córrego do Feijão iron ore mine, collapsed.

Solutions to avoid dam failures have been identified, including the retrieval of tailings, after draining most of the water from the slurry, for example, by using filter presses. The dried slurry can be reused as building material, for example, in bricks. If the tailings are laid over natural ground, phytoremediation can be used to clean up the new soil, anchor it on the stable substrate, and thus avoid mud slides.

Dam failures are one of the major environmental liabilities of the mining industry and therefore also of the production of iron, as mining is part of a value chain, which in effect shares negative and positive burdens.

These various issues ought to be included in an LCA, a method that aims at giving a full picture of the iron lifecycle, from cradle to grave. The dam failure issue, however, is not included in a standard LCA. Furthermore, the full upstream mass balance related to raw materials may or may not be included, depending on the scope of the study (gate or cradle) and on whether the overburden is properly taken on board or not. Therefore, some reexamination of LCA methodology or the development of some other metrics ought to be considered in the future.

3. Steel and Energy

The steel industry is considered as an *energy-intensive industry*, especially since *energy conservation* and *climate change* issues have created a drive towards the energy and ecological transitions. This can either be a tautological point, as making steel from ores requires a minimum amount of energy set by thermodynamics, or it can point to inefficiencies in industrial processes, which can and ought to be corrected. Since Roman times, the energy efficiency of carbon-based iron ore reduction was improved by a factor of 100, roughly. Today, the best performing steel mills are within 10% or 15% of a technical optimum [14], although not of the thermodynamic limit.

The steel sector, today, uses coal, natural gas (NG), and electricity mostly as energy sources, depending on the processing route, i.e., either the integrated route, the direct reduction route, or the electric arc furnace route, respectively.

Historically, however, iron was produced entirely from "renewables", either biomass in the form of *charcoal* for the bloomery, the muscles of blacksmiths, or hydraulic power (water wheels) to power the forge. The switch to coal and coke took place in the 18th century and is considered as one of the markers of the *first industrial revolution*. At the beginning of the 20th century, the generation of electricity moved iron production further away from renewables.

The *change from charcoal to coal* took place after a major environmental crisis, when the demand for charcoal contributed to forest depletion in industrialized countries [15]. Let us remember that the "discovery" of coal took place at a time when wood was becoming scarce, an early example of *material criticality* and *anthropogenically induced scarcity*! It also shows that *technology is a social construct:*; coal, which had been around forever, was "discovered" when it was needed by society.

Energy conservation in the steel sector was driven by the fact that energy costs account for roughly 20% of operating costs and therefore needed to be minimized for sound management. Steel therefore was one of the first industries to react to high energy prices, ever since the first energy crisis of 1974, cf. Figure 5. It is not simply a story of energy conservation, however, as the process chain for making steel matured and was perfected in the late 20th century and therefore made it easy to capitalize this cumulated improvement—although abruptly and this caused social pain in steel-intensive regions in the world. Last, this shows that energy conservation is one of the environmental constraints that has been internalized early in the market economy in which the steel sector functions, contrary to the usual paradigm, whereby the environment is considered as an externality.

An estimate of the energy consumption of a best-run integrated steel mill is 18.83 GJ/t_{HRC} (per ton of hot rolled coil). The corresponding mill is shown in Figure 6, complete with a detailed mass and energy balance at each process step. The energy balance refers to the steel mill, gate to gate (ore and coal in, hot rolled coil of steel out). An electric arc furnace (EAF) steel mill consumes 4.29 GJ/t_{HRC} (cf. Figure 7) and a direct-reduction and EAF-based mill 15.6 GJ/t_{HRC} (cf. Figure 8).

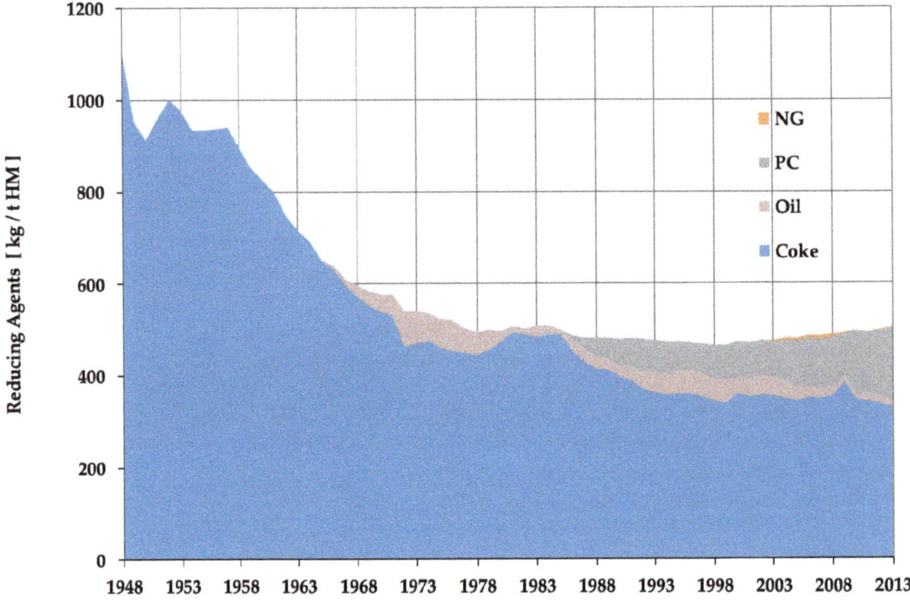

Figure 5. Historical evolution of reducing agent's consumption in the blast furnace (BF) in Europe (European Blast Furnace Committee, courtesy of VDEh).

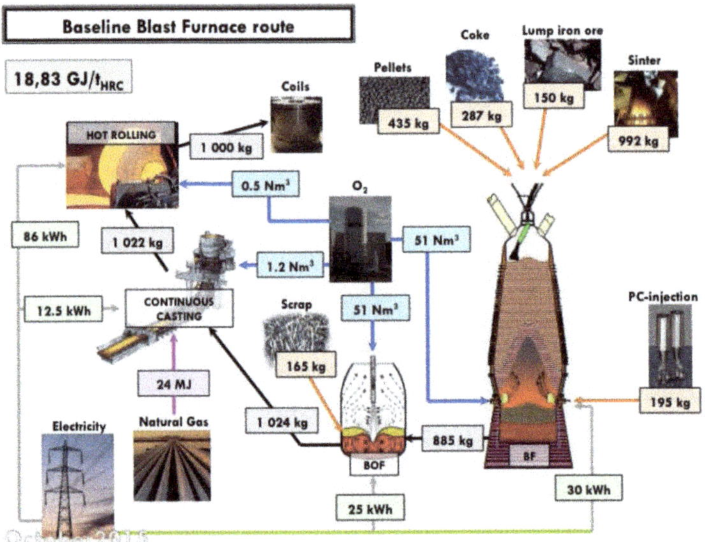

Figure 6. Mass and energy balance in an integrated steel mill (baseline blast furnace route, ULCOS simulation). Energy per ton of hot rolled coil (HRC).

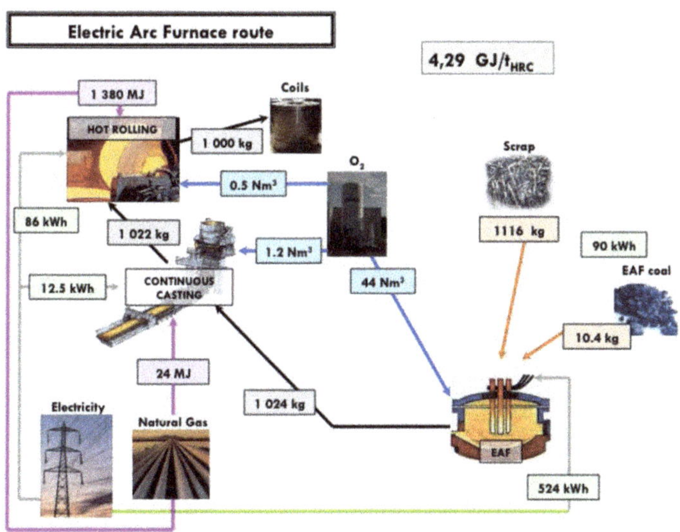

Figure 7. Mass and energy balance in an EAF steel mill fed with scrap.

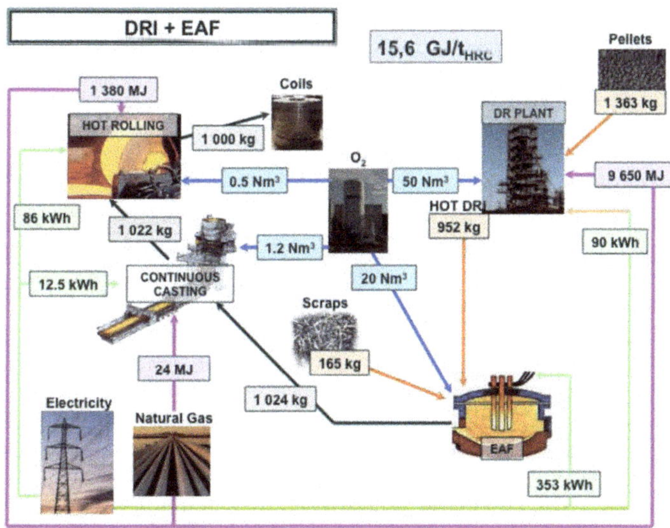

Figure 8. Mass and energy balance in an EAF-based steel mill fed with hot Direct Reduced Iron.

R&D into low-carbon steelmaking conducted as part of the ULCOS program (Ultra-LOw CO_2 Steelmaking) demonstrated that changing the operating point of the blast furnace and of most of the "ULCOS solutions" made it possible to improve energy use by roughly 20% to 25%, and not simply the 10–15% improvement stated before [14]. This is an extra benefit to be collected from switching to low-carbon process routes.

Hic et nunc, the energy transition, pushed by the cost of energy, by climate change policies, and by the perceived short life of the fossil energy resource (peak oil and peak gas), has pushed steelmaking processes to decarbonize. However, since energy consumption and GHG emissions have decoupled and will continue do so in the future, part of the story of upcoming progress in the area will be told in Section 5.

This, however, raises a number of interesting issues.

First, what role can electrification play in increasing the use of renewables in steel production?

Electricity is the simplest way to integrate renewables in the energy system. This is done either by injecting them directly at the level of the grid or by trading green certificates. In the short term, electricity is used in the steel sector by *electric arc furnaces* and all the electrical equipment used in a steel mill. In the future, it might be used to electrolyze water and *generate hydrogen*, to be used thereafter for *direct reduction of iron ore*, or to power the *direct electrolysis of iron ore*. Other processes, like the *reheating of steel* for hot rolling or heat treatments, can also be performed in electric heating furnaces (induction, conduction). A lot of technology is available but the high price of electricity, until now, curbed its broad use. There is therefore much leeway left to introduce more electricity in steelmaking and thus to decarbonize the sector at the same pace as electricity decarbonizes.

Note, however, that the energy needs of a steel mill are very large in terms of electrical power: A 5 Mt/y steel mill based on ore electrolysis would require a 1200 MW nuclear power plant or 240 recent wind turbines. This might require new investments in power generation [16]. On the other hand, this would open up opportunities in terms of *demand-side management* of the electricity grid. Indeed, if the steel mill can be turned on and off to accommodate electricity demand, this would alleviate or even suppress the complex matter of dealing with the intermittency of renewable electricity [17].

Second, are there other ways than green electricity to use renewables in steel production?

The short answer is no, as renewables are not meant to supply high-temperature energy of the kind that the steel industry needs. Heat reduction is out of reach, cf. Section 5.2. This should not keep inventors from trying to find new solutions, however.

Third, would not this decouple the search for less GHG emissions from the doxa of minimum energy consumption?

The implicit assumption today is that energy should be optimized first and then GHG emissions in a second step. If the price of carbon increased enough, these priorities would switch. The change could also be a matter of policy. This would release many constraints in the search for low-carbon solutions. Note, incidentally, that introducing renewables in any industrial system implicitly negates the energy optimization rule: Indeed, renewable energy is a rather inefficient way of generating electricity and, likewise, producing biomass by photosynthesis is also very inefficient, but this does not eliminate it from the search for solutions.

Finally, the point may not be so much to decrease the energy intensity (J/kg) of making steel, which may have reached a physical limit in the best-operated steel mills, but to decrease the lifecycle sector's or the value-chain's overall energy consumption. This will be achieved by *lean* and *frugal practices*, including reuse and recycling and *product-service systems (PSSs)*. Indeed, using less steel for the same services is a solution to cut energy consumption and GHG emissions at the same time. Sharing a car or a car ride, or an apartment in a short-term rental scheme are part of these solutions—this, in effect, increases the "productivity" of cars of or homes, minimizes the amount of steel engaged per unit of service, and should therefore decrease their environmental footprint, provided the extra management cost/footprint of these services remains small enough.

The future of energy is therefore deeply related to low-carbon practices at the level of large systems like the steel sector, on the one hand, and of individuals and their lifestyles, on the other hand.

4. Air Emissions and Pollution Related to Steel

Anthropogenic activities generate emissions to the atmosphere (air), the hydrosphere (water), or the geosphere (ground): The *emissions* evolve, interact, concatenate, sometimes move from one environmental compartment to another (e.g., ground to water), and eventually end up constituting what is called *overall pollution*. Thus, one should speak of the emissions of road traffic but of air pollution in a city and a region. A steel mill emits to the three environmental compartments and these emissions add up to the other local emissions. Another level of complexity is related to the geographical scale of the territory affected by that pollution: Emissions start as a local phenomenon, but pollution may propagate to other regions, countries, and sometimes continents. Thus, for example, *dioxin emissions* are local while *greenhouse gas emissions* are truly global, at the level of the whole planet. The trend, since the turn of the century, has been for more and more emissions to become more and more global.

Emissions are analyzed at a steel mill's or a single reactor's scale, either as statistical data on emissions or as process engineering analyses of how the emissions are generated. The global phenomenon of atmospheric pollution is only analyzed by meteorological tools. Emissions of the value chain, especially of its upstream part, should also be part of the discussion.

4.1. Air Emissions and Pollution Related to Steel

Data on the steel sector's air emissions are available from statistical data [18] and from publications about *the best available technologies*. They are expressed as emissions factors, either per ton of steel (g/t_{steel}) or per volume of smokestack emissions (mg/Nm^3)—they can be converted into each other, using the data published in [19].

Examples are given in Table 1 [19] and Table 2 [20], where different emissions parameters are presented: Heavy metals or HMs (Pb, Cd, Hg, Cr, Cu, NI, Se, Zn plus As, a non-metal), SO_x, NO_x, organic compounds (NMVOC, PCB, PCDD/F, PAHs, HCB), and particulate matter (PM, PM_{10}, $PM_{2.5}$, TSP); the first table shows global industry-wide figures while the second gives details of the emission limits of individual plants in the steel mill.

Table 1. Steelmill specific air emissions of heavy metals and organic compounds, sector averages and spreads [19].

Pollutant	Value	Unit	95% Confidence Interval	
			Lower	Upper
NMVOC	150	g/Mg_{steel}	55	440
TSP	300	g/Mg_{steel}	90	1300
PM_{10}	180	g/Mg_{steel}	60	700
$PM_{2.5}$	140	% of $PM_{2.5}$	40	500
BC	0.36	g/Mg_{steel}	0.18	0.72
Pb	4.6	g/Mg_{steel}	0.5	46
Cd	0.02	g/Mg_{steel}	0.003	0.1
Hg	0.1	g/Mg_{steel}	0.02	0.5
As	0.4	g/Mg_{steel}	0.08	2.0
Cr	4.5	g/Mg_{steel}	0.5	45.0
Cu	0.07	g/Mg_{steel}	0.01	0.3
Ni	0.14	g/Mg_{steel}	0.1	1.1
Se	0.02	g/Mg_{steel}	0.002	0.2
Zn	4	g/Mg_{steel}	0.4	43
PCB	2.5	mg/Mg_{steel}	0.01	5.0
PCDD/F	3	$\mu g\ I\text{-}TEQ/Mg_{steel}$	0.04	6.0
Total 4 PAHs	0.48	g/Mg_{steel}	0.009	0.97
HCB	0.03	mg/Mg_{steel}	0.003	0.3

Metals 2020, 10, 331

Table 2. Target values of various specific air emissions factors for steel production processes [20].

Pollutant	Process	Limit Values (mg/Nm³)				
		HM Protocol 1998	Gothenburg Protocol 1999/2005	Gothenburg Protocol 2012	HM Protocol 2012	POP Protocol 2012
SO$_x$	combustion of coke oven gas		new: 400 existing: 800	400	-	-
	combustion of blast furnace gas	-	new: 200 existing: 800	200	-	-
NO$_x$	combustion of other gaseous fuel	-	new: 200 existing: 350	new: 200 existing: 300	-	-
	sinter plant	-	400	400	-	-
Particulate matter	sinter plant	50	-	50		-
	pelletizing plant	25 40 g/t$_\text{pellets}$	-	crushing, grinding & drying: 20 all other process steps: 15		-
	blast furnace	50	-	hot stoves: 10		-
	basic oxygen steelmaking	-	-	30		-
	electric steelmaking	20	-	new: 5 - existing: 15		-
	hot & cold rolling	-	-	20 bag filters not applicable: 50		-
PCDD/F	sinter plant	-	-	-		0.5 ng/m³
	electric arc furnace plant	-	-	-		0.5 ng/m³

The list of pollutants mentioned in these tables is far from complete. Pollutants missing are, for example, silica (work on refractories), asbestos (not used any more, in most countries), and acid fumes (on pickling lines), etc. Moreover, categories inherited from toxicology (cf. Section 7) are not used in these inventories, e.g., carcinogenic or endocrine disruptors, although they would be relevant. Note also that some extremely important emissions could not be easily measured until rather recently (dioxins, 20 years ago, PM$_{2.5}$ and PM$_1$, 5 years ago and PM$_{0.1}$, presently).

The data on emissions are relative to industrial equipment, therefore to the combination of process reactors and exhaust gas-cleaning devices. Direct emissions from the processes alone are one or two orders of magnitude larger than what is presently dispersed to the atmosphere.

The ironmaking plant, comprising coke ovens, a sinter plant, and pelletizing plus a blast furnace, has the reputation of being the largest emitter in terms of particulate matters and combustion, although this "order of merit" can change depending on the actual technology used and on its age. Note also that emissions from ore piles, which are stocked inside the steel mill upstream of the ironmaking plant, are not usually accounted for in these emissions, even though this may raise serious issues (cf. Section 7).

4.2. Dust Generation

A more phenomenological approach can be adopted to identify the mechanisms that generate the dust [21], cf. Figure 9:

- *Saltation*, the basic mechanism for transporting powder from piles by the action of wind or of gas convection. It is the basic mechanism responsible for the airborne dust originating from sand deserts and transported thousands of kilometers away and for the displacement of sand dunes as well [22]. It is also effective in eroding piles of raw material, like ore, for example, and projecting dust in the atmosphere.
- *Erosion* can also generate dust, a common phenomenon in the biosphere/geosphere.
- *Volatilization (evaporation)* of volatile species (arrow 1 in the figure), such as some metals (e.g., zinc, lead, silicon—even though it is not a metal). This happens in the EAF and is the major mechanism in ferro-silicon furnaces (see further).
- *Projection by gas, electric arc, or powder injection* (arrows 2 and 2′), as in an EAF where coal or lime is transported with gas into the liquid steel bath through immersed lances while oxygen is injected in the same manner.

- *Bubble burst* (arrow 3 and 4) is active when large amounts of gas are generated in a liquid bath reactor, either by injection or as a result of a chemical reaction, like the decarburizing of the bath by "burning" dissolved carbon by injected oxygen.
- Probably more, like the *reemission of incoming* dust (arrow 5) or the *precipitation of graphite* when tapping a hot metal (pig iron) ladle or just letting it just sit while waiting for the converter.
- Note a variant of the bubble burst, called *droplet burst* [23], a phenomenon due to the oxidation of the droplet, its encapsulation as liquid metal inside a rigid shell, and then the explosion of that shell due to the buildup of inside pressure by evaporation, followed by the explosion of the droplet itself. Evidence of this mechanism was reported by Nedar Lotta in a steel BOF, but not by Huber et al. in the EAF. It is also reported in a different context, that of the explosion of fuel droplets during combustion.

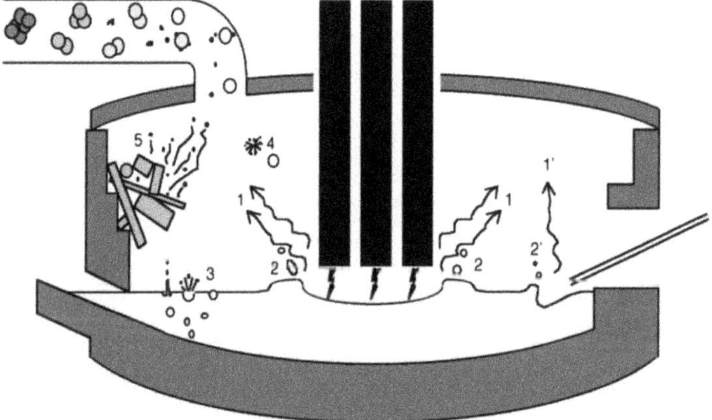

Figure 9. Mechanisms of dust generation in an electric arc furnace [21].

These various mechanisms lead to the formation of dust of a different nature in an electric arc furnace, as shown in Figure 10.

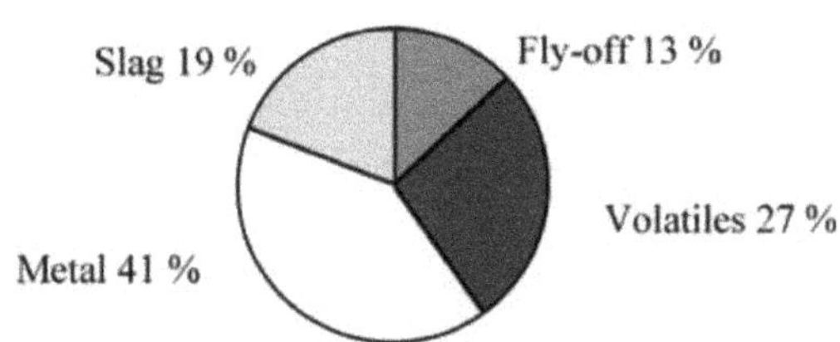

Figure 10. Nature of particulates generated inside an EAF [21].

Bubble burst was identified as the main mechanism responsible for the generation of dust in a steel electric arc furnace [24–36], cf. Figure 11. Experimental evidence in a steel bath is also shown at the top of that figure.

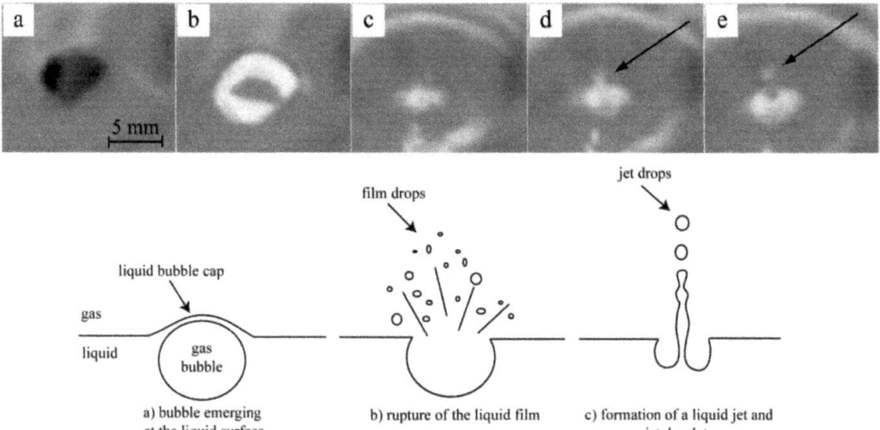

Figure 11. Top: frames taken from a high-speed video (emergence of a bubble (**a,b**), disruption of the bubble cap (**c**), formation of vertical liquid jet (**d**), emission of a drop from the jet (**e**); Bottom: bubble burst mechanism for dust generation [36].

Gas bubbles, generated in the melt by the nucleation of gaseous CO, rise to the free surface, where they explode and liberate small droplets of liquid iron, either from the film of the bubble or from the column of liquid metal that rises in the center of the bubble when the film collapses, the latter contributing more than the former; the droplets are then sucked into the gas exhaust system, where they solidify and oxidize in the draft alongside zinc vapor, which is the other major contributor to dust generation and originates from the evaporation mechanism.

The proportion of iron involved in an EAF due to this mechanism is 10.5 kg_{dust}/t_{steel}.

Dust is collected in a dry or a wet system, on the EAF itself (primary collection) or under the canopy of the steel shop (secondary collection), and generates the powdered dust or the sludge by-product. Some dust bypasses the collection systems and gets dispersed in the atmosphere to eventually drop to the ground (<5%).

Bubble burst is active in all liquid metal reactors, like converters or ladles. In combination with gas or powder injection, it is also essential in a liquid bath smelter.

Evaporation is the main mechanism in reactors that are "quiet", i.e., not stirred by powerful injections, decarburizing, or other mechanisms of in-bath gas generation, and which handle materials with a high vapor pressure. Metal is often the main substance that directly evaporates, like zinc in a steel EAF, or at the top of an imperial smelting furnace (ISF). Dust is only generated if at some point the vapor comes in contact with oxygen from air ingress; otherwise, the metal will condense and be recovered as in the ISF.

In the case of the production of *high-silicon alloys* including ferro-silicons, the volatile substance that escapes the furnace is SiO and not silicon itself [36]. Figure 12 shows the major mechanisms of dust generation in this context.

Evaporation is used to handle *dust from steel EAFs*, which are hazardous by-products and as such need to be treated separately. The Waelz kiln's process is based on the evaporation of zinc. Similarly, steel producers sometimes recycle the dust in the EAF itself, and thus enrich it in zinc by evaporation, up to 20% or 30%, in order to reduce the fee that they have to pay to the dust processors, which does not necessarily make economic sense from a total *cost of ownership* standpoint [37].

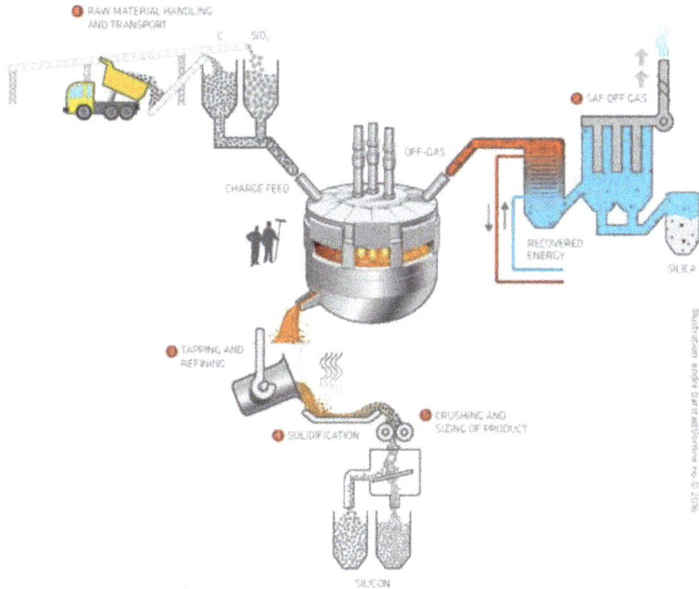

Figure 12. High silicon alloy production process and its primary emission sources [36].

4.3. Organic Compounds Genesis and Emissions

The *generation of organic emissions* is another area where process mechanisms have been described.

Studies of the emission of organic compounds by industrial processes are basically of two types: On the one hand, measurement campaigns are carried out to assess emissions both in terms of compounds formed and of their concentration [27]—this is a necessary step to analyze the compliance with regulatory limits on emissions; on the other hand, more basic work was carried out to understand the physico-chemistry mechanisms that drive the generation of some particular molecules from some particular sources [38–43].

For example, the transformation of materials added as contaminants to commercial scrap to the EAF charge, for example, bits of car tires, was analyzed in a laboratory simulator mimicking various conditions within the reactor. The degradation of the organic contaminant is due both to pyrolysis and to oxidation reactions; their relative importance depends on the temperature and on the concentration of oxygen.

Under pyrolytic conditions, hydrocarbons are thermodynamically unstable above 500 °C. Generally, their stability decreases when the molecular weight increases. Methane is the most stable compound below 1000 °C. A double bond is more stable than a single one. Because of resonance energy, aromatic compounds are very stable at high temperatures. Compared to other hydrocarbons excepting aromatic compounds, alkynes are more stable between 1000 and 1200 °C.

Therefore, under oxidizing conditions, both pyrolysis and oxidation take place, leading to the emission of numerous molecules more or less oxidized. Final products like carbon dioxide and water, inert gases but also unburnt compounds, such as H_2, CO, tars, VOCs, PAHs, etc.

More complex and harmful molecules can be generated in or outside of process reactors (for example, in their fume collection system), like *tetrachlorodibenzo-furans (PCDF)* and *tetrachlorodibenzo-dioxins (PCDD)*, both usually simply referred to as *dioxins* [38,39,43,44]. These compounds have a bad reputation in terms of health consequences (cf. Section 7) and therefore have been chased relentlessly by regulators. Industry has adapted accordingly.

Dioxins tend to form in an oxidizing atmosphere, thus rarely in the laboratory of the process reactor but most commonly in fume collection systems. Thus, in a steel blast furnace, which operates under reducing conditions, dioxins are never encountered. On the other hand, sinter plants, EAFs, and municipal waste incinerators (MWIs) have been well-known sources of dioxins.

Dioxins are generated by the degradation of organic compounds present in the charge of the reactor, where they are introduced either deliberately (fossil fuel, biomass) or as contaminants, but also by synthesis from inorganic substances, the so-called *de-novo mechanism*. Dioxins are gases at high process temperatures and then they condense as liquids in the dust collecting system (cf. Figure 13). Thus, for example, EAF dust is heavily contaminated by dioxins, which, in turn, might create problems at dust processors' plants.

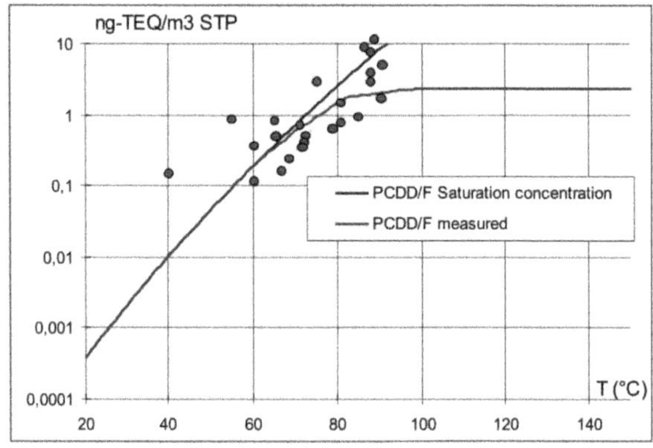

Figure 13. Comparison between the TEQ (toxic equivalent) saturation curve and experimental data from Arcelor Luxembourg.

Countermeasures were proposed to make sure that dioxins fully precipitate in the dust and are not carried away in the flue gases, heading for the smokestack and the atmosphere. The preferred abatement mechanism, however, is the post-combustion of the flue gas under conditions where dioxins are properly eliminated—i.e., combusting for at least 2 s and then quenching the gas very quickly, for example, between 400 and 250 °C—sometimes with the boost offered by a catalyzer.

4.4. Emissions and Pollution

The contribution of the emissions from a steel mill to pollution is case dependent. For example, the steel mill may be isolated, embedded in a city or part of an industrial region, where several other emitters coexist.

A case study [45] is shown in Figures 14 and 15. It relates to the Metropolitan Region of Vitória, Brazil, on the Atlantic coast, where dust pollution in terms of Total Suspended Particles (TSP, macroscopic dust) and PM_{10} (Particulate Matter less than 10 µm in size, breathable dust) were measured for 5 years (1995–1999) at four or two locations, respectively, for periods of 24 h every week. In total, 15 sources of dust were identified in the region and aggregated into 3 families: Industrial (integrated and EAF steel mills, mining terminal for overseas shipping of iron ore, pelletizing plant, coal, etc.), human activities (forest fires, soils, civil construction, quarries), and natural (marine aerosol). Dust in the Vitória region is overwhelmingly from anthropogenic origin, i.e., industrial and human activities' emissions. Human emissions are always more important than industrial ones, a clear-cut conclusion for large-particle dust and a weaker one for breathable dust.

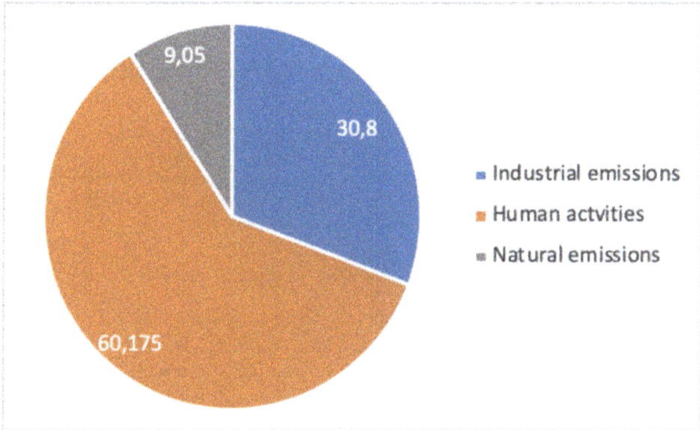

Figure 14. Local pollution in the Vitória metropolitan area: distribution of TSP according to sources of emissions (adapted from [45]).

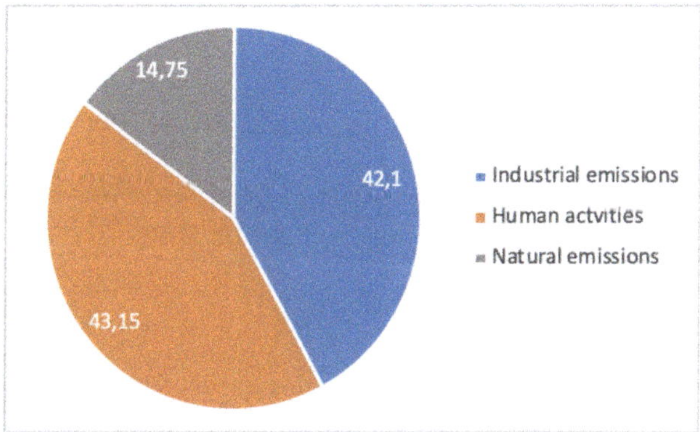

Figure 15. Local pollution in the Vitória metropolitan area: distribution of PM10 according to sources of emissions (adapted from [45]).

This shows clearly that pollution is multifactorial, and that industry is not the major emitter in this case, although it is an important one.

Dust pollution is a rather simple pollution phenomenon. Indeed, there are more complex situations where several pollutants add up to create new pollution, like in the case of *photochemical smog*. However, discussing these topics, which are not often analyzed in the literature in connection with steel mill emissions, goes beyond the scope of this article, although it points to interesting areas for further investigations.

Emissions to water and soil ought to be discussed in parallel with the previous discussion focused on air. We leave this for further reviewing work.

4.5. Lifecycle Emissions

Emissions from a steel mill, gate to gate, only encompass part of the emissions of the lifecycle of steel. Upstream emissions from mining were mentioned *en passant* in Sections 2 and 4.2 (saltation), relative to iron ore piles, solid waste and dust.

A Life Cycle Assessment (LCA) looks at emissions in a particular way related to *impact categories*. Table 3 from [46] examines seven mid-point impact categories for steel products (beams, coils of strip, etc.), related to: (1) Energy use (*primary energy demand*, PED), (2) GHG emissions (*global warming potential*, GWP), emission of acidic compounds that affect water quality either (3) by turning it acid (*acidification potential*, AP) or by providing nutriment for plant life (4) (*eutrophication potential*, EP), or generating emissions to air that cause *photochemical ozone creation* (POCP).

Table 3. Lifecycle significant flows, phases and processes in the worldsteel steel Life Cycle Inventory (excluding the end-of-life phase).

Impact Category	Main Input/Output	Main Phase	Main Processes
Primary energy demand	Hard coal (75–95%) Natural gas (0–15%)	Upstream (~100%)	Upstream energy: electricity and fuels
Global warming potential (100 years)	Carbon dioxide (90–95%) Methane (~6%)	Gate-to-gate (>60%) Upstream (20–30%)	
Acidification potential	Sulphur dioxide (50–60%) Nitrogen oxides (30–40%) Hydrogen sulfide (<10%)	Gate-to-gate (40–60%) Upstream (40–60%)	Gate-to-gate: steel production processes up to slab production
Eutrophication potential	Nitrogen oxides (>90%) Nitrous oxide (~2%) Ammonia (~2%) COD (~2%)	Gate-to-gate (>60%) Upstream (50–80%)	
Photochemical ozone creation potential	Carbon monoxide (60–70%) Sulphur dioxide (10–20%) NMVOCs (<10%) Nitrogen oxides (<10%)	Gate-to-gate (>80%) Upstream (~20%)	

The second column of Table 3 describes the inputs and the outputs (emissions) in/out of the lifecycle: energy comes from coal or natural gas, GWP is caused by CO_2 or CH_4 emissions, etc. The third column explains how much of the particular impact category is related to the steel mill itself (gate-to-gate) or to the upstream part of the lifecycle, thus to mining and raw materials' transportation. The upstream part of the lifecycle is significant, as it accounts for 20% to 100% of the midpoint indicator.

Of course, downstream emissions are related in a complex manner to the material. Moreover, beyond the negative impacts of materials, such as air emissions, they also have a positive impact, which has been described as the eco-socio-systemic value of that material [3]. This is, however, going beyond the scope of the present review paper.

There is much room left for more systematic work on emissions. Unifying the vocabulary used by the various disciplines that tackle these matters would be useful and this would make it easier to bring all the data into a single set. More work is needed on the phenomenology of emissions, at experimental and modeling levels. Connection with meteorological pollution and with health issues (see further, Section 7) would further flesh out the overall discussion. Extending the analysis to water and soil would also be necessary. Eventually, when this is all done, a more rigorous review than the present one would be in order.

5. Greenhouse Gas Emissions Related to Steel

5.1. Anthropogenic vs. Geologic Iron

The stock of iron available for anthropogenic use now and in the future is present under two forms:

- Metallic iron, constituting the stock of steel in the anthroposphere, which comprises steel-in-use engaged in various artifacts and discarded steel, known as scrap or as the anthropogenic mine.
- Iron compounds, mainly oxides, which constitute iron ores and are found in ore deposits (≈resources) and in geological mines (≈reserves).

The *circular economy*, a popular word today but a historical practice in the world of materials, should ensure that the anthropogenic mine takes over 80% of the iron supply by the end of this century [47]. This may read as somewhat paradoxical, as the recycling rate of steel is already above 80–90%. However, the residence time of steel in the economy is of the order of 40 years and the demand for steel is still increasing worldwide so that this complex dynamic will not deliver an intuitively circular economy until then.

Note also that an integrated steel mill, complete with coke oven batteries, generates coke oven gas (COG), rich in hydrogen (roughly 60%), which could serve as a source of hydrogen—and thus combine with low-carbon steelmaking. Such a scheme was indeed put into practice by Air Liquide at the Carcoke company in Zeebrugge, Belgium, in the 1980s. The gas was distributed to various customers through a regional system of pipes, which is still in operation today. The shutdown of the coke ovens brought this production of hydrogen to an end.

5.2. Reduction of Iron Ore

Until the advent of a steel production fully based on the circular economy, chemical reduction of iron ore will remain necessary—as physical reduction is out of reach, cf. Figure 16. Indeed, vacuum decomposition is impossible, even in the deep vacuum of outer space (10^{-11} Pa). The best laboratory vacuum (ultra-ultra-vacuum) is only 10^{-10} Pa. Thermal decomposition of iron oxides requires very high temperatures (3404 °C), which could be achieved, for example, in solar furnaces. However, as soon as the atoms (ions) would be separated, they would tend to recombine again, unless they are physically kept part, for example, in a kind of mass spectrometer. Needless to say, this does not correspond to a technology commensurate with the mass production of iron! The only physical option available for the decomposition of iron oxide is electrolysis, which will be discussed further in the report.

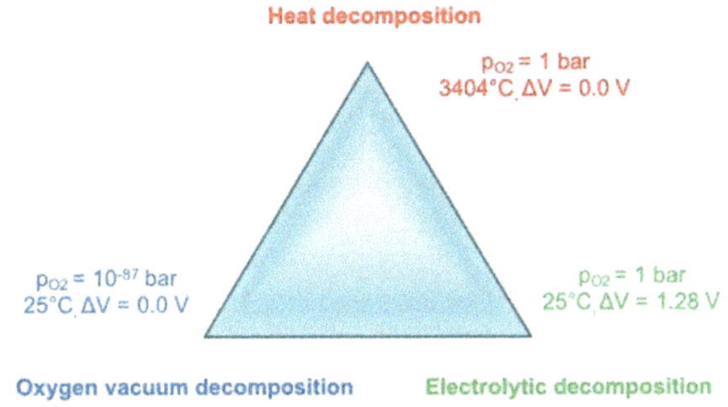

Figure 16. How to recover iron from oxides by pure physical means?

This is presently accomplished by smelting reduction based on the use of carbon, mainly from coal and, marginally, from natural gas, and other options include hydrogen and electricity, plus any combination of the three, as schematized in Figure 17, which also shows how they may contribute to low-carbon production.

Basically, there are three ways of reducing iron ore by chemical means:

- Historically, carbon has been used first, as charcoal initially and then as earth coal converted into coke (metallurgical coke). The blast furnace-based, integrated steel mill is the current state-of-the-art technological avatar of this solution.
- Hydrogen is the other option, although it requires the production of hydrogen first, whereas carbon stems directly from mining. Hydrogen reduction has been investigated at the laboratory extensively and some limited industrial operation took places at various scales years ago. See Section 5.9 for the revival of the concept in many on-going research projects.

- Electrons (e$^-$) can also behave as reducing agents, in the technological framework of electrolysis. The method is used extensively in non-ferrous metallurgy, most especially for the production of aluminum. However, several pieces of laboratory work have shown that the concept does work for iron as well (see further details in Sections 5.4 and 5.5).

- In addition to these "mainstream" solutions, there are more, like using metals as reducing agents or powerful reductants like hydrazine (N_2H_4). Due to the price of metals, especially non-ferrous metals that would have to be used in the case of steel, such as aluminum, this is not a practical solution, although it is used at the margin, e.g., aluminum is sometimes used in the steel shop to raise steel temperature in the ladle.

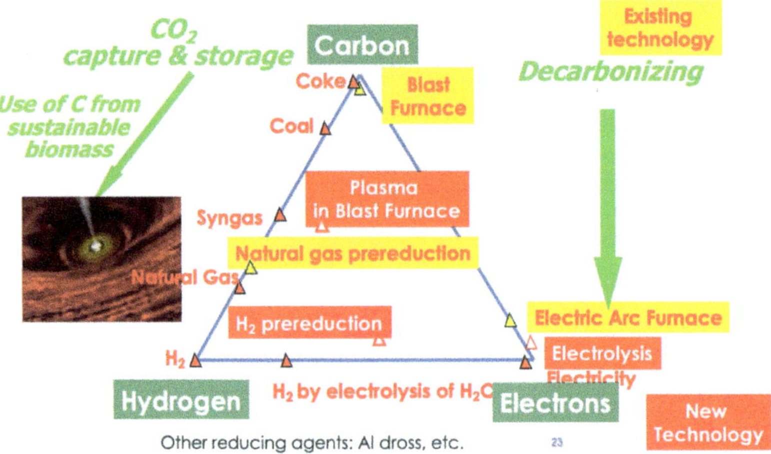

Figure 17. Practical ways to reduce iron ores: process route for iron and steelmaking.

Finally, for the sake of completion, hybrid methods have been imagined or tested in a long series of reactor concepts and sometimes pilot plants. An example is the combination of a nuclear power plant, preferably of the fourth generation, with an iron ore reduction plant: the former would provide the latter with part of the energy needed to bring the charge up to the reaction temperature. This work was never published but led to the concept of *high-temperature heat pumps* to boost the temperature above the level that waste heat from the nuclear reactor could deliver.

A more detailed analysis of many more possible ways to make low-carbon iron is available in [48]. It includes in particular a discussion of *thermochemical cycles*.

5.3. Benchmarking of Various Iron Reduction Routes

An exhaustive assessment of various existing or new process routes made possible through new technology developments (45 routes, altogether) was carried out in the framework of the ULCOS program [49]. Energy needs, CO_2 emissions, and total production costs were evaluated [14,50].

The routes differ slightly in terms of energy needs, but the differences are fairly small: all routes are either already fairly energy efficient (existing routes) or are extrapolated on the basis of high energy efficiency (new routes)—cf. Figure 18. Important details are in the original papers.

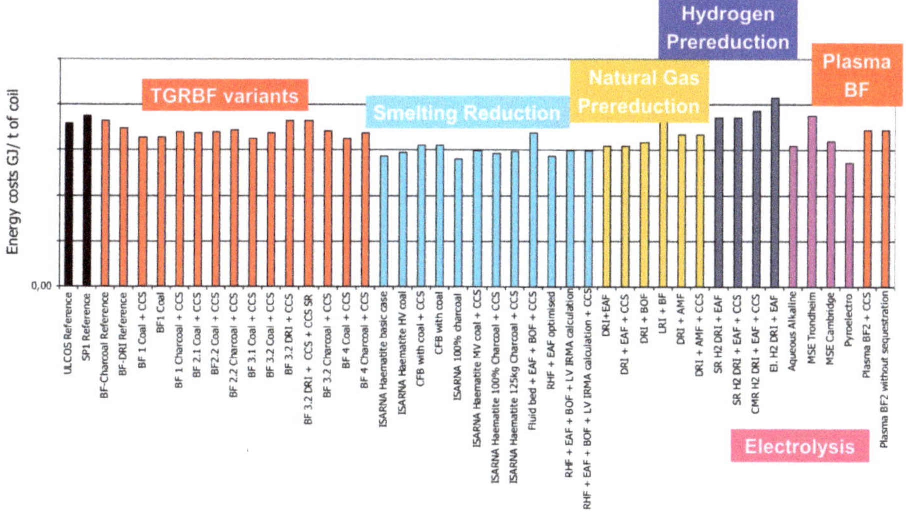

Figure 18. Specific energy needs (GJ/t$_{HRC}$).

CO_2 emissions, shown in Figure 19, are much more spread, because some of the routes adopt low-carbon technologies, such as CCS (carbon capture and storage) or some kind of green electricity. One route even exhibits *negative emissions*, because it uses biomass carbon (charcoal) from *sustainable plantations* and applies CCS, thus sequestering the CO_2 that was pumped from the atmosphere (in the plantation). The graph distinguishes between scope I and scope II emissions, i.e., between direct process emissions from steelmaking and indirect emissions (in green) related in particular to the production of electricity.

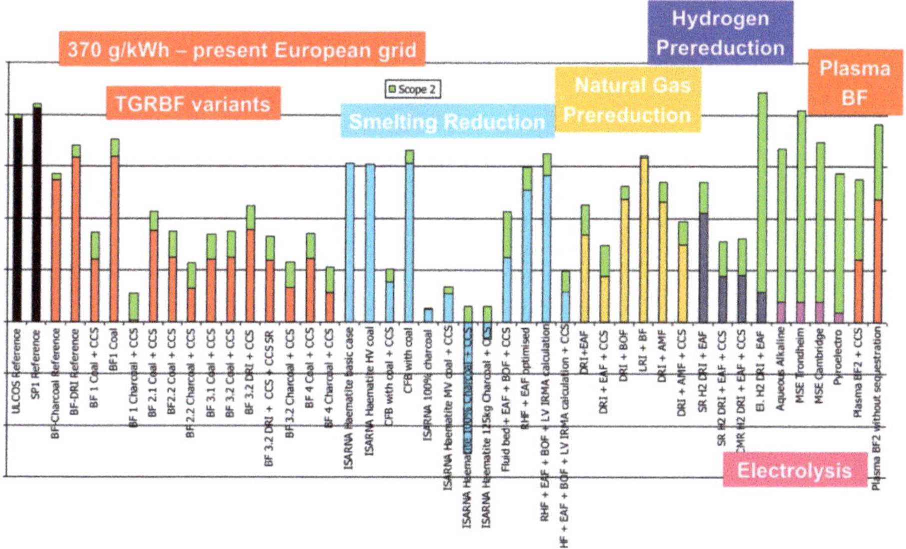

Figure 19. CO_2 specific emissions (t$_{CO2}$/t$_{HRC}$) for a 370 g/kWh grid.

Paying special attention to the hydrogen routes, they exhibit a slightly higher level of energy need than conventional routes and some other low-carbon solutions but do show improved emissions, except for the electrolysis Direct Reduction (DR) route when the carbon intensity of the electrical grid is 370 g/kWh, which was the European level at the time of the study (cf. Figure 19). The low-carbon grid in Figure 20 shows very low emissions.

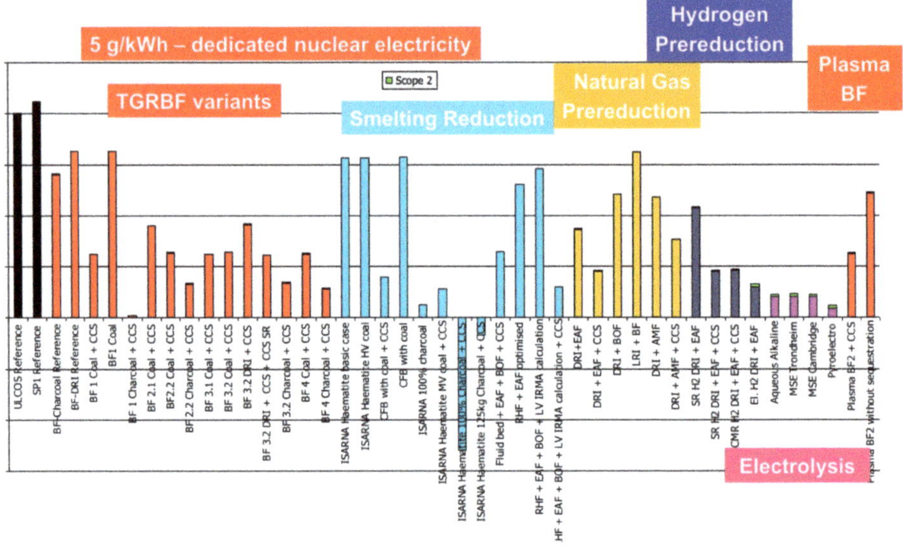

Figure 20. CO_2 specific emissions (t_{CO2}/t_{HRC}) for a 5 g/kWh grid.

The costs, not shown in this summary, are all significantly higher than the production costs of conventional BF ironmaking, which indicates that low-carbon technologies cannot "pay for themselves" and need some kind of extra funding, for example, through a carbon tax.

5.4. ULCOS Solutions and Their Present Evolutions

The *ULCOS solutions*, i.e., the short list of process routes that would be pursued at higher TRL after the end of the project, were defined in the second part of the ULCOS program in 2010. They are shown in Figure 21. The first three of them rely on CCS and the fourth one on the direct use of "green" electricity.

ULCOS-BF was to be built based on the existing Florange Blast Furnace of ArcelorMittal. It applied to the European NER 300 Program [51] and was to be awarded the required financing, when the blast furnace was shut down because of the economic crisis. The technology has since been frozen, but lower TRL research continued under the names of the VALORCO and LIS programs with French ADEME financing [52]. It is presently awaiting a blast furnace on which to resume the work.

Another ULCOS subproject ran in the first stage of the program (the injection of plasma at the tuyere of the blast furnace in order to reinject CO + CO_2 and reduce CO_2 with the electrical energy) was re-ignited and is now running as the IGAR project in Dunkerque plant, as schematized in Figure 22; it has received financing from French SGPI as part of the PIA program [53].

Coal & sustainable biomass		Natural gas	Electricity
Revamped BF	Greenfield	Revamped DR	Greenfield
ULCOS-BF	HIsarna	ULCORED	ULCOWIN ULCOLYSIS
tuyere tests in Dunkerque	demonstrator under experimentation	reformulated as HYBRIT	scale-up as SIDERWIN

Figure 21. The ULCOS solutions and their present status.

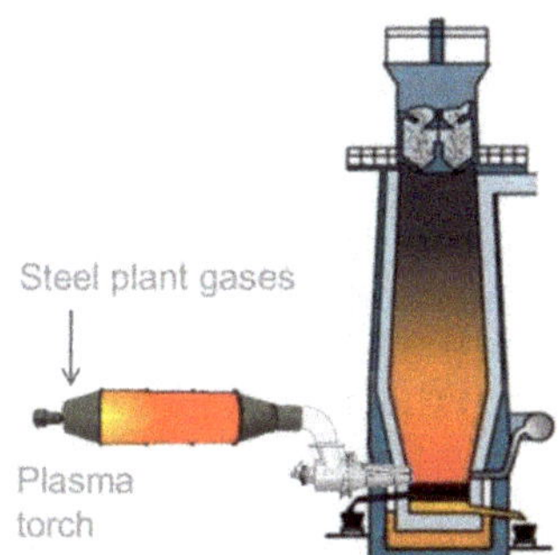

Figure 22. The explained I terms of what the acronym means project (source: ArcelorMittal).

The Smelting Reduction *HISARNA* process is continuing at a demonstrator scale in the Tata Steel mill of IJmuiden, Netherlands, with continuing financial support of the EC, through the RFCS and H2020 programs. It is presently conducting long-duration (weeks) trials.

The *ULCORED* process is now continuing as the HYBRIT project, thus as a synthesis of two ULCOS subprojects.

The electrolysis solutions, *ULCOWIN* and *ULCOLYSIS*, are presently continuing as the *SIDERWIN* project, a H2020 project [54,55], and as part of the VALERCO project, respectively, after various EU and National support schemes (EC's RFCS IERO, French ANR's ASCOPE, etc.), cf. Figure 23. SIDERWIN is presently designing a TRL 6 demonstrator that will be able to produce 100 kg of iron.

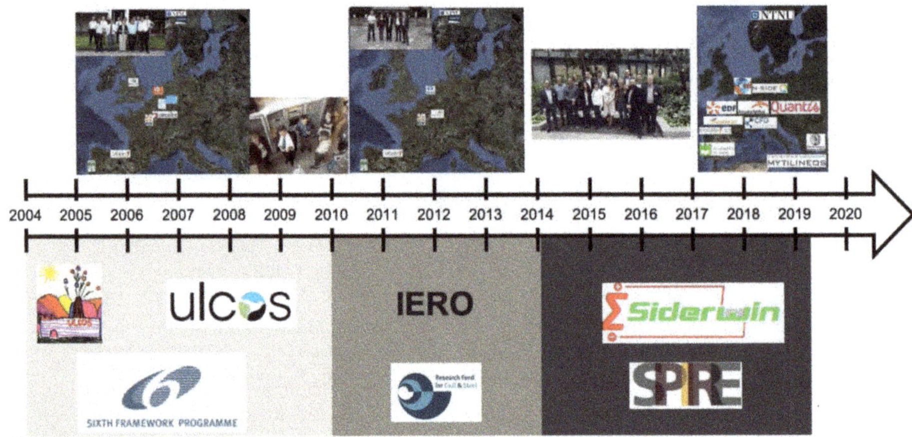

Figure 23. Evolution of the low-temperature electrolysis of iron ore, known as ULCOLYSIS and now as SIDERWIN (courtesy of ArcelorMittal).

A 48-month project called 3D-DMX (DMX Demonstration Dunkirk; DMX is the name of the amine-based capture technology) was launched in Dunkirk in May 2019 around ArcelorMittal's steel mill with EU financial support [56] in order to continue the development of a new technology for CO_2 capture carried out within the VALORCO project at IFP Energies Nouvelles's laboratory in Solaize, and demonstrate it on a stream of blast furnace gas (BFG) at the scale of the capture of 0.5 t_{CO2}/h [57,58]. The next step would be a large-scale demonstrator of 1 Mt/year and there are plans to gather a cluster of emitters around the North Sea and to use existing infrastructure for transport and storage, and thus to connect to the *Northern Lights project*, also funded under H2020 [59] and organized around the Sleipner Platform, where Statoil, now Equinor, developed know-how on the capture and storage of CO_2 extracted from the local natural gas stream.

Another new project is related to the decision of ArcelorMittal in Hamburg to build a demonstrator of *hydrogen-based reduction* with Midrex, based on the latter's technology of direct reduction [60]. The target is the production of 100,000 t/year production of direct reduced pellets. Larger scale-up would be envisaged depending on the technical outcome and political context. The other DRI technology providers are also working on their own concept of an H_2-based reduction furnace, for example the ENERGIRON ZR process [61].

The efforts carried out in Europe in terms of low-carbon steelmaking—now coined "zero-carbon steelmaking"—reflect the European Commission's commitment to transform Europe into the world's first climate-neutral continent by 2050, a policy known as the European Green Deal [62].

5.5. Low-Carbon Steelmaking Solutions in the US

In the US, two projects are aimed at low-carbon technology for steelmaking.

One is the Molten Oxide Electrolysis (MOE) project, a concept proposed by Donald Sadoway at MIT, which is in principle very similar to the ULCOLYSIS process, i.e., a transposition to iron of the Hall–Héroult process of making aluminum, without its drawbacks. The process takes place at liquid metal temperature (≥1538 °C) and metal and iron ore are both molten (cf. Figure 24). MIT has launched a spinoff company, Boston Metals, to develop the technology at a larger scale [63]. The challenge is to find an anode that is not made of carbon, which is what happens in aluminum cells, where it participates in the reduction by evolving CO_2 rather than simply O_2.

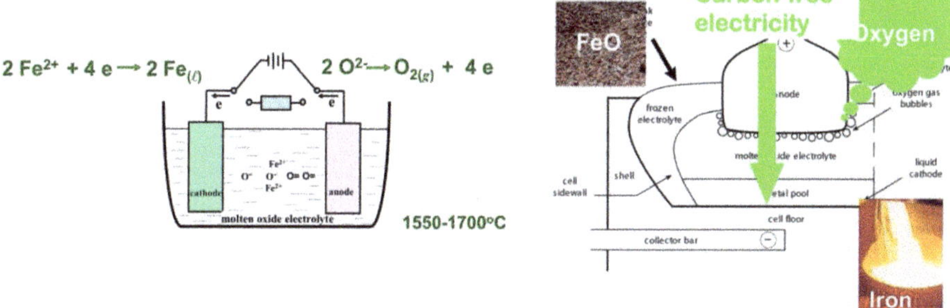

$$2\ Fe^{2+} + 4\ e \longrightarrow 2\ Fe_{(l)} \qquad 2\ O^{2-} \longrightarrow O_{2(g)} + 4\ e$$

Figure 24. The MOE process of D. Sadoway.

The second process originates from the University of Utah (Professor Hong Yong Sohn) and is somewhat similar to the SUSTEEL project already mentioned, except that the reactor concept here is a *flash smelter*, thus operating at very high temperatures in flight on particulate hematite, cf. Figure 25. It is called the new flash smelting process or the flash ironmaking technology (FIT). It receives support from AISI and US' DOE. The project operates a mini-pilot reactor, capable of 1550–1700 °C temperatures with a concentrate feeding rate of 2–5 kg/h and there are plans to scale it up [64].

Novel flash Iron Smelting

- University of Utah
- Gas-solid suspension reduction with hydrogen or natural gas
- Uses iron ore fines and eliminates coke making and pelletising/sintering
- Significant reduction in CO_2 and energy consumption
- Next step pilot plant construction.

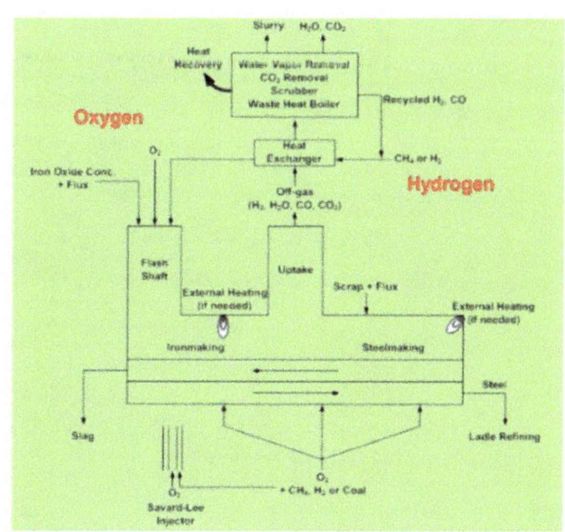

Figure 25. New flash melting process, University of Utah.

5.6. Low-Carbon Steelmaking Solutions in Japan

Japan has a national project, called COURSE 50 (CO_2 Ultimate Reduction in Steelmaking process by innovative technology for cool Earth 50) supported by NEDO, launched in 2007, and presently continuing in its third installment [65]. It collects a series of technology improvements and breakthroughs, the sum of which would add up to a significant reduction in emissions (30% by 2050), cf. Figure 26. It comprises work on CCS but has also a focus on hydrogen, cf. Figure 27.

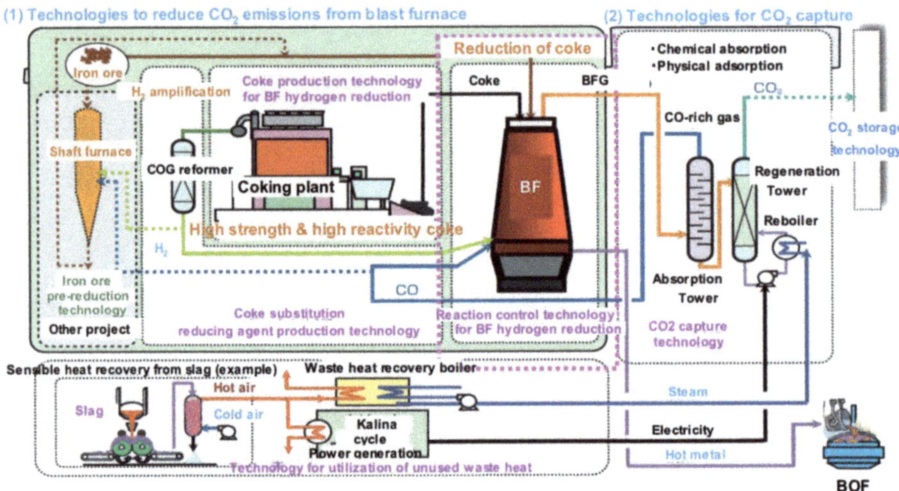

Figure 26. The Japanese COURSE50 program (source: ISIJ).

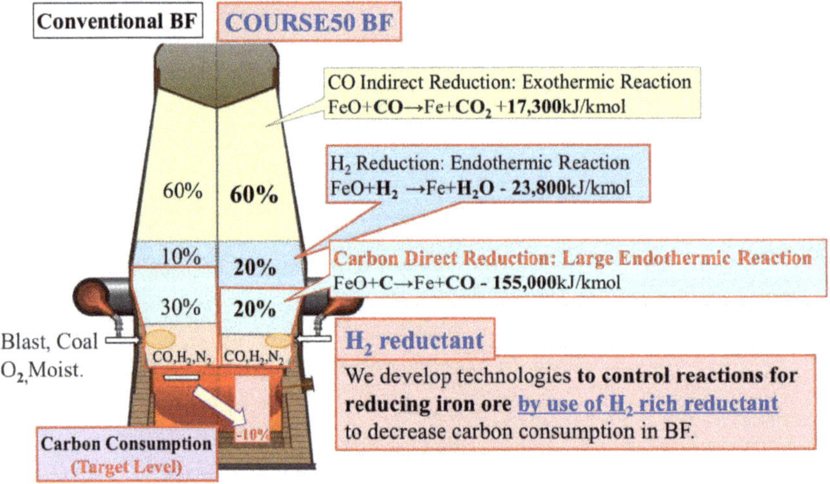

Figure 27. Concept of "iron ore hydrogen reduction reinforcement" in the Japanese COURSE50 program (source: ISIJ).

The idea of the "iron ore hydrogen reduction reinforcement" is to use hydrogen inside an otherwise classical BF by injecting a hydrogen-rich mixture at the tuyeres and then replacing the cohesive zone by two regions: one at the bottom where direct reduction of iron ore by carbon takes place and, on top of it, a zone in which hydrogen carries out the reduction. As both reactions are endothermic, energy is brought from the combustion at the nose of the tuyeres and from the descending burden where CO indirect reduction, which is exothermic, has been taking place. It is expected to reduce carbon consumption and therefore CO_2 emissions by 10% and a BF demonstrator, similar to the LKAB experimental furnace in Luleå, was built by NEDO to test the concept. Practical implementation is scheduled for beyond 2030.

To bridge the gap towards zero-emissions beyond the COURSE 50 program, Japan has developed a vision to reach carbon neutrality in the steel sector by 2100 [66]. It is based on the development and the implementation of a series of "super-innovative technologies", pushing COURSE50 targets towards more utilization of hydrogen in the blast furnace (super COURSE50 program), hydrogen reduction, and the systematic use of CCUS. The timeline is shown in Figure 28.

Development of technologies specific to iron & steel sector		2020	2030	2040	2050	2100
COURSE50	Raising ratio of H2 reduction in blast furnace using internal H2 (COG) Capturing CO2 from blast furnace gas for storage	R&D	Implementation			
Super COURSE50	Further H2 reduction in blast furnace by adding H2 from outside (assuming massive carbon-free H2 supply becomes available)	Stepping up R&D	Implementation			
H2 reduction iron making	H2 reduction iron making without using coal	Stepping up	R&D	Implementation		
CCS	Recovery of CO2 from byproduct gases.	R&D	Implementation			
CCU	Carbon recycling from byproduct gases		R&D	Implementation		

Development of common fundamental technologies for society		2020	2030	2040	2050	2100
Carbon-free Power	Carbon-free power sources (nuclear, renewables, fossil+CCS) Advanced transmission, power storage, etc.	R&D	Implementation			
Carbon-free H2	Technical development of low cost and massive amount of hydrogen production, transfer and storage	R&D	Implementation			
CCS/CCU	Technical development on CO2 capture and strage/usage Solving social issues (location, PA, etc.)	R&D	Implementation			

Figure 28. Long-term vision of the Japanese iron and steel industry for zero-carbon steelmaking [66].

5.7. Low-Carbon Steelmaking Solutions in China and Other Parts of the World

China, which is the largest steel producer in the world and therefore its prime sectoral CO_2 emitter, has been involved in low-carbon technologies since the onset of the CO_2 Breakthrough Program (CBP), cf. Section 5.9.

It communicates mostly on energy saving measures, CCU, and emphasizes the "social value" of steel, which helps other sectors and society in general control their emissions. Note that a CCUS center was created in Guangdong in cooperation with the UK and Australia, cf. Figure 29 [67].

Figure 29. The Canton CCUS center's logo.

India, especially Tata Steel, focuses on its energy efficiency accomplishments and the HISARNA demonstrator project running in its IJmuiden steel mill.

Australia is also involved in low-carbon technologies, especially focusing on the use of biomass, as a source of "green" carbon [68].

5.8. Carbon Capture and Usage (CCU)

Due to the alleged rejection of CCS by citizens in some parts of Europe, the idea of using CO_2 has been pushed forward, mainly by the chemical industry, which is wary of the announced peak oil and peak gas. Today, CO_2 is indeed used in the synthesis of urea and a broad array of reaction paths were explored to transform CO_2 into virtually any inorganic molecule, including fuels. Estimates of 5% to 15% of potential applications of CCU in terms of available industrial CO_2 has been put forward. Note that innovative uses of CO are also studied.

In Europe, the concept is being eagerly explored in various projects, including the VALERCO project already mentioned, but also in more recent ones, which have received public financial support:

- Carbon2Chem, proposed by ThyssenKrupp steel, Germany [69]. The concept is to make various chemicals, including ammonia, from steel mill gases (44% nitrogen, 23% carbon monoxide, 21% carbon dioxide, 10% hydrogen, and 2% methane) and with the application of renewable energy, which has the reputation of being carbon neutral. The project has support from the Federal Ministry of Education and Research, Germany. Cf. Figure 30.
- Carbon4PUR, involving ArcelorMittal and Dechema et al., is a SPIRE/H2020 Project [70]. The focus is on the production of polyurethane (PUR) foam and coatings from steel mill gases. Cf. Figure 31.
- STEELANOL, involving ArcelorMittal, Belgium (ethanol) and using Lanzatech technology, targets the transformation of CO from converter gas (BOG) by using fermentation driven by bacteria [71]. Cf. Figure 32.
- Other European projects focusing on CCU are FReSMe, ICO2CHEM, MefCO2, and RECO2DE [72].

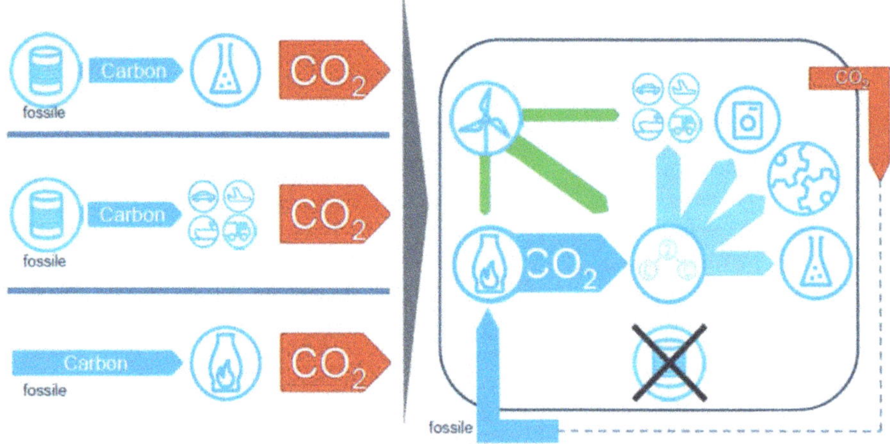

Figure 30. The Carbon2Chem project (source: ThyssenKrupp Steel).

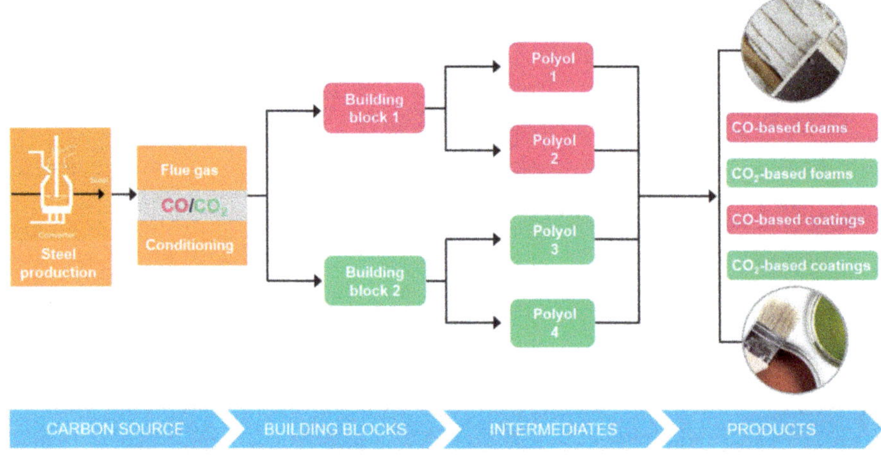

Figure 31. The Carbon4Pur project concept (source: ThyssenKrupp Steel).

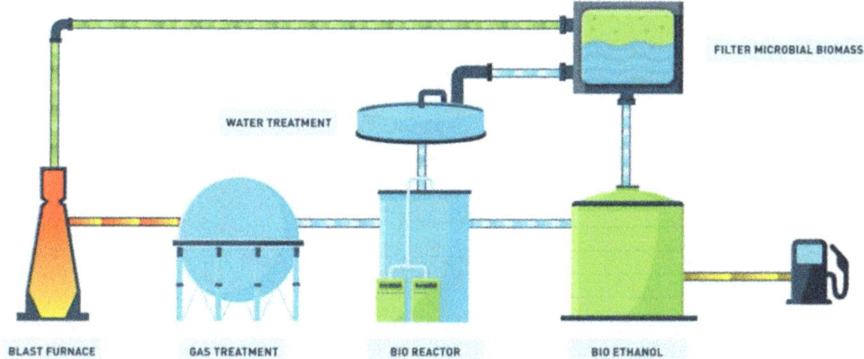

Figure 32. The STEELANOL project (source: ArcelorMittal).

These projects are presented as offering an opportunity for several energy-intensive industries to work together and, implicitly, to arrive at more efficient solutions than by acting as a single sector. Moreover, the claims that CCU reduces GHG emissions has to be substantiated by carrying out consequential LCA studies: is it indeed cutting emissions or simply postponing them by a few weeks or months? Does it avoid the use of fossil fuel or simply leaves it available for some other usage?

5.9. Institutional Support to Low-Carbon Steelmaking Technologies

Various institutions and organizations have been supporting the development of new low-carbon steelmaking technologies. The rationale is, on the one hand, that only breakthrough solutions could deliver the very low emissions that climate change requires, and, on the other hand, that this would call for a tremendous effort in terms of time, continuity of purpose, and financial support.

While the initial work on the topic goes back to the late 1980s and the early 1990s [73], the steel sector started to organize in the early 2000s and lobbied for financial support in the EU and in various countries from national funding agencies from that time on. This led to the *ULCOS program*, supported by the European Commission under the RFCS and Framework programs, and to the *CO_2 Breakthrough Program* (CBP) of Worldsteel, where participants from all over the world would seek support from their own countries. For a period of about 10 years, research on low-carbon steelmaking was conducted under a single banner, with programs in the 5–50 M€ range or slightly more.

ULCOS was about to launch a second stage, ULCOS II, with a budget size an order of magnitude larger than the previous generation of projects. Financing had been arranged through the *NER 300 program* until the ULCOS-BF project collapsed due to the shutdown of the Florange Blast Furnace, decided as a consequence of the 2008 economic crisis.

Since then, the various projects have been continuing individually and new directions are presently being explored (mainly hydrogen processes and carbon utilization). A so-called "big ticket" program is in the making in Europe, under the auspices of EUROFER and ESTEP and with preliminary support from the EC to an RFCS project (*LowCarbonFuture*) that should define its scope. The future ETS Innovation Fund, also called NER 400, is involved [73]. Work is continuing also in Japan, with a large size program (*COURSE50*) and the project to extend it further.

Institutions like the *IEA* and the *Steel Committee of OECD* have picked up an interest in the topic and the IEA is presently working on a *sectoral roadmap for steel* [74].

5.10. Production of Hydrogen

It is often taken for granted, in recent literature about hydrogen reduction, that hydrogen will be produced from electricity by *electrolysis of water* or, in the future, of steam (high-temperature electrolysis).

This would go along with a huge increase in electricity production capacity: a 5 Mt/year steel mill would require a 1200 MW nuclear power plant or 240 recent wind turbines. The electrolysis technology is not yet ready for immediate scale up to this large size, but R&D is ongoing.

One should therefore take on board the results shown in Figures 18–20, which indicate that the traditional technology based on the *steam reforming of methane* can continue to be used, especially if some technology improvements are introduced, such as CCS or the production of the less pure hydrogen that the steel sector is likely to need. The use of the hydrogen present in COG would also be of interest, cf. Section 5.1.

5.11. Conclusions on Zero-Carbon Steelmaking

There is no doubt that steel will continue to be produced *en masse* in the 21st century, and that production will keep increasing to accommodate the accession of more people to a decent standard of living across the world. Therefore, ways to alleviate the GHG emissions of the sector will need to be implemented and they will involve major changes in the way of making steel, thus they will need to be based on "breakthrough technologies".

There are myriad solutions for carrying out low-carbon steelmaking.

The future of *hydrogen reduction* depends on the future of hydrogen in general. The *hydrogen society* was brokered as a dream by people like Jeremy Rifkin [75], but the dream may be easier to implement in outer space than on Earth! Various sectors will compete for hydrogen, for example, transport and steel and transport today would be ready to pay more for hydrogen than steel. The value-in-use in each sector is quite different, an order of magnitude at least!

Moreover, if hydrogen use does not take off for transport, i.e., if fuel cell electric vehicles do not compete seriously with electric battery vehicles, then it is unlikely that it will be used in the steel sector.

Additionally, hydrogen reduction will compete with the direct electrolysis of iron ore. This process is developing briskly, and it could seriously be a credible alternative to "green" hydrogen reduction in 2030 and beyond. On paper, it looks like it would be more energy efficient than green hydrogen reduction, but this needs experimental confirmation to be completely believable.

Additionally, the continuing use of coal and natural gas, associated with CCS or CCU will probably remain large in the steel sector.

Thus, research and development have to continue, in parallel, exploring the many possible low-carbon routes to steelmaking.

6. Steel, Biodiversity, and Ecosystem Damages

The loss of biodiversity has today reached a crisis level, becoming the sixth major extinction since the advent of life on earth [76]. This is certainly one of the major threats to "the environment", caused by a combination of factors, including climate change and the growing urbanization of the world. Biodiversity loss in addition to the loss of million species also means the loss of ecosystem services that biodiversity usually brings. Any human activity, individual, collective or industrial has a responsibility in this erosion of biodiversity, as it influences its causes. However, the connection with steel and steel production is not specific and therefore will not be discussed further here [77].

7. Steel, Health of People, Animals and Ecosystems

In biological matter (*biochemistry*), iron is an essential element in the fabric of life [78]. Steel, i.e., iron and its alloying elements, may also act as a toxicant, and as such is studied by *toxicology*. Steel production also raises health issues in the workplace (*occupational health*) or around steel mills (*public health*): both subjects are covered by *epidemiology*. The connection between steel mills and people's contamination is an environmental issue related to the phenomena of *emissions and pollution*. For example, the creation of an environmental department at IRSID in the 1970s was related to an industrial accident, the intoxication of cattle near the steel mill of Le Breuil in Le Creusot in France, by molybdenum emissions from the electric arc furnace. Therefore, there has been a strong connection between emissions, pollution, and "public" health, since the early beginnings of the discipline. Beyond effects on human health, the environment also affects life in general and, more broadly, ecosystems, studied by specific disciplines, *ecotoxicology* and *ecoepidemiology*. Moreover, the field, globally, is covered by an emerging discipline called *environmental health*. It is defined by WHO in its 1989 conference in Frankfurt as being "related to aspects of human health and diseases, which are driven by the environment. This also refers to the theory and practice of controlling and measuring environmental factors that may potentially affect health" [79].

The role of iron in the biochemistry of animals, from microorganisms to human beings, is related to the toggle between the two redox states, Fe^{2+} and Fe^{3+}, a mechanism used to transfer electrons inside cells and thus participate to its *metabolism*. Important enzymes (ferritin, transferrin), the hemoglobin of blood cells (aptly called *hematies*), and other biomolecules contain iron. Fe is therefore an essential trace element in the human diet, with health issues if there is too much or too little present in the daily input (7–11 g/day) [78].

This, however, can be considered as unrelated to the iron and steel sector, because *the geobiochemical cycle of iron* is mainly disconnected from the *anthropogenic iron cycle* [80].

Toxic metals, usually referred to as *heavy metals*, include the following [81]:

- Plant toxicity: Al, As, Cd, Cr, Cu, Pb, Mn, Hg, Ni, Pt, Se, Ag, Th, W, U, V, Zn. Animal toxicity: Al, As, Cd, Cr, Cu, Pb, Li, Mn, Hg, Se, Th, Sn, W, U, V, Zn.

While the following metals are essential to life (thus, some are both toxic and essential at different levels):

- Essential to plants: Cu, Fe, Mg, Mn, V.
- Essential to animals: Cr, Co, Cu, Fe, Mg, Mn, Ni, Se, Sn, V, Zn.

Iron is both *essential* and *non-toxic* to any of the life kingdoms, but many of the elements used in alloys do raise health concerns (Al, Cr, Cu, Pb, Mn, Ni, Se, W, Sn, Zn). Note that animal and human toxicity are not identical and that there are many levels of toxicity (acute vs. chronic, lethal or not, confirmed vs. alleged, etc.); see a specialized source for details [83]. Moreover, there are important ongoing controversies regarding the toxicity of common metals like aluminum, chromium, or nickel. Lastly, other authors have published slightly different lists of toxicant metals (for example, Mo is not in the lists). In a practical way, toxic problems are most acute in the working environment of steel mills and, therefore, are handled as part of *occupational health and safety*, mostly satisfactorily nowadays in world-class steel mills [84].

Regarding air emissions caused by the production of steel (cf. Section 4), there are various factors that raise concern:

- Particulate matter or dust: TPM, PM_{10}, and now $PM_{2.5}$ are routinely measured nowadays [85], but critics point out the fact that smaller particles than 2.5 μm (or 1 for PM 1) and particularly nanoparticles are still unaccounted for, even though they are very likely to penetrate deeply into the body of living organisms and, thus, to be raising the most serious health risks;
- PAHs;

- PCBs;
- PCDDs and PCDFs;
- HCB;
- More generally, VOCs;
- POPs; and
- Inorganic pollutants (SO_2, H_2S, NH_3, HCl, HCN, HNO_3, O_3, CO, CO_2, black carbon, etc.) and products of incomplete combustion (NOx). Note that inorganic and organic pollutants can combine, such as dust particles made of a core of black carbon and a coating of PAHs.

Steel processes are direct sources of all of these emissions and the standard way to control them is to capture the fumes and "treat" them with a technology that is deemed to deliver the level of purification required by regulations. Discussion of the best *abatement technologies* lies outside of the scope of this paper [20]. Direct process emissions have the reputation of being properly captured. Emissions that escape to the atmosphere (like canopy emissions or other uncaptured emissions) are dispersed in the environment. Sometimes, the issue disappears because of *dilution*, but, in other cases *uncaptured emissions* aggregate with emissions from other sources and contribute to *large-scale pollution* and to their associated health problems.

Most uncaptured emissions are related to particulate matter, cf. Section 4.2. Figure 33 shows the evolution of PM_{10} emissions in EEA-33 Europe [86]. Industry-related emissions (green and beige), part of which come from the steel sector, account for more than 1/3 of total emissions and their absolute amount has only slightly decreased over the past 30 years. The data for $PM_{2.5}$ are not available for this extended time period. In 2017, regarding PM_{10} and $PM_{2.5}$, industry was responsible, respectively, for 17% and 10% of the total emissions [87]. The EEA reports on the largest polluters in Europe. In 2015, 7 of the 10 top polluters in PM_{10} were integrated steel mills [88].

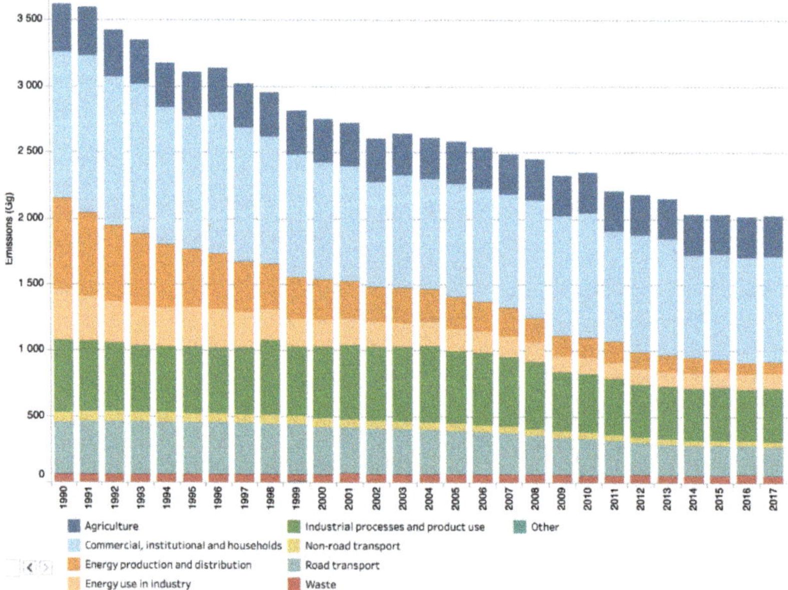

Figure 33. PM_{10} emissions by economic sector in Europe (EU-28). Source: European Environment Agency [82].

Air pollution has serious health effects:

- WHO reports 4.2 million premature deaths ("death that occurs before the average age of death in a certain population", according to the NIH) due to outdoor pollution (worldwide ambient air pollution accounts for 29% of all deaths and disease from lung cancer, 17% from acute lower respiratory infection, 24% from stroke, 25% from ischemic heart disease, 43% from chronic obstructive pulmonary disease. Pollutants with the strongest evidence for public health concern, include particulate matter (PM), ozone (O_3), nitrogen dioxide (NO_2), and sulphur dioxide (SO_2)).
- 3.8 million premature deaths are due to indoor pollution worldwide, as shown in Figure 34 [89].

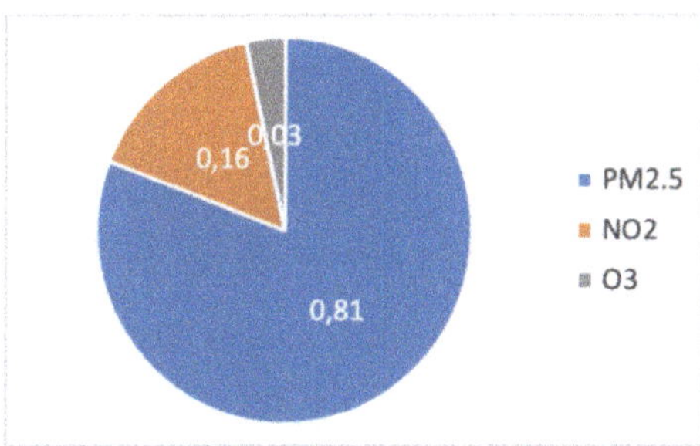

Figure 34. Repartition of air pollution in terms of premature deaths due to $PM_{2.5}$, NO_2, and O_3. for EU-28 in 2015.

Premature deaths were essentially due to $PM_{2.5}$, NO_2, and O_3, in the proportions of 81%, 16%, and 3% in EU-28 in 2015, i.e., 483,400 deaths, of which 391,000 are attributed to microscopic particulate matter [90].

Allocating part of these premature deaths to the steel sector is difficult due to a lack of fine-grained data and of relevant literature. More generally, this matter of premature deaths as related to air pollution should be examined at the scale of industrial sectors or activities like transport, whereas WHO has been carrying it out at an aggregated world level.

There is, however, a troth of literature devoted to emissions from steel mills, people's impregnation by pollutants, and health. However, even though all contribute some relevant information, none establish causal relationships between emissions and health, beyond undocumented statements, militant guesses, rumors, and even plain fake news. The difficulty is that a steel mill is usually part of an industrial complex, which generates a variety of emissions that aggregate to cause local pollution. This was analyzed in detail, for example, in the case of the Vitória metropolitan area (cf. Section 4.5). The literature comprises reports of militant organizations (NGOs) [91], their counterpart from steel business, i.e., CSR reports [92], special reports on local case studies in Taranto, Italy [93,94] and Fos-sur-Mer, France [95,96], and more.

In the future, when all society stakeholders will continue to seek reduced emissions of particulate matter, the steel sector will have to participate in that effort [97]. This will call on new investment and, probably, on the development of new measuring devices and new abatement technologies.

Reviews will be needed, but probably also new studies, to clarify whatever data and knowledge are available today.

8. Conclusions

This review paper has looked at steel, and particularly steel production, as it is related to the whole spectrum of environmental issues. This was done in two steps.

The first step consisted in taking stock of the fact that the main issues have been properly studied, understood, and kept under control:

- Iron resources are abundant, both primary and secondary raw materials: indeed, iron ore resources and reserves are plentiful, and steel is fully active in the circular economy, being the most recycled material.
- Energy consumption is already lean, an evolution driven by the high energy intensity of the sector and the high prices of energy. Change will take place by incorporating more renewables in the sector's energy mix, implementing the energy transition in an original way (electrification, CCUS, and/or use of green hydrogen), and codeveloping lean, frugal, and more durable product solutions with steel users, as well as switching to PSSs.
- Emissions to air, water, and soil have been curbed and most of them have started to decrease, at least in countries with advanced state-of-the-art steel mills. Among all air emissions, rogue emissions of particulate matter (coarse PM_{10}, fine $PM_{2.5}$, and ultrafine nanoparticles) probably still need to be better measured and better captured than it is yet the case today.
- We did not linger on the looming major biodiversity extinction, because the connection of this phenomenon with a particular metal like steel is only indirect, through its contribution to climate change and to urbanization of the anthroposphere.
- Steel is responsible for a sizeable chunk of greenhouse gas emissions, generating roughly twice as much CO_2 as the quantity of steel produced. However, the sector started to explore radical solutions for cutting emissions early, and a large number of them have been identified and tested at some intermediary scale, laboratory, pilot, or demonstrator.

The second step consisted in exploring the remaining open and unsolved issues:

- Regarding raw materials, mine tailings are still mostly out of control, with too many dams failing regularly across the world, thus creating major industrial disasters. All metals bear a similar responsibility.
- Emissions, particularly air emissions, still constitute a major problem in and around steel mills. The main remaining issue is particulate matter, which is generated by most human activities but also and significantly by steel: out of the 10 largest polluters in Europe, 7 are steel mills! The key reason for worrying about PM is the number of premature deaths that it causes.
- The third major task is to arrive at practical solutions to reach carbon neutrality by 2050, a commitment that all stakeholders in Europe have made and that they have to materialize.

Today, in the first quarter of the 21st century, the world has abandoned the irenic view that progress can easily improve the standard of living of most people on Earth and, at the same time, preserve nature and leave the environment intact for future generations. In the case of steel, this means aggressively addressing the major issues that are still open, and finding practical and working process solutions. This will mean radically redesigning steel production with even more demanding environmental targets in mind.

It is no longer possible today to treat environmental issues at the margin, like in an extra chapter of a metallurgy treatise. They need to be taken on board at the onset of any major technical action.

Funding: This research received no external funding.

Conflicts of Interest: The author declares no conflict of interest.

Glossary

AP	acidification potential
BC	Black Carbon
CCS	Carbon capture and storage
CCU	Carbon capture and usage
CCUS	Carbon capture, usage and storage
COURSE 50	CO_2 Ultimate Reduction in Steelmaking process by innovative technology for cool Earth 50 (Japan)
EAF	Electric Arc Furnace
EEA	European Environment Agency (EU)
EP	eutrophication potential
GHG	Greenhouse gases
GWP	Global Warming potential
HCB	Hexachlorobenzene
HM	Heavy Metal
HRC	Hot-rolled coil
IRSID	Institut de recherches de la sidérurgie française (France)
IPBES	Intergovernmental Science-Policy Platform on Biodiversity and Ecosystem Services (UN)
ISF	Imperial Smelting Furnace (zinc blast furnace)
LCA	Life Cycle Assessment or Life Cycle Analysis
MIT	Massachusetts Institute of Technology (US)
Mg	megagram = ton
NEDO	New Energy and Industrial Technology Development Organization (Japan)
NIH	National Cancer Institute (US)
NMVOC	Non-methane volatile organic compound
PAHs	Poly-Aromatic Hydrocarbon
PIA	Programme d'investissements d'avenir (France)
PCB	Polychlorinated biphenyl
PCDD	polychlorodibenzodioxin
PCDF	polychlorodibenzofuran
PCDF	polychlorodibenzofuran
PED	Primary energy demand
PM	Particulate Matter
PM_{10}	Particulate Matter with dimensions less than 10 μm
$PM_{2.5}$	Particulate Matter with dimensions less than 2.5 μm
POCP	photochemical ozone creation
POP	persistent organic pollutant
PSS	Product Service System
PUR	polyurethane
SGPI	Secrétariat général pour l'investissement (France)
TSP	Total Suspended Particles (<100 μm)
ULCOS	Ultra-LOw CO_2 Steelmaking
USGS	United States Geological Survey
VOC	Volatile Organic Compound
WHO	World Health Organization, a specialized agency of the UN (*OMS* in French)

References

1. Birat, J.-P.; Declich, A.; Belboom, S.; Fick, G.; Thomas, J.-S.; Chiappini, M. Society and Materials, a series of regular seminars based on a dialogue between soft and hard sciences. *Metall. Res. Technol.* **2015**, *112*, 501.
2. Birat, J.-P. Musica Universalis or the Music of the Spheres, 2018. *Matériaux Tech.* **2017**, *105*, 509. [CrossRef]
3. Birat, J.-P. Product innovations of key economic importance for the steel industry. *Metall. Res. Technol.* **2018**, *115*, 420. [CrossRef]
4. *Mineral Commodity Summaries 2019*; National Minerals Information Center, USGS, February 2019; p. 89. Available online: https://prd-wret.s3-us-west-2.amazonaws.com/assets/palladium/production/atoms/files/mcs2019_all.pdf (accessed on 29 February 2020).

5. Birat, J.-P.; Thomas, J.-S.; Brimacombe, L.; Colla, V.; Schmidt, D.; Verniory, F.; Linsley, K.; Oliver, S.; Lopes, E.M.D.; Pistelli, I.; et al. *The Sustainable Use of Resources in the European Steel Industry: A Roadmap PrePared by WP4/Planet of ESTEP*; ESTEP: Brussels, Belgium, 2008.

6. Birat, J.-P. Resources, materials and primary raw materials, chapter 7. In *Sustainable Materials Science; Environmental Metallurgy, Origins, Basics, Resources & Energy Needs*; EDP Sciences: Les Ulis, France, 2020; Volume 1, p. 459.

7. Source: Worldsteel, WORLD STEEL IN FIGURES 2019. Available online: https://www.worldsteel.org/en/dam/jcr:96d7a585-e6b2-4d63-b943-4cd9ab621a91/World%2520Steel%2520in%2520Figures%25202019.pdf (accessed on 29 February 2020).

8. CRM, Critical Raw Materials. Available online: http://criticalrawmaterials.org/coking-coal/ (accessed on 29 February 2020).

9. Mancini, L.; de Camillis, C.; Pennington, D. *Security of Supply and Scarcity of Raw Materials—Towards a Methodological Framework for Sustainability Assessment*; EU publication, JRC81762, EUR 26086 EN; Publications Office of the European Union: Luxembourg, 2013.

10. Rundman, K. Material Flow in Industry, 42 Pages, 2004. Available online: https://www.academia.edu/5976780/Steel (accessed on 29 February 2020).

11. Paulina Concha Larrauri, ID and Upmanu, Chronology of major tailings dam failures, from 1960. *Environments* **2018**, *5*, 28. [CrossRef]

12. World Mine Tailings Failures—From 1915, Supporting Global Research in Tailings Failure Root Cause, Loss Prevention and Trend Analysis. Available online: https://worldminetailingsfailures.org/ (accessed on 29 February 2020).

13. Benjamin, K.; Saleem, S.; Ali, H.; Bazilian, M.; Radley, B.; Nemery, B.; Okatz, J.; Mulvaney, D. Sustainable minerals and metals for a low-carbon future. *Science* **2020**, *367*, 30–33.

14. Birat, J.-P.; Lorrain, J.-P.; de Lassat, Y. The "CO_2 Tool": CO_2 emissions & energy consumption of existing & breakthrough steelmaking routes. *Rev. Métall. Int. J. Metall.* **2009**, *106*, 325–336.

15. Iles, L. The Role of Metallurgy in Transforming Global Forests. *J. Archaeol. Method Theory* **2016**, *23*, 1219–1241. [CrossRef]

16. Birat, J.-P.; Lavelaine, H. *The Eco-Techno-Systemic Services that Steel Plays in the Energy System*; SCANMET V 12–15 June 2016; Swerea MEFOS: Luleå, Sweden, 2016.

17. Birat, J.-P. The special roles that metals and copper play in the energy system, chapter 1. In *Sustainable Materials Science; Environmental Metallurgy, Origins, Basics, Resources & Energy Needs*; EDP Sciences: Les Ulis, France, 2020; Volume 2, p. 459.

18. Kuenen, J.; Berdowski, J.; van der Most, P.; Slager, J.M.; Mulder, W.; Hlawiczka, S.; Fudala, J.; Bloos, J.P.; Verhoeve, P.; Quass, U.; et al. *EMEP/EEA Air Pollutant Emission Inventory Guidebook 2016*; European Environment Agency, Publication Office of the European Union: Luxembourg, 2016.

19. Remus, R.; Miguel; Monsonet, A.A.; Roudier, S.; Sancho, L.D. *Best Available Techniques (BAT). Reference Document for Iron and Steel Production Industrial Emissions Directive 2010/75/EU Integrated Pollution Prevention and Control*; EUR 25521EN; Publication Office of the European Union: Luxembourg, 2013.

20. Plickert, S. BAT in the Iron and Steel Industry. In Proceedings of the BAT Workshop, Berlin, Germany, 21 April 2015.

21. Huber, J.C.; Rocabois, P.; Faral, M.; Birat, J.P.; Patisson, F.; Ablitzer, D. La formation des poussières dans un réacteur sidérurgique. *Rev. Métall* **2001**, *98*, 399–410. [CrossRef]

22. De Wever, P.; Duranthon, F. *Histoire d'un Grain De Sable, EDP Sciences—Collection: Terre à Portée De Main*; EDP Sciences: Paris, France, 2015.

23. Nedar, L. *Dust Formation in a BOF Converter*; Steel Research: Weinheim, Germany, 1996; Volume 67, pp. 320–327. [CrossRef]

24. Huber, J.C.; Patisson, F.; Rocabois, P.; Birat, J.P.; Ablitzer, D. Some means to reduce emissions and improve the recovery of electric arc furnace dust by controlling the formation mechanisms. In Proceedings of the REWAS'99 Global Symposium on Recycling, Waste Treatment and Clean Technologies Proceedings, San Sebastian, Spain, 5–9 September 1999; pp. 1483–1492.

25. Huber, J.C.; Patisson, F.; Rocabois, P.; Faral, M.; Ablitzer, D. Formation of electric arc furnace dust and influence of the process on the recoverable part. In Proceedings of the ECCE'2, 2nd European Conference on Chemical Engineering Proceedings, Montpellier, France, 5–7 October 1999.

26. Huber, J.C.; Rocabois, P.; Faral, M.; Birat, J.P.; Patisson, F.; Ablitzer, D. The formation of EAF dust. In Proceedings of the 58th Electric Furnace Conference Proceedings, Orlando, FL, USA, 12–15 November 2000; pp. 171–181.

27. Arion, A.; Baronnet, F.; Birat, J.-P.; Faral, M.; Gonthier, S.; Huber, J.-C.; Marquaire, P.-M. Dust and organic compounds in the flue gas of a steelmaking furnace. In Proceedings of the (新製鋼プロセスフォーラムSSE研究成果-18), ISIJ Autumn Meeting, Nagoya, Japan, 1–3 October 2000; Research of SSE program in Shinseiko Process Forum-18, CAMP-ISIJ. Volume 13, p. 789.

28. Patisson, F.; Huber, J.C.; Faral, M.; Birat, J.P.; Sessiecq, P.; Ablitzer, D. Emission de Poussières Par éclatement de Bulles de Gaz Dans un Four électrique d'aciérie. In Proceedings of the 8ème Congrès de Génie des Procédés, Nancy, France, 17–19 October 2001; Lavoisier Technique et Documentation: Paris, France, 2001. Société française de Génie des Procédés n°88. Volume 15, pp. 43–50.

29. Guézennec, A.G.; Huber, J.C.; Patisson, F.; Sessiecq, P.; Birat, J.P.; Ablitzer, D. *Etude Expérimentale de L'éclatement de Bulles de Gaz à la Surface D'un Bain D'acier Liquid*; Matériaux: Tours, France, 2002.

30. Guezennec, A.G.; Huber, J.C.; Patisson, F.; Sessiecq, P.; Birat, J.P.; Ablitzer, D. Experimental Study of the Bubble Burst Phenomenon at the Surface of a Liquid Steel Bath. In Proceedings of the TMS Annual Meeting, San Diego, CA, USA, 2–6 March 2003; Schlesinger, M., Ed.; EPD Congress: Warrendale, PA, USA, 2003.

31. Guézennec, A.G.; Huber, J.C.; Patisson, F.; Sessiecq, P.; Birat, J.P.; Ablitzer, D. Etude de la formation de poussières dans les réacteurs sidérurgiques. In Proceedings of the Journées d'Automne de la SF2M, Paris, France, 28–29 October 2003.

32. Huber, J.C.; Patisson, F.; Sessiecq, P.; Birat, J.P.; Ablitzer, D. Dust Formation by Bubble-Burst Phenomenon at the Surface of a Liquid steel Bath. *ISIJ Int.* **2004**, *44*, 1328–1333.

33. Guézennec, A.G. La formation des Poussières de Four Électrique d'aciérie: De la Genèse des Particules à Leur Évolution Morphologique. Génie des Procédés. Ph.D. Thesis, Institut National Polytechnique de Lorraine—INPL, Nancy, France, 2004. Available online: https://tel.archives-ouvertes.fr/file/index/docid/370527/filename/Th_AGG.pdf (accessed on 29 February 2020).

34. Guézennec, A.; Huber, J.; Patisson, F.; Sessiecq, P.; Birat, J.; Ablitzer, D. Dust formation in electric arc furnace: Birth of the particles. *Powder Technol.* **2005**, *157*, 2–11. [CrossRef]

35. Guézennec, A.G.; Patisson, F.; Sessiecq, P.; Huber, J.C.; Ablitzer, D. Modelling Dust Emissions in Electric Arc Furnace (EAF) Fume Extraction System Process. In Proceedings of the 2006 Sohn Int'l Symposium, San Diego, CA, USA, 27–31 August 2006.

36. Kero, I.; Gradahl, S.; Tranell, G. Airborne Emissions from Si/FeSi Production. *JOM* **2017**, *69*. [CrossRef]

37. Birat, J.P.; Schneider, M.; Gros, B. Recycling of steel and recycling of zinc: A steel producer's viewpoint. *Rev. Met. Paris.* **1998**, *95*, 721–730. [CrossRef]

38. Jackson, K.; Aries, E.; Fisher, R.; Anderson, D.R.; Parris, A. Assessment of exposure to PCDD/F, PCB, and PAH at a basic oxygen steelmaking (BOS) and an iron ore sintering plant in the UK. *Ann. Occup. Hyg.* **2012**, *56*, 12.

39. Oberlé, C. Etude de la Dégradation de Matériaux Organiques Associés Aux Ferrailles Automobiles. Ph.D. Thesis, Université de Haute Alsace, Mulhouse, France, 1 December 1995.

40. Arion, A. Etude de la Dégradation Thermique des Comatériaux Organiques Associés aux Ferrailles Automobiles. Ph.D. Thesis, INPL, Nancy, France, 11 December 1998.

41. Birat, J.-P.; Arion, A.; Faral, M.; Baronnet, F.; Marquaire, P.-M.; Rambaud, P. Maîtrise des émissions de Produits Organiques Dans Les Fumées du Four électrique à arc. *La Revue de Métallurgie-CIT* **2001**, *98*, 839–854. [CrossRef]

42. Arion, A.; Baronnet, F.; Lartiges, S.; Birat, J.P. Characterization of emissions during the heating of tyre-contamined scrap. *Chemosphere* **2001**, *42*, 853–859. [CrossRef]

43. Arion, A.; Birat, J.P.; Marquaire, P.-M.; Baronnet, F. Abatement of dioxin emissions in electric arc furnace exhaust flue gas. *Organohalogen Compd.* **2002**, *56*, 345–348.

44. Dioxin Facts, a Website Representing Chlorine Producers in the US. Available online: http://www.dioxinfacts.org/sources_trends/the_way.html (accessed on 31 July 2017).

45. De Souza, P.A.; de Quieroz, R.S.; Morimoto, T.; Guimarães, A.F.; Garg, V.K. Air pollution investigation in Vitória Metropolitan Region, ES, Brazil. *J. Radioanal. Nucl. Chem.* **2000**, *246*, 85–90. [CrossRef]

46. Birat, J.P. Life Cycle Assessment (LCA) and related methodologies, part of chapter XII. In *Treatise on Process Metallurgy*; Elsevier: Amsterdam, The Netherlands, 2013; p. 25.

47. Birat, J.-P. Materials, Secondary Raw Materials and the Circular Economy, chapter 8. In *Sustainable Materials Science*; Environmental Metallurgy; EDP Science: Les Ulis, France, 2020; Volume 1, p. 459.

48. Birat, J.-P.; Antoine, M.; Dubs, A.; Gaye, H.; de Lassat, Y.; Nicolle, R.; Roth, J.L. Vers une sidérurgie sans carbone? *Rev. Métall.* **1993**, *90*, 411–421. [CrossRef]

49. Birat J, P. Selected papers from 4th ULCOS SEMINAR (Part 1 and 2). *Metall. Res. Technol.* **2009**, *106*, 317.

50. Birat, J.-P.; Lorrain, J.-P. The "Cost Tool": Operating and capital costs of existing and breakthrough routes in a future studies framework. *La Rev. Métall.* **2009**, *106*, 337–349. [CrossRef]

51. Available online: https://ec.europa.eu/clima/policies/lowcarbon/ner300_en (accessed on 29 February 2020).

52. Available online: https://www.ademe.fr/sites/default/files/assets/documents/valorco.pdf (accessed on 29 February 2020).

53. Available online: https://www.usinenouvelle.com/article/europlasma-aide-arcelormittal-a-arreter-de-fumer. N665359 (accessed on 29 February 2020).

54. Development of New MethodologieS for InDustrial CO_2-FreE Steel PRoduction by ElectroWINning, H2020 SPIRE Project. Available online: https://www.siderwin-spire.eu (accessed on 29 February 2020).

55. Lavelaine, H.; ΣIDERWIN partners. ΣIDERWIN Project: Electrification of Primary Steel Production for Direct CO_2 Emission Avoidance. In Proceedings of the METEC & 4th ESTAD, Düsseldorf, Germany, 24–28 June 2019.

56. H2020, "Advanced CO_2 Capture Technologies"—Call ID: LC-SC3-NZE-1-2018. Available online: https://ec.europa.eu/info/funding-tenders/opportunities/portal/screen/opportunities/topic-details/lc-sc3-nze-1-2018 (accessed on 28 February 2020).

57. Broutin, P.; Carlier, V.; Grancher, F.; Bertucci, S.; Allemand, J.P.; Coulombet, D.; Valette, P.; Baron, H.; Laborie, G.; Normand, L. Pilot Plant in Dunkirk for Demonstrating the DMX^{TM} Process. In Proceedings of the 14th International Conference on Greenhouse Gas Control Technologies, GHGT-14, Melbourne, Australia, 21–25 October 2018.

58. Broutin, P.; Petetin, B.; Courtial, X.; Lacroix, M. 3D: A H2020 Project for the Demonstration of the DMX Process. In Proceedings of the 5th Post Combustion Capture Conference PCCC-5, Kyoto, Japan, 17–19 September 2019. organized by IEA-GHG and hosted by RITE.

59. Available online: https://northernlightsccs.eu (accessed on 28 February 2020).

60. Available online: https://hamburg.arcelormittal.com/icc/arcelor-hamburg-de/broker.jsp?uMen=f0e10ffc-365a-0e51-a18f-7ff407d7b2f2&uCon=3a860d3f-5001-3d61-ca98-3d0632940ec7&uTem=aaaaaaaa-aaaa-aaaa-aaaa-000000000042 (accessed on 28 February 2020).

61. Duarte, J. *Trends in Hydrogen Steelmaking*; Steel Times International: Surrey, UK, 2020; Volume 44, No 1; pp. 35–39.

62. The European Green Deal. *Communication from the Commission*; EUROPEAN COMMISSION: Brussels, Belgium, 2019; COM (2019) 640 Final.

63. Available online: https://mrl.mit.edu/index.php/156-doe-awards-2-15-million-to-boston-metal (accessed on 28 February 2020).

64. Sohn, H.Y.; Mohassab, Y.; Elzohiery, M.; Fan, D.Q.; Abdelghany, A. Status of the Development of Flash Ironmaking Technology. In *Applications of Process Engineering Principles in Materials Processing, Energy and Environmental Technologies*; Wang, S., Free, M., Alam, S., Zhang, M., Taylor, P., Eds.; The Minerals, Metals & Materials Series; Springer: Cham, Switzerland, 2017. [CrossRef]

65. *Course50, Breaking the CO_2 Reduction Deadlock*; Highlight Japan, 12–13 September 2016. Available online: http://dwl.gov-online.go.jp/video/cao/dl/public_html/gov/pdf/hlj/20160901/12-13.pdf (accessed on 28 February 2020).

66. *JISF Long-Term Vision for Climate Change Mitigation A Challenge Towards Zero-Carbon Steel*; JISF Press Release, September 2019. Available online: https://www.jisf.or.jp/en/activity/climate/documents/JISFLong-termvisionforclimatechangemitigation_text.pdf (accessed on 28 February 2020).

67. Available online: http://www.gdccus.org/en/ (accessed on 28 February 2020).

68. Jahanshahi, S.; Mathieson, J.G.; Somerville, M.A.; Haque, N.; Norgate, T.E.; Deev, A.; Pan, Y.; Xie, D.; Ridgeway, P.; Zulli, P. Development of Low-Emission Integrated Steelmaking Process. *J. Sustain. Metall.* **2015**, *1*, 94–114. [CrossRef]

69. Available online: https://www.thyssenkrupp.com/en/carbon2chem/ (accessed on 28 February 2019).

70. Available online: https://www.carbon4pur.eu (accessed on 28 February 2020).

71. Available online: http://www.steelanol.eu/en (accessed on 28 February 2020).

72. Available online: https://www.carbon4pur.eu/news-and-events/spire-projects-on-the-utilisation-of-co2-and-co/ (accessed on 28 February 2020).

73. Available online: http://ner400.com (accessed on 28 February 2020).

74. Kick-off Workshop for the IEA Global Iron & Steel Technology Roadmap. 20 November 2017. Available online: https://www.iea.org/workshops/kick-off-workshop-for-the-iea-global-iron--steel-technology-roadmap-.html (accessed on 28 February 2020).

75. Rifkin, J. *The Hydrogen Economy: The Creation of the Worldwide Energy Web and the Redistribution of Power on Earth*; Penguin Publishing Group: Oxford, UK, 2002; Volume 304.

76. Bongaarts, J. *IPBES (2019): Summary for policymakers of the global assessment report on biodiversity and ecosystem services of the Intergovernmental Science-Policy Platform on Biodiversity and Ecosystem Services*; Díaz, S., Settele, J., Brondízio, E.S., Ngo, H.T., Guèze, M., Agard, J., Arneth, A., Balvanera, P., Brauman, K.A., Butchart, S.H.M., et al., Eds.; IPBES Secretariat: Bonn, Germany, 2019.

77. Peters, K.; Colla, V.; Moonen, A.C.; Branca, T.A.; del Moretto, D.; Ragaglini, G.; Delmiro, V.M.M.; Romaniello, L.; Carler, S.; Hodges, J.; et al. Steel and Biodiversity: A Promising Alliance. *Matériaux Tech.* **2017**, *105*, 5–6. [CrossRef]

78. Emsley, J. *Nature's Building Blocks, an A-Z Guide to the Elements*; Oxford University Press: Oxford, UK, 2001.

79. *See for Example the Journal Environmental Health*; Springer Nature: London, UK. Available online: https://ehjournal.biomedcentral.com (accessed on 28 February 2020).

80. Birat, J.-P.; Zaoui, A. Le Cycle du Fer ou le recyclage durable de l'acier. *Rev. Métall. Cah. Inf. Tech.* **2002**, *99*, 795–807. [CrossRef]

81. Merian, E.; Anke, M.; Ihnat, M.; Stoeppler, M. *Elements and their Ccompounds in the Environment: Occurrence, Analysis and Biological Relevance*, 2nd ed.; Wiley-VCH: Berlin, Germany, 2008.

82. *Air Pollutant Emissions Data Viewer (Gothenburg Protocol, LRTAP Convention) 1990–2017*; European Environment Agency, EEA, Publications Office of the European Union: Luxembourg. Available online: https://www.eea.europa.eu/data-and-maps/dashboards/air-pollutant-emissions-data-viewer-2#tab-based-on-data (accessed on 28 February 2020).

83. Gunnar, F.; Bruce, N.; Fowler, A.; Nordberg, M. *Handbook on the Toxicology of Metals*, 4th ed.; Elsevier: Amsterdam, The Netherlands, 2015; Volumes 1–2.

84. The ILO Encyclopaedia of Occupational Health and Safety, Workplace Health and Safety Information. Available online: http://www.ilocis.org/en/contilo.html (accessed on 28 February 2020).

85. Klimont, Z.; Kupiainen, K.; Heyes, C.; Purohit, P.; Cofala, J.; Rafaj, P.; Borken-Kleefeld, J.; Schöpp, W. Contributions to cities' ambient particulate matter (PM): A systematic review of local source contributions at global level. *Atmos. Chem. Phys. Discuss.* **2016**. [CrossRef]

86. *Air Quality in Europe—2018 Report*; European Environment Agency, EEA, Publications Office of the European Union: Luxembourg, 2018; ISSN 1977–8449.

87. *Air Quality in Europe—2017 Report*; European Environment Agency, EEA, Publications Office of the European Union: Luxembourg, 2017; p. 22.

88. *Releases of Pollutants to the Environment from Europe's Industrial Sector—2015*; Briefing Published; EEA: Luxembourg, 2017; Last modified 10 August 2018.

89. World Health Organization. *Ambient Air Pollution: A Global Assessment of Exposure and Burden of Disease*; World Health Organization (WHO), Publications of WHO: Geneva, Switzerland, 2016; ISBN 978-92-4-151-135-3.

90. Mortalité Prématurée Imputable à L'exposition Aux Particules Fines (PM2.5), à l'ozone (O_3) et au Dioxyde d'azote (NO_2) en 2012 Dans 40 Pays Européens et Dans L'Ensemble de l'UE, 21/04/2016, European Environnement Agency, Publications Office of the European Union, Luxembourg. Available online: https://www.eea.europa.eu/fr/pressroom/newsreleases/de-nombreux-europeens-restent-exposes/mortalite-prematuree-imputable-a-la (accessed on 28 February 2020).

91. Peek, B.; Sadykova, D.; Urbaniak, D.; Jan Haverkamp; Jan Šrytr; Gallop, P.; Dubey, S. ArcelorMittal: Going Nowhere Slowly, A Review of the Global Steel Giant's Environmental and Social Impacts in 2008–2009, May 2009. Friends of the Earth Europe, CEE Bankwatch Network and Global Action on ArcelorMittal; Available online: https://bankwatch.org/publication/arcelormittal-going-nowhere-slowly-a-review-of-the-global-steel-giants-environmental-and-social-impacts-in-2008-2009 (accessed on 28 February 2020).

92. Integrated Annual Review 2018, Smarter, Circular, Lower-Emissions. This Is How We Are Inventing the Steels of the Future. Pamphlet, ArcelorMittal, 2019. Available online: https://annualreview2018.arcelormittal. com/~{}/media/Files/A/Arcelormittal-AR-2018/AM_IntegratedAnnualReview_2018.pdf (accessed on 28 February 2020).

93. Tonelli, F.; Short, S.W.; Taticchi, P. Case study of ILVA, Italy: The impact of failing to consider sustainability as a driver of business model evolution. In Proceedings of the 11th Global Conference on Sustainable Manufacturing—Innovative Solutions, Berlin, Germany, 23–25 September 2013; Seliger, G., Ed.; Univ.-Verl. der Technische Universität: Berlin, Germany, 2013.

94. Vagliasindi, G.M.; Gerstetter, C. *ILVA Industrial Site in Taranto, in-Depth Analysis*; European Parliament, October 2015; IP/A/ENVI/2015-13 PE 563.471; European Parliament: Brussels, Belgium, 2015. Available online: https: //www.europarl.europa.eu/RegData/etudes/IDAN/2015/563471/IPOL_IDA(2015)563471_EN.pdf (accessed on 28 February 2020).

95. Allen, B.L.; Lees, J.; Jeanjean, M.; Ferrier, Y.; Lecour, M.; Cohen, A.K. Synthèse du Rapport FOS EPSEAL—Étude Participative en Santé Environnement Ancrée Localement sur le Front Industriel de Fos-sur-Mer et Port-Saint-Louis-du-Rhône (volet 1)—La Commune de Saint-Martin-de-Crau (volet 2), juin. 2019. Available online: https://reporterre.net/IMG/pdf/synthe_se_finale_volet_2.pdf (accessed on 28 February 2020).

96. Goix, S.; Periot, M.; Douib, K. INDEX—Rapport d'étude. Etude d'imprégnation de la Population Aux Polluants Atmosphériques de la Zone Industrialo-Portuaire de Fos-sur-Mer. 2018. Available online: https:// www.researchgate.net/publication/325545879_INDEX_-_Rapport_d%27etude_Etude_d%27impregnation_ de_la_population_aux_polluants_atmospheriques_de_la_zone_industrialo-portuaire_de_Fos-sur-Mer (accessed on 28 February 2020). [CrossRef]

97. Yang, H.; Liu, J.; Jiang, K.; Meng, J.; Guan, D.; Xu, Y.; Tao, S. Multi-objective analysis of the co-mitigation of CO_2 and $PM_{2.5}$ pollution by China's iron and steel industry. *J. Clean. Prod.* **2018**, *185*, 331–341. [CrossRef]

Article

Review of the Energy Consumption and Production Structure of China's Steel Industry: Current Situation and Future Development

Kun He [1,2], Li Wang [1,3,*] and Xiaoyan Li [2]

[1] School of Energy and Environmental Engineering, University of Science and Technology Beijing, Beijing 100083, China; hekun0805@163.com
[2] China ENFI Engineering Corporation, Beijing 100038, China; lixiaoyan@enfi.com.cn
[3] Beijing Engineering Research Center for Energy Saving & Environmental Protection, Beijing 100083, China
*   Correspondence: liwang@me.ustb.edu.cn; Tel.: +10-6233-2566

Received: 17 January 2020; Accepted: 11 February 2020; Published: 26 February 2020

Abstract: China produced 49.2% of the world's total steel production in 2017. From 1990 to 2017, the world's total steel production increased by 850 Mt, of which 87% came from China. After 30 years of rapid expansion, China's steel industry is not expected to increase its production in the medium and long term. In fact, the industry is currently in the stage of industrial restructuring, and great changes will arise in production structure and technical level to solve pressing issues, such as overcapacity, high energy intensity (EI), and carbon emission. These changes will directly affect the global energy consumption and carbon emissions Thus, a review of China's steel industry is necessary to introduce its current situation and development plan. Therefore, this paper presents an overview of the Chinese steel industry, and factors involved include steel production, production structure, energy consumption, technical level, EI, carbon emission, scrap consumption, etc. In addition, four determinants are analyzed to explain the EI gap between China and the world's advanced level. In addition, comparison of steel industries between China and the world, development plans for energy savings, and emission reduction are also included in this paper to give readers a clear understanding of China's steel industry.

Keywords: steel industry; industrial restructuring; energy consumption; carbon emission; technology upgrade

1. Introduction

Steel production is an energy-, resource-, and pollution-intensive process [1,2]. China is currently the world's largest steel producer; indeed, the country's steel production accounted for 49.2% of the world's total steel production in 2017 [3]. The energy consumption of China's steel industry accounted for over 20% of the national industry energy consumption in 2017, and the CO_2 emissions from steel enterprises also accounted for over 10% of the country's total CO_2 emissions [4–6]. As such, improving the energy efficiency of steel production should be a primary concern for China, especially in times of high energy price volatility.

The rapid development of China's steel industry began in the 1990s. From 1990 to 2017, the world's total steel production increased by 850 Mt, of which 87% came from China [7]. Rapid expansion of production capacity has had generally positive effects on the energy efficiency of the industry, and the energy intensity (EI) of China's steel industry decreased by 11.5% from 2006 to 2017 [8]. However, there is still a gap between the EI of China's steel industry and the world's advanced level.

In 2018, the production of China's steel industry increased to 928 Mt. After years of rapid expansion, China's steel industry is currently at the end of the stage of production growth. In fact, the industry is facing numerous pressures, such as overcapacity and resources, energy, and environmental issues. These issues must be solved under the condition of ensuring an adequate steel supply, and this is a challenge for the Chinese government.

There are various energy-efficiency opportunities that exist in China, and many existing research works on the steel industry from a technical point of view are available in the literature, and the energy saving potential of China has been assessed in scientific papers [9–15]. However, currently, a comprehensive review of China's steel industry is still necessary to give readers a clear understanding of the present situation and development plan to realize the production structure adjustment and technical level upgrade.

The work presented in this paper is a unique study for the steel industry, as an extensive review of China's steel industry was conducted in this study. This paper specifically discusses (1) the development and present situation of steel production and consumption in China; (2) the implementation rate of major energy-saving technologies, gas recovery and utilization, and secondary energy generation in key enterprises in the country; (3) the development of the overall energy intensity (EI), specific process EIs, and the EI gap in key steel enterprises between China and the world; and (4) the carbon emissions of China's steel production and its main sources.

This paper also presents an analysis of the reasons behind the EI gap between China's and the world's steel industries. The factors considered include scrap ratio (SR), production structure, energy structure, and industrial concentration. In addition, development plans for major energy conservation in China's steel industry are also introduced and analyzed. These plans include eliminating backward production capacity, developing and implementing energy-saving technologies, and adjusting production structures by increasing scrap consumption in steel production.

We hope this study could be a useful reference for global policy makers, researchers, and industrial energy users, and be helpful for energy conservation and emission reduction work of China's steel industry.

2. Development and Present Situation of China's Steel Industry

2.1. Steel Production and Consumption

The rapid development of China's steel industry began in the 1990s, and the rise of this industry has had an important impact on the development of the global steel production (Figure 1). China became the world's largest steel producer in 1996 and has retained this status thus far. From 1990 to 2017, the world's total steel production increased by 850 Mt, of which 87% came from China [7].

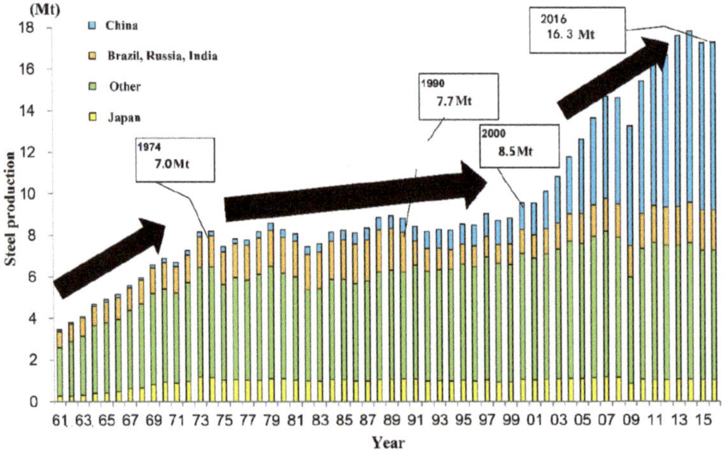

Figure 1. Development of global steel production.

In 2017, the world's total steel production was 1689.4 Mt (Figure 2), and the top 10 steel producers included China (831.7 Mt), Japan (104.7 Mt), India (101.4 Mt), the United States, (81.6 Mt), Russia (71.3 Mt), South Korea (71.0 Mt), Germany (43.4 Mt), Turkey (37.5 Mt), Brazil (34.4 Mt), and Italy (24.1 Mt) [16].

Country	2017	
	Rank	Poduction(Mt)
China	1	831.7
Japan	2	104.7
India	3	101.4
United States	4	81.6
Russia	5	71.3
South Korea	6	71.0
Germany	7	43.4
Turkey	8	37.5
Brazil	9	34.4
Italy	10	24.1
Taiwan, China	11	22.4
Ukraine	12	21.3
Iran	13	21.2
Mexico	14	19.9
France	15	15.5
World		1689.4

Figure 2. Top 15 global steel producers.

As the world's largest steel producer, China is also the world's largest steel consumer. In 2017, China's steel consumption accounted for 46.4% of the world's total steel consumption; by comparison, European Union (EU) countries and North American Free Trade Agreement (NAFTA) countries accounted for 10.2% and 8.9% of the world's total steel consumption, respectively [16]. In addition, Japan accounted for 4.1% of the world's total steel consumption, and other Asian countries consumed 15.9% of the world's total steel (Figure 3).

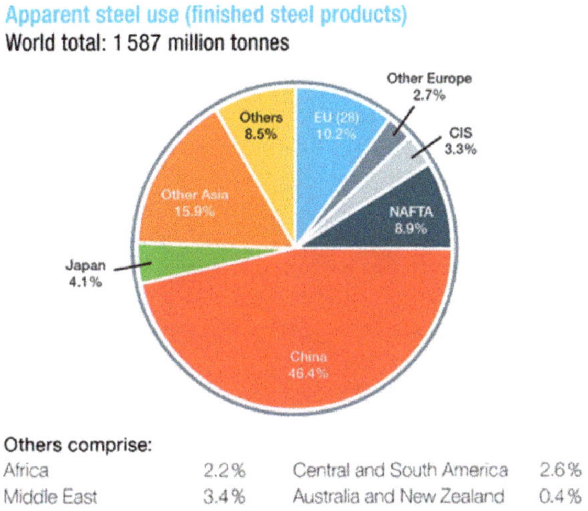

Figure 3. Global steel consumption in 2017 [16], with permission from World Steel Association 2019.

2.2. Production Route

Iron and steel industry has a complex structure. However, only a limited number of routes are applied worldwide, and these production routes use similar energy resources and raw materials (Figure 4). Globally, steel is produced via two main routes, namely, the blast furnace–basic oxygen furnace (BF–BOF) route and the electric arc furnace (EAF) route.

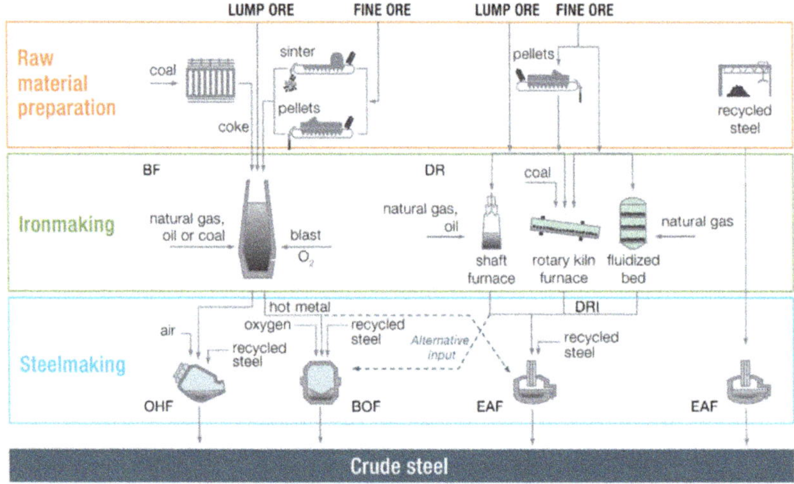

Figure 4. Steel production routes [17], with permission from World Steel Association 2012.

In the BF–BOF route, iron ore is first processed into iron, also known as molten or pig iron. Then, the molten iron is converted into steel in a converter. After refining, casting, and rolling, the steel is delivered in the form of steel plates, section steel, or steel bars.

The EAF route uses electricity to melt scrap steel in an electric arc furnace. Additives, such as alloys, can be added during steelmaking to adjust the required chemical composition of the steel, and oxygen can be injected into the EAF. The downstream processing stages of this route, such as casting, reheating, and rolling, are similar to those of the BF–BOF route.

The key difference between the BF–BOF and EAF routes is the type of raw materials consumed. The main raw material of the BF–BOF route is iron ore (generally accounting for 70%–100% of the total raw material); scrap, pig iron, and hot-pressed iron are also added. By comparison, the EAF route produces steel using mainly recycled steel (generally accounting for over 70% of the total raw material consumed) [18]. Depending on the plant configuration and availability of recycled steel, other sources of metallic iron, such as direct reduction iron (DRI) or hot metal, can also be used in the EAF route.

In 2017, the BF–BOF and EAF routes respectively accounted for 71.6% and 28.0% of the world's total steel production; another 0.4% of the world's steel production was derived from the open-hearth route [3]. China, Japan, Russia, Korea, Germany, Brazil, and Ukraine, as some of the world's major steel producers, use the BF–BOF route as their main route of steel production; the EAF route is used as the main mode of steel production in the United States, India, and Turkey (Table 1).

Table 1. Crude steel production by process in 2017.

Country	BF–BOF (%)	EAF (%)	Open-Hearth (%)
China	90.7	9.3	-
Japan	75.8	24.2	-
U.S.	31.6	68.4	-
India	44.2	55.8	-
Russia	66.9	30.7	2.4
South Korea	67.1	32.9	-
Germany	70.0	30.0	-
Turkey	30.8	69.2	-
Brazil	77.6	21.0	-
Ukraine	71.8	6.8	21.5
World	71.6	28.0	0.4

2.3. Production Technology Development

Given the rapid growth of China's steel production began in the 1990s, the country's energy consumption has also increased dramatically. Therefore, China attaches great importance to energy conservation in the steel industry. Implementing energy-saving technology is an effective way to reduce energy consumption in steel production. Over the past 30 years, China has made remarkable progress in the development of energy-saving technologies in steel production, and the technical indicators of China's steel industry have considerably improved.

2.3.1. Implementation Rates of Coke Dry Quenching and Top Pressure Recovery Turbine Technologies

Coking and blast furnace (BF) are the highest energy-consuming processes in steel production. Using coke dry quenching (CDQ) and top-pressure recovery turbine (TRT) technologies can effectively reduce the EI of coking and BF. In 2000, the implementation rates of CQD and TRT technologies of China' steel industry were only 12% and 14%, respectively (Table 2). After years of development, the implementation rates of CQD and TRT technologies increased to 90% and 99% in 2015, respectively, and the total number of CDQ units in China now exceeds 200 sets (processing capacity 25,000 t/h). Approximately 700 TRT-equipped BFs exist in China, of which 597 are gas dry dedusting equipment [19,20].

Table 2. Implementation rates of CDQ and TRT in China.

Year	CDQ	TRT
2000	12%	14%
2005	26%	74%
2010	85%	95%
2015	90%	99%

2.3.2. By-Product Gas Recovery and Utilization

By-product gas resources are the most important secondary energy resource in steel production. A large amount of energy can be saved by recycling and utilizing by-product gas resources. The recovery and utilization rates of China's key steel enterprises are relatively high (Table 3), and over 98% of the BF gas and coke oven gas produced was recycled, and converter gas recovery was 114 m^3/t in 2017 [19,20].

Table 3. Recovery and utilization of by-product gas in China's key steel enterprises.

Year	Utilization Rate of BF Gas (%)	Utilization Rate of Coke Oven Gas (%)	Converter Gas Recovery (m³/t)
2016	98.26	98.16	115
2017	98.34	98.77	114

2.3.3. Power Generation from Secondary Energy

Energy consumption for steel production only accounts for 30% of the total energy consumption, and the remaining 70% of the energy consumed is converted into various forms of waste heat and residual energy, such as by-product gas resources, the sensible heat of slag, and the waste heat of products [20,21]. These waste heat and residual energy resources can be used to preheat materials, generate steam or self-contained power plants, and generate power.

In 2017, power generation from secondary energy resources in China's key steel enterprises accounted for approximately 41.3% of the total electricity consumption (Figure 5), of which 57.8% originated from by-product gas, 16.2% from TRT, 11.7% from CDQ, 5.0% from sintering waste heat, and 9.3% from other secondary energy sources [19].

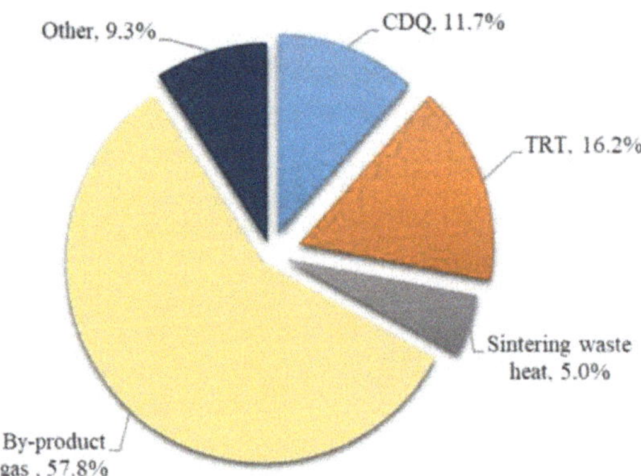

Figure 5. Power generation from secondary energy resources in China's key steel enterprises.

2.4. Energy Consumption

Steel enterprises use numerous energy evaluation indicators, such as comprehensive EI and comparable EI, as well as EI indicators for coking, sintering, pelletizing, ironmaking, steelmaking, and rolling process. In this section, the EI of China's steel industry is comprehensively described through an exhaustive analysis of various indicators.

2.4.1. Overall Energy Consumption

The comprehensive EI includes all forms of energy directly consumed by steel enterprises and their auxiliary production systems and the total amount of energy actually consumed by subsidiary production systems directly serving the production of steel enterprises [22]. The comprehensive EI is calculated as follows:

$$e_{\text{Comprehensive}} = \frac{E_i}{P},$$ (1)

where E_i is the energy consumption of the i category energy, kgce; P is the steel production, t.

From 2006 to 2017, the comprehensive EI of China's steel industry decreased from 645 kgce/t to 571 kgce/t (decrease by 11.5%) [19,23,24]. This reduction shows that China's steel enterprises have made remarkable progress in increasing their overall energy efficiency (Figure 6).

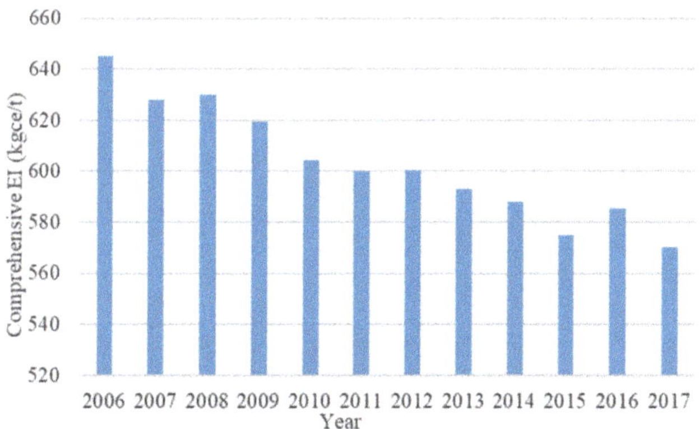

Figure 6. Comprehensive EI of China's steel industry.

2.4.2. EI of the Production Process

The EI of the production process reflects the energy consumption of the main production processes in steel production. From 2006 to 2017, the EI of the production process of China's key steel enterprises decreased dramatically: The EI of sintering, pelletizing, coking, BF, and steel processing respectively decreased by 12.8%, 20.1%, 19.0%, 9.6%, and 5.0%, respectively (Table 4) [19,23–25].

Table 4. The EI of production process of China's key steel enterprises. (Unit: kgce/t).

Year	Sintering	Pelletizing	Coking	BF	Converter	EAF	Processing
2006	55.61	33.08	123.11	433.08	9.09	81.26	64.98
2007	55.21	30.12	121.72	426.84	6.03	81.34	63.08
2008	55.49	30.49	119.97	427.72	5.74	81.52	59.58
2009	54.52	29.96	113.97	410.55	2.78	73.44	57.66
2010	52.65	29.39	105.89	407.76	−0.16	73.98	61.69
2011	54.34	29.60	106.65	404.07	−3.21	69.00	60.93
2012	50.60	28.75	102.72	401.82	−6.08	67.53	57.31
2013	49.76	28.58	99.87	399.88	−7.81	62.38	60.32
2014	49.48	27.12	98.15	388.70	−8.73	66.06	63.30
2015	48.53	26.72	99.66	384.43	−11.89	60.38	63.44
2016	47.78	26.16	97.46	387.75	−12.24	65.90	61.78
2017	48.49	26.17	99.67	391.37	−14.26	60.22	61.73

2.4.3. Comparison of EIs between Steel Industries in China and the World

Comparable EI is used to compare the energy consumption of steel production in different enterprises or countries; this parameter represents the sum of energy consumption in each production process [26]. The Comparable EI is calculated as:

$$e_{Comparable} = (1/P)\left(\sum P_i \times e_i + I + J + K\right),\qquad(2)$$

where P_i is the production of i process, t;

e_i is the average energy consumption of the i process, kgce/t product;

I, J, and K are the energy consumption for processing and transportation of fuel, energy consumption for locomotive transportation, and changes in enterprise energy stock, respectively.

According to the International Energy Agency (IEA) and the Research Institute of Innovative Technology for the Earth [27,28], Japan possesses the world's most energy-efficient steel industry. All steel mills in Japan use existing technologies with minimal potential for further energy-conservation measures.

China's steel industry has achieved good results in reducing energy consumption over the last 20 years (Figure 7). However, a wide gap remains between the steel industries of China and Japan [8].

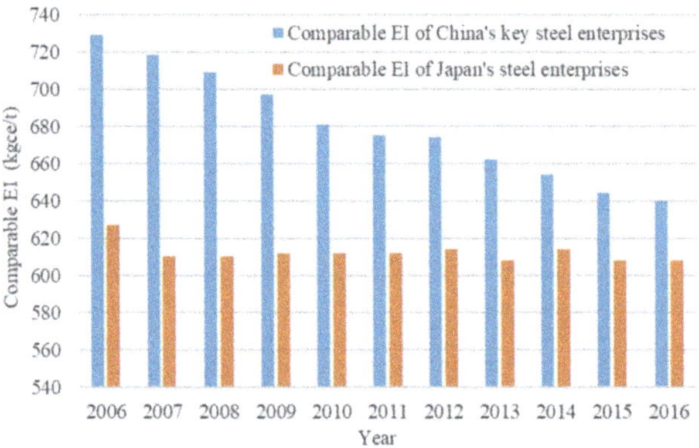

Figure 7. Comparable EI of steel enterprises in China and Japan.

2.5. CO_2 Emissions

Traditional steel production relies heavily on fossil fuels, such as coal and coke. Therefore, China's steel industry has become the second largest national emitter of CO_2 after the power industry. In 2015, CO_2 emitted by China's steel enterprises accounted for over 15% of the country's total CO_2 emissions [5,6].

CO_2 emission from steel enterprises is mainly caused by coal combustion. Coal accounts for approximately 80% of the energy consumption structure in China's steel enterprises [4]. According to statistics, CO_2 emissions from China's iron-making system (sintering, pelletizing, coking, and blast furnace) account for approximately 85% of the total emissions from the steel industry [29,30]. Therefore, reducing CO_2 emissions from the iron front process is imperative.

In 2015, at the Paris Climate Conference, the Chinese government proposed reducing CO_2 emissions per unit GDP by 65% compared with 2005 levels by 2030 and establishing a national carbon emissions trading market in 2017 [30]. The latter proposal will integrate eight industries, including electricity, steel, and cement, into one system to promote overall carbon emission reduction. Therefore, China's steel enterprises are facing the severe situation of CO_2 emission reduction.

Using scrap instead of iron ore can directly reduce the production of the iron-making system and drastically reduce energy consumption and CO_2 emissions. The EI of direct steel-making with scrap is only 30% that of the BF–BOF route. This finding indicates that using one ton of scrap in China's steel production can save 350 kgce and reduce CO_2 emissions by 1.4 tons [31].

At present, scrap consumption per ton of steel in China is far below the world's average level (specific description is in Section 3.1). In the future, China's recyclable scrap resources will increase substantially as the scrap consumed in previous decades gradually reaches their recycling cycle, and increasing the use of scrap steel in steel production will become an inevitable trend.

3. Comparison of Steel Industries between China and the World

Over the last two decades, China's steel industry has made remarkable achievements in reducing its EI by improving technology levels and promoting energy-saving technologies; these efforts have resulted in considerable reductions in the energy consumption of steel production. However, a gap still remains between the EIs of key enterprises between China and the world's advanced level. Therefore, in the future, reducing energy consumption will remain a key issue for China's steel industry. In this section, the main reasons behind this EI gap are analyzed by comparing steel production in China and other countries.

3.1. Difference in Scrap Ratio

Iron ore and scrap steel are the two main raw materials for steel production. Compared with that of iron ore, using scrap in steel production can save energy and resources by reducing the production of iron-making systems. *SR* is used to define scrap consumption in steel production:

$$\text{Scrap ratio} = \frac{\text{Scrap consumption}}{\text{steel production}} \tag{3}$$

Over the last 10 years, the SR of the global steel industry remained at 35%–40%, and is approximately 37% on average. Among the world's major steel-producing countries, the United States has the highest SR of 75%, which fluctuates considerably. The EU's SR is also high at approximately 55%–60%, South Korea's average is approximately 50%, and Japan's average is approximately 35% [32]. The SR of China's steel industry was only 11.2% in 2016. The consumption of scrap resources per ton of steel in China is considerably lower than the global average, and this gap has a negative impact on energy conservation because processing of iron ore into hot metal requires a large amount of energy and resources.

A low SR leads to the dependence of China's steel enterprises on iron ore as the main raw material. The SR and iron–steel ratios of key steel enterprises in China from 2006 to 2016 are calculated according to China Steel Yearbook [33]. As shown in this Figure 8, the scrap ratio of China's key enterprises declined from 2006 to 2016, and the corresponding iron–steel ratio showed an opposite increasing trend. This dependence explains why the iron–steel ratio of the country's iron-making system is higher than that of other countries.

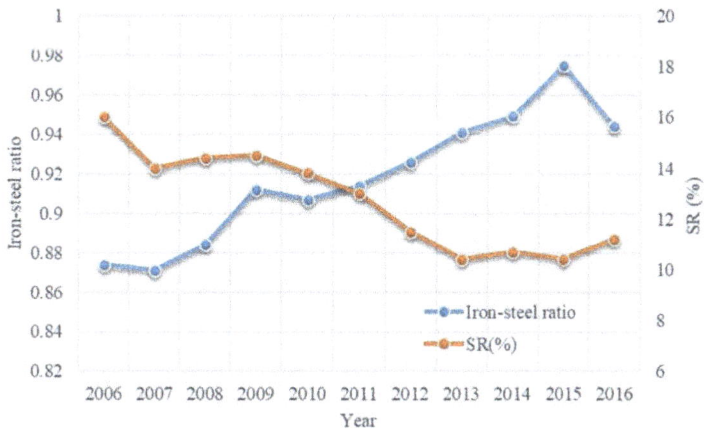

Figure 8. Scrap ratio and iron–steel ratio of China's key steel enterprises.

In 2016, China's iron–steel ratio was 0.867; by comparison, the global average iron–steel ratio was 0.734. The global average was only 0.573 after deducting China's iron–steel ratio [20]. The iron–steel ratios of the United States and Germany are 0.333 and 0.646, respectively. Generally, if the iron-steel ratio increases by 0.1, the comprehensive EI of steel production will increase by about 50 kgce/t, so the comprehensive EI of China is about 110–250 kgce/t higher than the advanced level just because of the high iron-steel ratio [20].

3.2. Differences in Production Structure

Energy consumption and pollutant emissions in steel enterprises are mainly concentrated in iron-making systems (from the iron ore entering the plant to coking, sintering, pelletizing, and ironmaking). Therefore, the energy consumption of the BF–BOF route is generally higher than that of the EAF route (Figure 9). Compared with the BF–BOF route, direct steelmaking with scrap via the EAF route can save approximately 60% of the energy expenditure to produce steel and reduce CO_2 emissions by 80% [34–37].

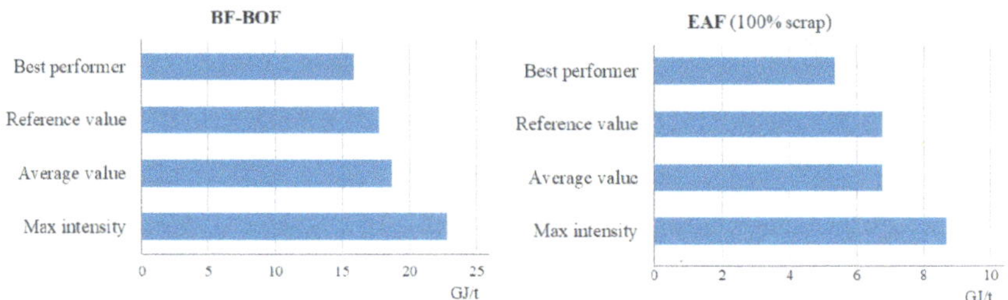

Figure 9. Energy intensity of the BF–BOF and EAF routes.

Increasing the use of the EAF route can reduce energy consumption in steel production. However, production through EAF is limited by the availability of scrap steel resources. The actual situation of scrap resources varies greatly in different countries and regions around the world. Some countries (regions) have abundant scrap resources and low prices; in this case, additional EAF steel plants can be built and additional scrap steel can be consumed in converters. For example, in the United States, the proportion of EAF steel accounts for over 60% of the total crude steel production. In some developing countries with insufficient scrap resources, the BF–BOF route remains the main mode of steel production. For example, in China, the EAF steel ratio has hovered at approximately 10% over the last few years [7].

Insufficient scrap storage is the main reason behind the low SR in China. Scrap recycling has a certain cycle, and China began to use a large number of steel products in 2000. Thus, a large gap in scrap resources exists in China. Therefore, over the last 30 years, China's steel production growth has mainly originated from the BF–BOF route, and the production of EAF steel has been stable due to the limitation of scrap quantity.

From 2000 to 2017, the production of China's BF–BOF route rose by over 800%; thus, the proportion of the EAF route in China has continuously declined (Figure 10). In 2017, China's BF–BOF production accounted for 90.7% of total steel production, while its EAF production accounted for only 9.3%, which is far below the world average level (28%) [3].

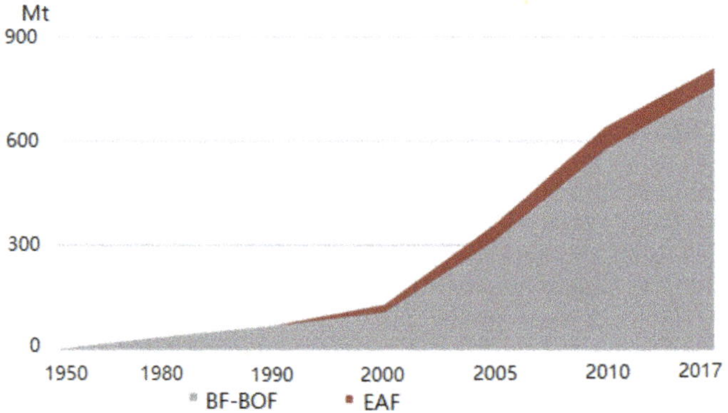

Figure 10. Development of the BF–BOF and EAF routes in China.

3.3. Differences in Energy Structure

According to IEA statistics, the energy consumption of the steel industry in 2017 accounted for 17% of the world's total industrial energy consumption. In terms of total energy consumption, coal is the main energy source (64% of the total energy consumption), followed by electricity (20% of the total energy consumption) and natural gas (11% of the total energy consumption) [4]. Oil contributes only 1% of the energy consumption. The remaining energy consumption is provided by other types of energy, such as biomass (Figure 11).

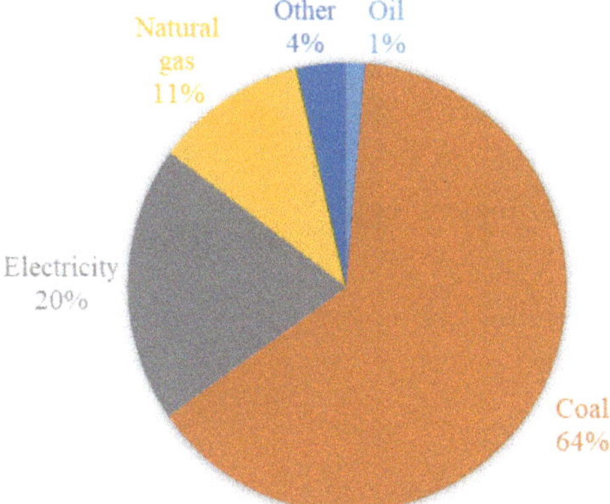

Figure 11. Energy structure of the world's steel industry.

The energy structures of different countries vary remarkably. For example, coal and natural gas account for 76% and 2% of energy consumption in China's steel industry, respectively [4]. By contrast, only 24% of the energy consumption of the United States comes from coal; 47% comes from natural gas (Figure 12). The energy structures of different steel-producing countries differ, and the industrial conversion efficiencies of different kinds of energy vary. Thus, differences in energy structure will have a certain impact on the energy efficiency of steel production.

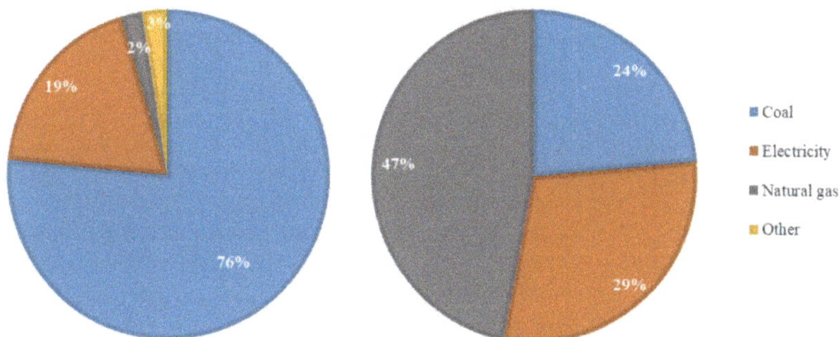

Figure 12. Energy structures of the steel industries of China (**left**) and the United States (**right**) in 2017.

In industrial production, the energy efficiency of natural gas is higher than that of coal regardless of their use in fuel or power generation, and the carbon emission of natural gas is lower than that of coal. Using 1 m^3 of natural gas can save 0.76–1.19 kgce and reduce carbon emissions by 3.33–5.01 g compared with the using coal [38]. At present, the main energy source of steel production in China is coal, and the proportion of natural gas in the energy consumption of China's steel industry is drastically lower than the world average; this situation is unfavorable for energy savings and carbon emission reduction.

3.4. Differences in Industrial Concentration

In China, EI varies among steel enterprises of different scales, and the EI of small steel enterprises is generally higher than that of key steel enterprises, due to small size production equipment being mostly used in small steel enterprises, which are of high production energy consumption [39]. In addition, the management and technological advantages of large-scale steel enterprises allow them to consume less energy than small-scale enterprises.

Over the past decade, the concentration of China's steel industry has shown a downward trend (Figure 13), with the concentration of the top 10 enterprises declining from 45% in 2001 to 36% in 2016. By contrast, in Japan, the concentration of the top five enterprises accounted for over 80% of the country's total steel production, and the large-scale production of steel industry has been achieved [7,40,41].

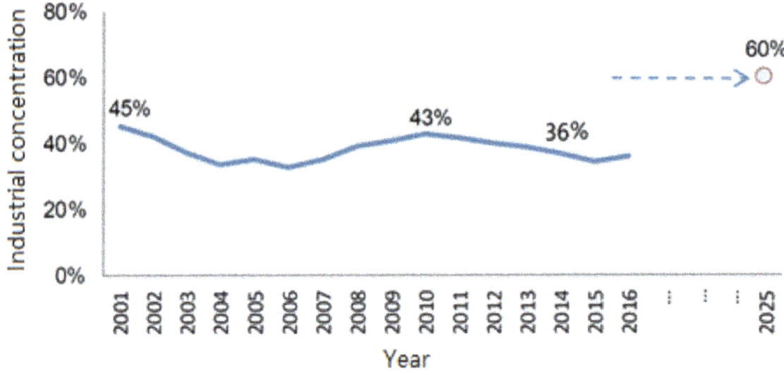

Figure 13. Industrial concentration of China's steel industry.

Increasing the production proportion of large-scale enterprises is a development trend in China's steel industry. According to "Adjustment Policy of Iron and Steel Industry" published in 2015, the concentration of the top 10 steel enterprises in China should not be less than 60% by 2025, and three or five super-large steel enterprise groups with strong competitiveness in the global scope should be formed.

4. Development Directions for Energy Savings and Emission Reduction in China's Steel Industry

After decades of rapid development, China's steel production has entered the peak arc region, and, in the medium and long term, overall steel production is not expected to increase. According to relevant plans published by the Chinese government, in the future, China's steel industry will focus on industrial restructuring to solve problems arising from previous rapid development stage. Eliminating backward production capacity (technological upgrading), promoting energy-saving technologies, and restructuring production are key directions for energy savings and emission reduction.

4.1. Eliminating Backward Production Capacity

Overcapacity, a common problem currently faced by the global steel industry, presents a very serious challenge to China. In 2015, the excess capacity of China's steel industry was 336.2 Mt, accounting for 46% of the global excess capacity [42]. At the same time, China also retains backward capacity, which affects the total energy consumption of steel production. Against this background, the Chinese government published "Opinions on the Iron and Steel Industry to Eliminate Overcapacity and Realize Development from Difficulties," which demands the following: the crude steel production capacity should be reduced by 10–150 Mt within five years from 2016, and future development should aim at industry merging and reorganization, industrial structure optimization, and resource utilization efficiency improvement.

China strictly enforces the energy conservation law and defines backward production capacity in accordance with process energy consumption. Steel production capacity that fails to meet mandatory standards, such as "Energy Consumption Limit for Products of Major Processing Units in Crude Steel Production", should be reformed and upgraded within six months (Table 5). The steel production involved in this effort is estimated to range from 10 Mt and 150 Mt; this amount will effectively improve the energy efficiency of China's steel industry.

Table 5. Energy consumption requirements for new and reformed steel enterprises [43].

Types of Enterprises	Coking	Sintering	Blast Furnace	Converter	Common Steel EAF	Special Steel EAF
New construction and transformation enterprises	≤122 (Top loading) ≤127 (Tamping)	≤50	≤370	≤−25	≤90	≤159
Existing enterprises	≤150 (Top loading) ≤155 (Tamping)	≤55	≤435 ≤485 (Vanadic titanomagnetite)	≤−10	≤92	≤171

China's steel industry eliminated backward production capacity by 65 Mt in 2016 and by 55 Mt in 2017. The productivity utilization rate has increased from approximately 70% in 2015 to over 85% in 2017 [44], as shown in Table 6.

Table 6. Effect of the elimination of backward productivity.

Parameter	2014	2015	2016	2017
Capacity (Mt)	1151	1134	1069	1019
Capacity reduction (Mt)	31.1	17.1	65	50
Capacity utilization rate (%)	71.5	70.9	75.6	86.65

4.2. Research and Promotion of Energy-Saving and Emission-Reduction Technologies

The promotion and application of energy-saving and secondary energy-recovery technologies in steel production has always been the focus of the energy-saving work of China's steel industry. After years of development, some major energy-saving technologies have been widely implemented in China. These technologies include sintering waste heat recovery, TRT power generation capacity, and converter gas recovery.

In the future, the progress of research and development of energy-saving technologies will further decrease the EI of steel production in China. Several domestic steel industry researchers have begun to actively pay attention to research on topics such as high-temperature and pressure dry quenching, heat recovery from coke oven riser waste, coal humidification, heat recycling from sintering waste gas waste, heat recovery from slag washing water waste, heat recovery from converter flue gas waste, comprehensive utilization of pure dry dust removal, and high-parameter gas-generating units [44] (Table 7).

Table 7. Major technological progress and their energy-saving effect in the future.

No.	Technology	Energy Saving Effect	Development Plan	Coal Saving (Mtce)
1	High temperature and pressure CDQ	40kgce/t coke	add more than 30 new equipments, involving 57 Mt of coke production capacity	2.28
2	Waste heat recovery of coke oven riser	90 kg steam (0.6 MPa)/t coke (equivalent to 9 kgce/t)	add more than 30 new waste heat recovery equipments, involving 57 Mt of coke production capacity	0.51
3	Coal moisture control	6 kgce/t coke	add more than 10 new equipments, involving 20 Mt of coke production capacity	0.12
4	Recycling of waste heat from sintering	4 kgce/t sinter	about 40 new units, involving 80 Mt of sintering production capacity	0.32
5	Waste heat recovery of slag washing water	40 ktce/heating cycle (1 million m^2 of heating area)	add more than 7000 m^2 heating area, of which 2300 m^2 have been added in 2016	2.80
6	Comprehensive utilization of waste heat from converter flue gas- pure dry dedusting	8 kgce/t steel	applied to 200 converters, involving 300 Mt of steel production	2.40
7	High parameter gas power generator unit	thermal efficiency increased by over 5%, 40 ktce compared with medium temperature and medium pressure unit (enterprises with steel production of 10 Mt/year)	applied to 135/65/54 MW high parameter gas generator units in 30 enterprises (steel production over 5 Mt/year), involving 250 Mt of steel production	1.00
		Total		9.43

4.3. Production Structure Transformation

A large amount of scrap steel resources accumulated by China is expected to reach their recycling cycle in recent years. Under these conditions, utilization of scrap steel could be expected to increase over the next few years.

According to estimations of recyclable period of steel products in China, during the 13th Five-Year Plan period, in addition to scrap vehicles, many bridges, houses and military equipment have reached the scrap period. This phenomenon will further increase China's scrap stock. The amount of recyclable scrap steel is expected to reach approximately 200 Mt in 2020, 272.2 Mt in 2025, and 346 Mt in 2030 [37] (Figure 14). Therefore, increasing the scrap ratio is an inevitable trend in the development of China's iron and steel industry.

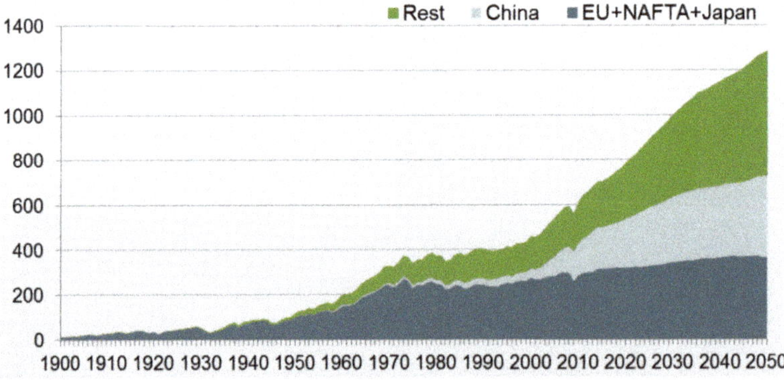

Figure 14. Amount of recyclable scrap resources [37], with permission from World Steel Association 2017.

The 13th Five-Year Plan is anticipated to be a major turning point for the scrap industry. Efforts to rationally utilize scrap steel resources, increase the proportion of EAF steel, and realize structural and energy savings will have tremendous potential for growth and development.

China's crude steel production has been in the downward zone of the peak arc, and steel production is not expected to increase in the medium and long term. Pig iron production shows the same characteristics, and, given the gradual increase in scrap resources, the average decline rate of pig iron production over the long run will be higher than that of crude steel. At present, over 65% of China's total iron production is produced by imported iron ore [33]. From a long-term perspective, the demand for iron ore will continue to decline [44] (Figure 15).

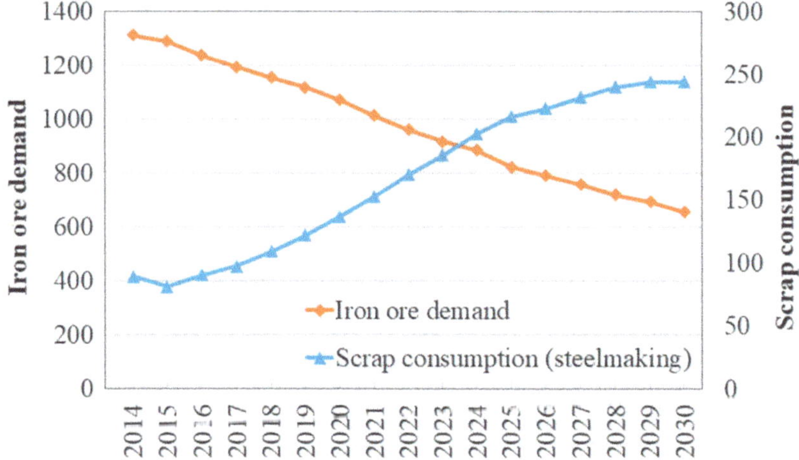

Figure 15. Trend of iron ore and scrap consumption by China's steel industry.

At present, 70% of energy consumption and 85% of CO_2 emissions of China's steel industry are concentrated in converting iron ore into molten iron [29,30]. The SR of China's steel industry hovering between 10% and 15% over the last decade, and it will be not less than 25% in 2025 according to the development plan of China's steel industry. This will make a significant contribution to reducing the energy consumption of China's steel industry.

5. Results and Discussion

5.1. Conclusions

This paper presents a comprehensive overview of the development and current situation of China's steel industry, including steel production, steel consumption, production structure development, energy-saving technology development, overall energy consumption, process energy consumption, and carbon emissions. The future development of the energy-saving work of China's steel industry is also analyzed.

At present, China's steel production is not expected to increase in the medium and long term. Pig iron production shows the same characteristics, and the demand for coke and iron ore will also continue to decline. Over the next few years, China can reduce its energy consumption and carbon emissions by eliminating backward production capacity (technological upgrading), implementing energy-saving technologies, increasing scrap consumption, and reducing the production of iron-making systems.

5.2. Policy Implications

5.2.1. Scientifically and Rationally Elimination of Backward Production Capacity

Scientific and reasonable methods should be used to achieve the capacity removal of China's steel industry. First, uneconomical and low value-added production capacity should be reduced. Second, illegal and irregular steel enterprises should be investigated and punished in accordance with the law. Third, industry regulations should be conscientiously implemented to eliminate backward production capacity. These regulations include "Standard Conditions for Iron and Steel Industry" (revised in 2015) and "Standardized Enterprise Management Measures for Iron and Steel Industry."

5.2.2. Improvement of the Production Technology Level

Special funds should be allocated to support the research and development of key energy-saving technologies. Energy savings and emission reduction in the steel industry should be treated as key aspects of national technological transformation. The government should increase their support for key special projects related to energy saving and emission reduction. For example, technology research should focus on key issues, such as insufficient power self-generation, low power generation efficiency, taxes and fees of power network, and the poor utilization of low-temperature waste heat in summer.

5.2.3. Normalization of the Scrap Steel Market

The consumption of scrap resources in steel production will dramatically increase as the availability of recyclable scrap in China increases. At present, China's scrap industry has four main problems: (1) the capacity of scrap processing only accounts for 30% of the social scrap volume; this value is far from the demand of the steel industry; (2) the quality of equipment available for scrap processing is low, and dismantling lines for automobiles has not yet been extensively established; (3) reliable statistical data for the classification of scrap resources and technical standards for the scrap processing products industry are lacking; and (4) the taxation of scrap import and distribution enterprises is not conducive to scrap recycling. Therefore, management of the scrap steel industry should be improved, and scientific processing and classified sales should be implemented to meet the needs of various users and provide strong support for the transformation of the production structure of the steel industry.

Author Contributions: K.H. and X.L. prepared and revised the manuscript under the supervision of L.W. All authors contributed to the general discussion. All authors have read and agreed to the published version of the manuscript.

Funding: This work is financially supported by the National Key R&D Program of China (Grant No. 2016YFB0601101).

Conflicts of Interest: The authors declare no conflict of interest.

References

1. Quadera, M.; Ahmed, S.; Ghazillaa, R.; Ahmed, S.; Dahari, M. A comprehensive review on energy efficient CO_2 breakthrough technologies for sustainable green iron and steel manufacturing. *Renew. Sustain. Energy Rev.* **2015**, *50*, 594–614. [CrossRef]
2. Mousa, E.; Wang, C.; Riesbeck, J.; Larsson, M. Biomass applications in iron and steel industry: An overview of challenges and opportunities. *Renew. Sustain. Energy Rev.* **2016**, *65*, 1247–1266. [CrossRef]
3. World Steel Association. *Steel Statistics Yearbook 2018*; World Steel Association: Belgium, Brussels, 2019.
4. International Energy Agency (IEA). Energy Balance Flows 2015. Available online: http://www.iea.org/Sankey/index.html (accessed on 11 January 2020).
5. Zhang, Q.; Xu, J.; Wang, Y.; Hasanbeigi, A.; Zhang, W.; Lu, H.; Arens, M. Comprehensive assessment of energy conservation and CO_2 emissions mitigation in China's iron and steel industry based on dynamic material flows. *Appl. Energy* **2018**, *209*, 251–265. [CrossRef]
6. Liu, H.; Fu, J.; Liu, S.; Xie, X.; Yang, X. Calculation methods and application of carbon dioxide emission during steel-making process. *Iron Steel* **2016**, *51*, 74–82.

7. World Steel Association. Statistics. 2018. Available online: https://www.worldsteel.org/steel-by-topic/statistics.html (accessed on 11 January 2020).

8. National Bureau of Statistics of China. *China Energy Statistics Yearbook 2016–2017*; China Statistics Press: Beijing, China, 2018.

9. Tan, X.; Li, H.; Guo, J.; Gu, B.; Zeng, Y. Energy-saving and emission-reduction technology selection and CO_2 emission reduction potential of China's iron and steel industry under energy substitution policy. *J. Clean. Prod.* **2019**, *222*, 823–834. [CrossRef]

10. Wang, X.; Wen, X.; Xie, C. An evaluation of technical progress and energy rebound effects in China's iron & steel industry. *Energy Policy* **2018**, *123*, 259–265.

11. Hasanbeigi, A.; Price, L.; Zhang, C.; Nathaniel, A.; Li, X.; Shangguan, F. Comparison of iron and steel production energy use and energy intensity in China and the U.S. *J. Clean. Prod.* **2014**, *65*, 108–119. [CrossRef]

12. Li, Y.; Zhu, L. Cost of energy saving and CO_2 emissions reduction in China's iron and steel sector. *Appl. Energy* **2014**, *130*, 603–616. [CrossRef]

13. Zhang, S.; Worrell, E.; Crijns, G.; Wagner, F.; Cofala, J. Co-benefits of energy efficiency improvement and air pollution abatement in the Chinese iron and steel industry. *Energy* **2014**, *78*, 333–345. [CrossRef]

14. Wen, Z.; Meng, F.; Chen, M. Estimates of the potential for energy conservation and CO_2 emissions mitigation based on Asian-Pacific Integrated Model (AIM): The case of the iron and steel industry in China. *J. Clean. Prod.* **2014**, *65*, 120–130. [CrossRef]

15. He, F.; Zhang, Q.; Lei, J.; Fu, W.; Xu, X. Energy efficiency and productivity change of China's iron and steel industry: Accounting for undesirable outputs. *Energy Policy* **2013**, *54*, 204–213. [CrossRef]

16. World Steel Association. *World Steel in Figures 2018*; World Steel Association: Belgium, Brussels, 2019.

17. World Steel Association. *Sustainable Steel at the Core of a Green Economy*; World Steel Association: Belgium, Brussels, 2012.

18. International Energy Agency. *Energy Technology Perspectives 2012*; International Energy Agency: Paris, France, 2012.

19. The Editorial Board of China Steel Yearbook. *China Steel Yearbook 2018*; Metallurgical Industry Press: Beijing, China, 2019.

20. Wang, W. Energy consumption and energy saving potential analysis of China's iron and steel industry. *Metall. Manag.* **2017**, *8*, 50–58.

21. Ding, Y.; Deming, S. High-efficiency utilization of waste heat at fully integrated steel plant. *Iron Steel* **2011**, *46*, 88–98.

22. Zhang, G. *Research and Application of Index System for Evaluating Energy Consumption in Iron and Steel Enterprises*; Northeastern University: Shenyang, China, 2013.

23. China Iron and Steel Industry Association. Energy-Saving and Emission-Reduction. 2018. Available online: http://www.chinaisa.org.cn/gxportal/login.jsp (accessed on 11 January 2020).

24. He, K.; Wang, L.; Zhu, H.; Ding, Y. Energy-Saving Potential of China's Steel Industry According to Its Development Plan. *Energies* **2018**, *11*, 948. [CrossRef]

25. Hao, Y. Study on Machanism and Model of Circular Economy in China's Iron and Steel Industry. Ph.D. Thesis, University of Science and Technology, Beijing, China, 2014.

26. National Bureau of Statistics. *Handbook of Energy Statistics*; China Statistics Press: Beijing, China, 2010.

27. The Research Institute of Innovative Technology for the Earth. Estimated Energy Unit Consumption in 2010. Available online: https://www.rite.or.jp/system/en/global-warming-ouyou/download-data/E-Comparison_EnergyEfficiency2010steel.pdf (accessed on 11 January 2020).

28. The Japanese Iron and Steel Federation. Activities of Japanese steel industry to Combat Global Warming. 2017. Available online: http://www.jisf.or.jp/en/activity/climate/documents/ActivitiesofJapanesesteelindustrytoCombatGlobalWarming.pdf (accessed on 11 January 2020).

29. Cai, J. Air consumption and waste gas emission of steel industry. *Iron Steel* **2019**, *54*, 7–14.

30. Wang, H.; Zhang, J.; Wang, G.; Jiang, X. Analysis of carbon emission reduction before ironmaking. *China Metall.* **2018**, *28*, 1–6.

31. Wang, P.; Jiang, Z.; Zhang, X.; Geng, X.; Hao, S. Long-term scenario forecast of production routes, energy consumption and emissions for Chinese steel industry. *J. Univ. Sci. Technol. Beijing* **2014**, *36*, 1683–1693.

32. Bureau of International Recycling. *World steel recycling in Figure 2017*; Bureau of International Recycling: Belgium, Brussels, 2018.

33. The Editorial Board of China Steel Yearbook. *China Steel Yearbook 2010–2017*; Metallurgical Industry Press: Beijing, China, 2018.
34. International Energy Agency (IEA). Energy Use in the Steel Industry. Available online: https://www.iea.org/media/workshops/2014/industryreviewworkshopoct/8_Session2_B_WorldSteel_231014.pdf? (accessed on 11 January 2020).
35. United States Environment Protection Agency (EPA). *Available and Emerging Technologies for Reducing Greenhouse Gas Emissions from the Iron and Steel Industry*; EPA: Research Triangle Park, NC, USA, 2012.
36. Zang, Y.; Liu, W.; Zhang, J. Accelerating the utilization of scrap to promote energy-saving and emission-reduction of China's iron and steel industry. *Res. Iron Steel* **2010**, *38*, 43–46.
37. World Steel Association. Global Steel Industry Outlook, Challenges and Opportunities. 2017. Available online: https://www.worldsteel.org/en/dam/jcr:d9e6a3df-ff19-47ff-9e8f-f8c136429fc4/International+Steel+Industry+and+Sector+Relations+Conference+Istanbul_170420.pdf.? (accessed on 11 January 2020).
38. Sun, H.; Li, W. Natural Gas Play an Important Role in the Energy Conservation and Emissions Reduction. *Pet. Plan. Eng.* **2009**, *20*, 7–9.
39. Zhang, C. *Potential Analysis and Synergy Approaches of Energy Saving and Pollution Reduction: Case in Iron Industry*; Tsinghua University: Beijing, China, 2015.
40. Editorial Department of Metallurgical Management. Ranking of crude steel production of global steel enterprises in 2017. *Metall. Manag.* **2018**, *6*, 4–15.
41. Lin, B.; Wu, Y.; Zhang, L. Estimates of the potential for energy conservation in the Chinese steel industry. *Energy Policy* **2011**, *39*, 3680–3689. [CrossRef]
42. Lukas, B. *Overcapacity in Steel-China's Role in a Global Problem 2016*; Alliance for American Manufacturing: Washington, DC, USA, 2016.
43. Ministry of Industry and Information Technology of the People's Republic of China. Steel Industry Standard Conditions 2015. Available online: http://www.miit.gov.cn/n1146285/n1146352/n3054355/n3057569/n3057577/c3571380/content.html (accessed on 11 January 2020).
44. Research group of China Iron and Steel Industry Association. *Supply Side Structural Reform of Iron and Steel Industry*; China Iron and Steel Industry Association: Beijing, China, 2018.

metals MDPI

Review

Challenges and Outlines of Steelmaking toward the Year 2030 and Beyond—Indian Perspective

Sethu Prasanth Shanmugam [1], Viswanathan N. Nurni [2,*], Sambandam Manjini [1], Sanjay Chandra [2] and Lauri E. K. Holappa [3]

1   Research and Development, JSW Steel Limited, Salem 636453, India; ssprasanth47@gmail.com (S.P.S.); manjini.sambandam@jsw.in (S.M.)
2   Centre of Excellence in Steel Technology (CoEST), Department of Metallurgical Engineering and Materials Science, Indian Institute of Technology Bombay, Mumbai 400076, India; sanjaychandra@iitb.ac.in
3   Department of Chemical and Metallurgical Engineering, School of Chemical Engineering Aalto University, 02150 Espoo, Finland; lauri.holappa@aalto.fi
*   Correspondence: vichu@iitb.ac.in; Tel.: +91-22-25767611

Abstract: In FY-20, India's steel production was 109 MT, and it is the second-largest steel producer on the planet, after China. India's per capita consumption of steel was around 75 kg, which has risen from 59 kg in FY-14. Despite the increase in consumption, it is much lower than the average global consumption of 230 kg. The per capita consumption of steel is one of the strongest indicators of economic development across the nation. Thus, India has an ambitious plan of increasing steel production to around 250 MT and per capita consumption to around 160 kg by the year 2030. Steel manufacturers in India can be classified based on production routes as (a) oxygen route (BF/BOF route) and (b) electric route (electric arc furnace and induction furnace). One of the major issues for manufacturers of both routes is the availability of raw materials such as iron ore, direct reduced iron (DRI), and scrap. To achieve the level of 250 MT, steel manufacturers have to focus on improving the current process and product scenario as well as on research and development activities. The challenge to stop global warming has forced the global steel industry to strongly cut its CO_2 emissions. In the case of India, this target will be extremely difficult by ruling in the production duplication planned by the year 2030. This work focuses on the recent developments of various processes and challenges associated with them. Possibilities and opportunities for improving the current processes such as top gas recycling, increasing pulverized coal injection, and hydrogenation as well as the implementation of new processes such as HIsarna and other CO_2-lean iron production technologies are discussed. In addition, the eventual transition to hydrogen ironmaking and "green" electricity in smelting are considered. By fast-acting improvements in current facilities and brave investments in new carbon-lean technologies, the CO_2 emissions of the Indian steel industry can peak and turn downward toward carbon-neutral production.

Keywords: iron ore; coking coal; DRI; scrap; blue dust; natural gas; energy saving; decarbonization

Citation: Shanmugam, S.P.; Nurni, V.N.; Manjini, S.; Chandra, S.; Holappa, L.E.K. Challenges and Outlines of Steelmaking toward the Year 2030 and Beyond—Indian Perspective. *Metals* **2021**, *11*, 1654. https://doi.org/10.3390/met11101654

Academic Editors: Geoffrey Brooks and Alexander McLean

Received: 9 September 2021
Accepted: 15 October 2021
Published: 19 October 2021

Publisher's Note: MDPI stays neutral with regard to jurisdictional claims in published maps and institutional affiliations.

1. Introduction

Steel manufacturing is a technologically complex industry having subsequent linkages in terms of material flow and plays a vital role in determining infrastructure and the overall development of a country. The global steel industry and its supply chain constitute 40 million jobs across the world. In 2019, India established itself as the second-largest steel producer with 111.3 million tons [1], constituting 5.9% of total crude steel production on the planet for the respective year, and it has ambitious plans to produce 250 million tons by 2030 with a per capita consumption aim of 160 kg. This is understandable, as the current per capita steel consumption in India is only 74 kg, which is much lower than the average global consumption of 229 kg [2]. India's crude steel production and per capita consumption with projection for the year 2030–2031 is given in Figures 1 and 2. Moreover, the steel industries in India have challenges in terms of the non-availability of metallurgical grade coal within the country, leaner grades of ore, solid waste utilization, and water scarcity. Environmentally, the steel sector also faces challenges in terms of bringing down the carbon emission by 20% of 2005 levels within 2030. Indian steel manufacturers and associated metallurgical and mining industries are taking various measures and also need to take various measures in the future in terms of process improvements and developments to cater to the above needs. This paper attempts to elaborate on these challenges and speculate as to how the steel industries shall evolve in India in 2030 and beyond.

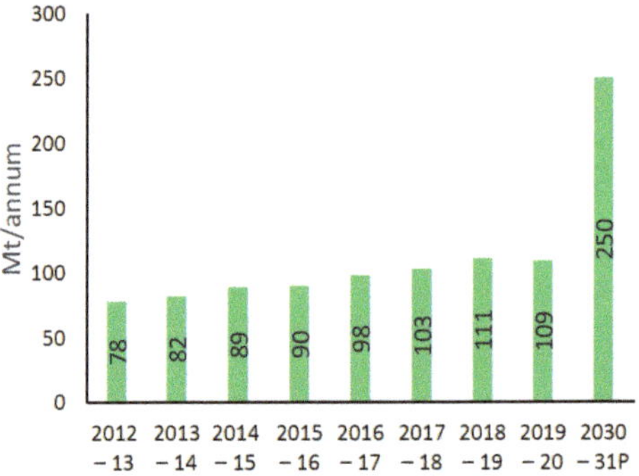

Figure 1. Crude steel production; (Graph plotted using data from ministry of steel [3]).

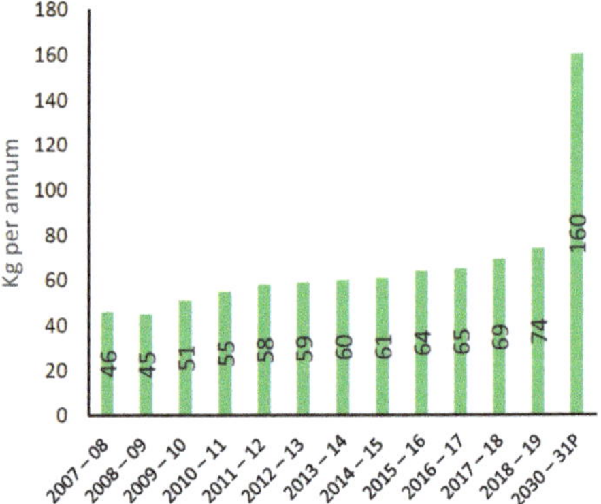

Figure 2. Per capita consumption of steel in India; Graph plotted using data from India Brand Equity Foundation [4]).

2. Organization of Indian Steel Sector

Based on production routes, Indian steel industries are divided into two types: industries producing steel primarily through a hot metal route using large integrated steel manufacturing units and others through primarily scrap and DRI in relatively smaller manufacturing units. The former ones are denoted as oxygen route manufacturers and the latter are denoted as electric route manufacturers henceforth in this paper.

Operations of the oxygen route manufacturers start from coke making, reducing iron ore primarily through blast furnace (BF) for the production of hot metal, and subsequently producing crude steel with standard specifications. There are also steel plants that use direct reduction (MIDREX), smelting reduction (COREX), and BF combination followed by the converter-arcing (CONARC) process. They all are ore-based production routes. Typically, these integrated manufacturers produce more than 5 million ton per annum from a single work location.

Electric route manufacturers primarily use scrap (they also produce DRI through a rotary kiln process), direct reduced iron (DRI), or hot briquetted iron (HBI) as raw material for the induction furnace route, and a few of them use the electric arc furnace route (EAF). Apart from these two, there are also independent hot and cold rolling units as well as sponge iron producers who use rotary kiln processes and pig iron producers who use mini blast furnaces. Typically, these manufacturers produce less than 1 million tons per annum from a single unit. Almost half of the Indian steel production is made from the electric route steel producers. Detailed process maps followed by Indian manufacturers are shown in Figure 3. The production capacity of various steelmaking processes in India is shown in Table 1. Crude steel production by process with its share percentage in Indian steel industry over the years is shown in Figure 4.

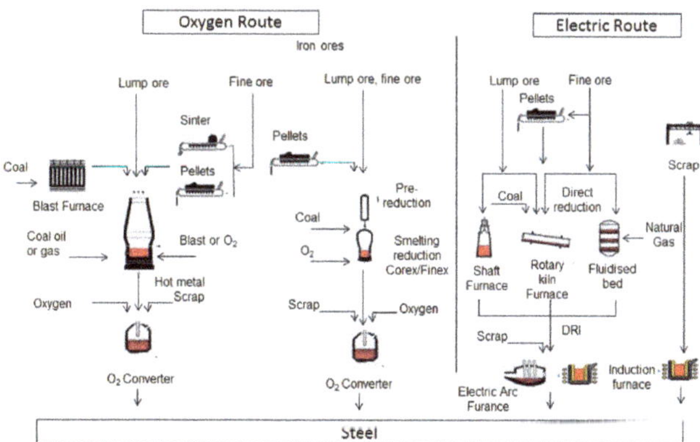

Figure 3. Indian steelmaking routes process map.

Table 1. Capacity of iron and steel industry in India as of 2018–2019; Source: The Energy and Resources Institute based on data from Joint Plant Committee.

Segment	No of Units	Capacity (Million Ton)	Production in 2018–2019
Blast Furnace	59	80	72.5
COREX	2	1.65	1.4
Sponge Iron	318	49	37
BOF	17	57	49
IF	1174	49	33
EAF	56	42	28

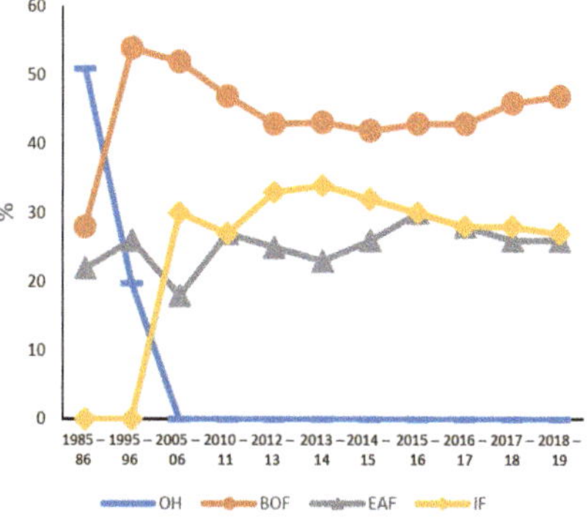

Figure 4. Crude steel production by process; share (%) OH—Open Hearth Process; BOF—Basic Oxygen Furnace; EAF—Electric Arc Furnace; IF—Induction Furnace. Graph plotted using data from India Brand Equity Foundation [4].

3. Current Trends and Challenges

3.1. Raw Materials

India is the 4th largest iron ore producer and the 3rd largest coal producer in the world. Coal is also identified as one of the major sectors of "Make in India", which is an initiative by the Government of India launched by the prime minister [5]. India is also the world's largest producer of sponge iron: about 37 million tons per annum.

3.1.1. Iron Ore

Proven hematite resources in India are around 29 billion ton, of which only 13% is high grade (>65% Fe), 47% are medium grade (62–65% Fe), and the remaining are low-grade ores [6]. During mining, lumps ($-40 + 10$ mm), fines ($-10 + 0.15$ mm), and slimes (<0.15 mm) are generated. For the efficient use of all these size ranges of ores, suitable ore agglomeration techniques are followed across the country. Over a while, sinter and pellet processes have improved and gained a significant place. Ore fines, -6 to 100 mm mesh size are being used for making sinter, and below 100 mesh fines are used for manufacturing pellets.

Indian iron ores have a major problem with high alumina and phosphorus. Highly friable hematite ore generates a huge quantity of fines during mining and crushing, which are rich in Al_2O_3. These fines are generally in the form of goethite (hydrated iron oxide) in which alumina is present in the matrix. Alumina exists in the form of gibbsite and kaolinite [7]. The high alumina content in these fines make them less amenable to physical separation and creates a problem in getting the final concentrate grade. In addition to magnetic and gravity separation, there is a need to explore the possibility of efficient beneficiation through froth flotation and selective dispersion with chemical aids. Some research efforts in this direction are being conducted at the National Metallurgical Laboratory, Jamshedpur, and Institute of Minerals and Materials Technology, Bhubaneswar.

As a result of the highly reducing atmosphere at the ironmaking stage, all the phosphorus in iron ore ends up in hot metal. Furthermore, if phosphorous is not removed in subsequent processes, it causes cold brittleness in steel. Efforts to remove phosphorous during mineral processing techniques are being explored using techniques such as thermal treatment, bioleaching, froth flotation, etc. [8]. The high P content in hot metal results in high P-containing slag in the steelmaking process and further poses difficulty in recycling this steelmaking slag back into the ironmaking process.

Blue dust is fine, powdery, soft, and friable ore rich in Fe content (65–67%) present in mines. Due to its fineness, it is not used directly in the furnaces. An estimated reserve of around 550 million ton of blue dust is available in India. India needs to find an appropriate technology to utilize this high-grade fine iron ore.

3.1.2. Coal

Although India is the third-largest coal producer after China and the USA, coking coal is only 15% out of the total coal mined [9]. In addition, Indian coking coal is not suitable for blast furnace due to its high ash content and hence requires extensive crushing and washing. About 85% of the coking coal requirements of Indian steel industries are met through imports [10]. Nevertheless, it invariably increases the coke rate in Indian blast furnaces compared to other countries. The cost of coke in the international market is expected to remain high in the foreseeable future and has forced the Indian manufacturers to go toward decreasing coke consumption by injecting pulverized coal. Increased usage of these injectants also paves the way for high production rates resulting from more oxygen input but decreases permeability, resulting in a larger pressure drop in blast furnaces [11]. Unfortunately, coal grades available in India are also found to be mostly not suitable for pulverized coal injection due to its high ash content. Thus, both for coke making as well as pulverized coal injection, Indian steel industries rely primarily on imported coal. Manufacturers in the electric route who use a rotary kiln process for the production of DRI/sponge iron also use primarily imported coal for their processes.

In summary, India has almost no reserves of coal that can be used in blast furnaces as well as other DRI-making units. This is one of the major issues that will shape the future of Indian steel industries and also should encourage efforts to decarbonize the Indian steel industry.

3.1.3. Alternative Fuels

Natural gas is the cleanest of fossil fuels, causing smallest CO_2 emissions. It is available in oil and gas fields located at the Hazira basin, Assam, Tripura, and Mumbai offshore regions [12]. A total of 90 Million Metric Standard Cubic Meters Per Day (MMSCMDs) of domestic gas production was achieved in FY 2018–2019. Apart from this, India imports Liquified Natural Gas (LNG) through six operational LNG regasification terminals with a combined capacity of $\approx$140 MMSCMD. Utilizing hydrogen-rich tuyere injectant such as natural gas helps to decrease the coke quantity as well as CO_2 emission of the blast furnace.

Other hydrocarbons such as waste plastics when heated up in the absence of oxygen, produce CO and hydrogen, which can be utilized in blast furnaces. However, before the injection of plastics through tuyeres, it needs to be crushed and pelletized (if necessary). Not all types of plastics can be injected; they should be segregated and sized. Indian manufacturers have yet to explore these advantages from plastics. Plastics can also be used in a coke oven as chemical feedstock [13].

3.1.4. Scrap

Major raw materials for electric route manufacturers are steel scrap and DRI. The import quantity of scrap and DRI production in India are shown in Figures 5 and 6. Domestic scrap supply from unorganized industry is around 25 MT (around 78% of the demand) in the year 2017–2018. [14]. However, the gap is expected to be reduced due to the increased production foreseen in the future, and it is likely to increase the scrap quantity considerably. In addition, the national scrap policy of 2017 envisages setting up a system that improves the processing and recycling of ferrous scraps through organized and scientific metal scrapping centers to minimize imports as well as to become self-sufficient in availability. In recent years, there has been an increasing trend in manufacturing steel through the scrap route, as it reduces greenhouse gas (GHG) emissions. Every ton of scrap utilized for steel production avoids the emission of 1.5 tons of CO_2 and the consumption of 1.4 tons of iron ore, 740 kg of coal, and 120 kg of limestone [15]. As of March 2019, 47 electric arc furnaces and 1128 induction furnaces are in operation across the country, and their main source of raw material is scrap.

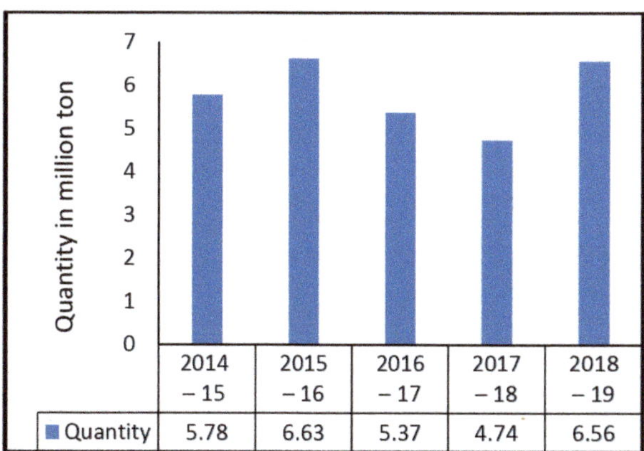

Figure 5. Steel scrap import over the years(Graph plotted using data from: Ministry of Steel).

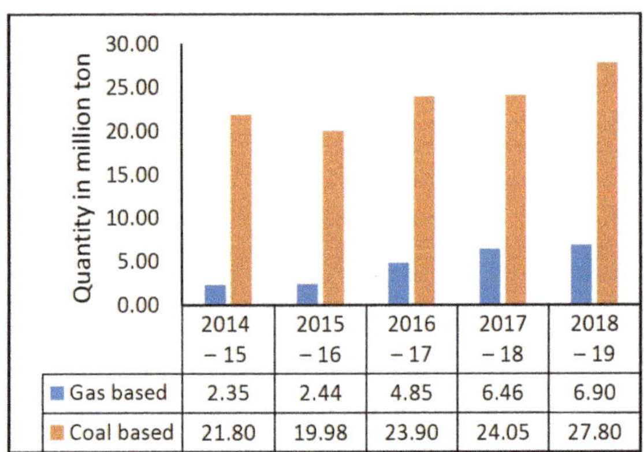

Figure 6. DRI Production (Graph plotted using data from: Ministry of Steel).

3.2. Processes

From the viewpoint of blast furnace operation, sinters and pellets in burden material give better bed permeability, decreasing the thickness of the cohesive zone due to narrower softening and melting ranges compared to lump ore. They also save energy since the calcination of limestone is avoided by making fluxed pellets/sinters, and we reduce alkali content in the blast furnace. As a result of these advantages, the use of lump ore is drastically reduced to around 10% [16].

The high alumina content in the iron ore poses a major challenge in blast furnace operation. To maintain appropriate slag viscosity, plants practice the addition of quartz and correspondingly the addition of lime to take care of basicity [17]. This results in a significant increase in the slag rate for Indian blast furnaces (375–420 kg/ton of hot metal as compared to 350–375 kg/ton of hot metal in China). High slag rates demanding more heat also result in higher coke rates. It also poses challenges toward increased pulverized coal injection in Indian blast furnaces.

High phosphorus in ore directly results in higher P% in the hot metal. To remove P, an excessive quantity of flux is necessitated in the converter process, resulting in significant temperature loss. If an external pre-treatment for hot metal dephosphorization is used, the process should be fast enough to match the BOF sequence.

Rotary kiln processes employ non-coking coal for converting iron ore to metallic iron and have the advantage of low investment cost. The energy balance of rotary kiln with theoretical and actual consumption showed only about 55% energy efficiency. A significant factor is that the waste gas liberated during the process remains unutilized at the moment [18]. It decreases the efficiency of the rotary kiln process. In addition, the utilization of this waste gas by waste heat recovery boiler experiences difficulty due to the accretion formation in the kiln and dust particles in the gas. These conditions make the rotary kiln highly inefficient compared to gas-based processes.

Typically, electric route manufacturers in India work with capacity less than 1 million tons per annum with the DRI-EAF/Induction Furnace route. Over the years, such small units have evolved closer to the market. It is also interesting to note that the electric route industries that are located in the western parts of the country use a higher proportion of scrap due to the accessibility of imported scrap by the sea route. On the other hand, on the eastern parts where ore reserves are abundant, they use rotary kiln units in combination with an induction or an electric arc furnace. Among electric arc furnace and induction furnace routes, the induction furnace is preferred because of its high yield (95–96% for IF and 92–93% for EAF), better electrical efficiency, and low investment cost [19,20]. The IF also has the characteristic of not using other forms of energy than electricity (O_2, fuel, carbon injection). However, unlike electric arc furnaces, at present, the maximum induction furnace capacity available in India is 50 tons only.

Although gas-based DRI is an environmentally effective process, the final carbon in the DRI is high compared to coal-based DRI from Indian plants (gas based 1.5–1.8%, coal based 0.2–0.25%). Thus, it makes it difficult for electric route manufacturers to reduce carbon in the product to the required range, especially when they use Induction Furnaces with acidic lining. However, manufacturers who use EAF prefer DRI from gas-based units, as the higher carbon content gives higher energy efficiency [21]. In addition, the high porosity, the low thermal, and the electrical conductivity of DRI are some of the other problems faced by electric route manufacturers. The proper selection of scrap and separation is much needed to have control over tramp elements in the final steel. Since most of the electric route manufacturers do not have extensive ladle-refining facilities, control over the nitrogen, hydrogen, and total oxygen contents are not present. These shortcomings are the reason why most of the induction-based manufacturers are making rebars and construction-quality steel rather than engineering and automotive-grade steel. In addition, the furnace lining of induction-based electric route industries is generally silica-based, and it results in ineffective dephosphorization compared to oxygen route products. The typical furnace lining life is 18 to 20 heats.

As the production is increasing, the demand for making clean steel is also increasing. It is the steelmaker's responsibility to fulfill the stringent quality requirements of the end-user. For example, the fatigue life of bearing steel greatly depends on the non-metallic inclusions. Bearing customers would want to have an inclusion size as low as possible (hard aluminum oxide and other oxide inclusions larger than 30 μm should be avoided). In that regard, reducing the total oxygen content in the steel becomes a necessary criterion. Reducing the total oxygen level from 15 ppm to less than 10 ppm would reduce the non-metallic inclusions as well. Achieving the low level of oxygen requires stringent control of the secondary steelmaking practice.

Steel manufacturers with large capacity and that have blast furnaces produce a versatile range of steel grades right from construction-quality steel grades to automobile, defense, railways, and aerospace grades. Larger capacity manufacturers have their rolling mills inbuilt in their facilities, and they produce a plethora of products in the form of bars, flats, wire rods, hexagon bars, etc.

Steel manufacturers with induction and arc furnaces produce mostly construction-quality grades in the form of rebars due to the factors mentioned earlier. Some manufacturers have a vacuum degassing facility and argon oxygen decarburization facility (AOD) and produce high-quality alloy steels for automotive applications and stainless steel grades with a smaller heat size. In Indian steel market conditions, highly critical steel grades for applications in space and aerospace equipment are imported. India needs to grow in terms of high-value niche special steels.

3.3. Environment and Energy

Steel industries generate around 30% of the total CO_2 emissions from all industrial sectors on the planet. The reason for this huge quantity is due to the usage of coke and the high consumption of energy factory-wide. Nearly 65% of the emissions of oxygen route industries come from ironmaking processes, out of which 90% is contributed by coke and coal. Steel industries all around the world are striving hard to improve the energy efficiency of blast furnaces by approaching the theoretical limits of production and carbon consumption.

The iron and steel industry is one of the major energy consumers in India as well. By using the best available technologies (BAT), the specific energy consumptions are BF-BOF route-16.4 GJ/TCS (ton of crude steel), COREX-BOF-19.3 GJ/TCS, DRI-EAF (coal-based) route-19 GJ/TCS, and DRI-EAF (gas-based)-15.9 GJ/TCS [14]. However, major steel plants in India have a specific energy consumption of 27.3 GJ/TCS [14]. Although there exists a substantial potential to save energy by adopting the best practices and newest innovations for reducing energy consumption, reaching this target can be quite challenging considering the quality of raw materials in India. Thus, the authors opine that arriving at a target-specific energy consumption considering the local raw material quality can be quite fruitful in defining the road map for steel technology for India.

A major challenge for the electric route with arc furnaces or induction furnaces is the electrical energy consumption. The melting process starts from room temperature, and there is no heat recovery from off-gases. Typically, Indian manufacturers have electricity consumption 600 kWh/T compared to the world average of 416 kWh/T. However, DRI-EAF based routes generally consume less energy than companies that use EAF alone. The specific energy consumption of gas-based DRI is 10.46 to 14.43 GJ/T, and coal-based plants have 15.9 to 20.9 GJ/T [22].

Waste disposal and treatment is a big challenge for steel manufacturers. The steel industry generates solid, liquid, and gaseous wastes, and the most common are iron and steel slag, scrap, sludge, effluent, flue gases, etc. Approximately 0.4 to 0.8 tons of solid waste is generated for a ton of liquid steel (also, 0.5 tons of effluent water and 8 tons of moist laden gases) [22]. The waste management systems practiced by steel industries involve processing the wastes for a recycling in-plant process or disposing by appropriate methods. Ironmaking slag such as BF or COREX slag is used as a raw material for slag cement making and for producing slag sand, which can replace river sand. Due to the presence of free lime, BOF slag is not used in construction applications and generally goes for landfilling. Steel plants are developing and adopting methods for accelerated weathering of the steel slag and use it in construction activities, tide breakers in the coastal areas, and soil ameliorants. Some of the other applications for weathered BOF slag are rail ballast, cement making, road and paver blocks, and brick and tiles manufacturing [23].

On the contrary, electric arc slag is granulated and used as agglomerates in road construction and similar applications. Fine granulated EAF slag is also used in shot blasting and industrial water filter applications. Ladle-refining slags have high CaO, MgO, and SiO_2 contents and could be used as a raw material in cement production [24]. However, the powdery nature of the LF slag makes it difficult to handle. In addition, the presence of high sulfur, heavy metals such as Cr, V should be avoided and thus necessitates proper screening and care. One of the important solid wastes to be recycled is the slag from BOF, as it cannot be dumped due to its high phosphorus content.

4. Possibilities and Opportunities

4.1. Raw Materials

a.   Iron Ore

Demand for iron ore has been increasing with the increased production of iron and steel in developing countries such as India and China. However, the quality of iron ore has deteriorated over the years globally due to long-term mining. The low-grade iron requires beneficiation before agglomerating for use in the iron-making process. The iron ore interlocked with silica and alumina has to be liberated for efficient beneficiation [25]. This requires finer communition, resulting in iron ore fines often finer than 300 microns. It means extra energy consumption, too. The pelletization process is better suited as an agglomeration process for finer fraction iron ores. It is envisaged that the use of pellets in the Indian blast furnaces will increase the replacement of the usage of sinter. For pellet induration, natural gas can be used, which is less polluting than the coke fines used in sinter making.

A large quantity of fine iron ore in the form of slime arises after beneficiation having high Al_2O_3, making it unsuitable for blast furnace operation. Slime beneficiation techniques such as hydro-cyclone and magnetic separation techniques are being explored to produce an iron concentrate with ~63% of Fe [25]. Bhagwan Singh et al. from NMDC utilized the blue dust concentrate to produce ultra-pure and high-quality ferric oxide, premium-grade sponge iron powder using hydrogen as a reducing agent [26]. This kind of process is expected in the near future for full-scale production.

b.   Coal substitutes

Plastics can be used as chemical feedstock in the coke oven and in blast furnaces. Waste plastics contain carbon and hydrogen, and if it is thermally heated along with coal in a coke oven, it generates coke, coke oven gas, and hydrocarbon oil. In the latter case, the plastics shall be dechlorinated, pulverized, and injected through the tuyeres of the blast furnace as a reducing agent. Nomura et al. have analyzed the different types of plastics used in the coke oven and out of which only PE (polyethylene) and PVC (polyvinyl chloride)-type plastics have a small effect on properties, and particularly, PE increased the strength of coke. In contrast, PS (polystyrene) and PET (polyethylene terephthalate) inhibited the strength [13]. The size of the granules also requires optimization for different varieties of plastics. India can explore the possibility of using waste plastics in coke ovens, as it ensures a sustainable social system. For every ton of plastic used, coke can be reduced by 750 kg.

c.   Alternate Fuels

India is expecting 15% of the energy mix from natural gas utilization by 2030 from 6% today. Currently, India has six operating LNG terminals. India started building an LNG port at Kakinada by 2019, and the government has plans to build 11 more terminals, particularly on the east coast. Apart from this, India is working on constructing a national gas grid, increasing to around 14,000 km of additional pipeline networks across the country. These plans would increase the availability of natural gas for steel manufacturers. With the increased availability of natural gas, the projections for sponge iron production by the process route in India by 2030 is given in Figure 7. Although the share of the gas-based production will increase from 25% to 33%, also the coal-based production is planned to increase, thus ruling out the efforts to mitigate CO_2 emissions.

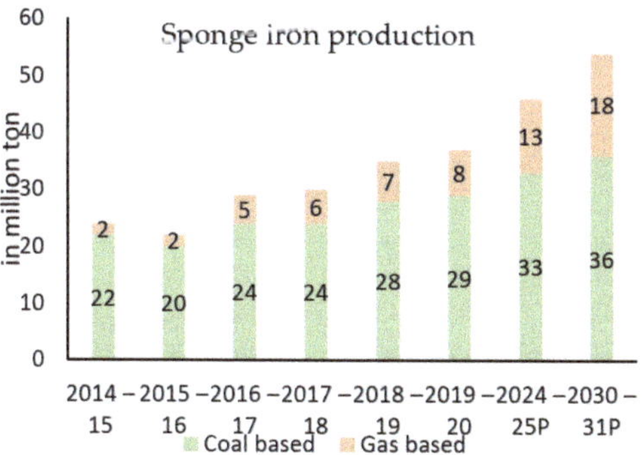

Figure 7. Sponge iron production forecast (Graph plotted using data from Ministry of Steel).

d.   Scrap

The Ministry of Steel has issued the steel scrap recycling policy [14]. It aims to promote the metal scrapping centers and also ensures the scrap processing and recycling from various sources. It envisages a structure to provide standard guidelines for the collection, dismantling, and shredding of scrap. The scrap requirement of India in 2030 is expected to more than double from the current level 32 MT/2019 to 70–80 MT [12]. The Indian government expects that the increased production of vehicles for the last two decades would generate a continuous flow of steel scrap for recycling to steel production. Once scrap becomes more available, then DRI + scrap utilization through an induction furnace or electric arc furnace may increase in the future.

With the increased availability of natural gas and scrap, the recent history and the projected process-wise crude steel production for the year 2030 is given in Figure 8. Principally, the trend is positive, but again, the projected strong production growth will inevitably lead to increasing emissions.

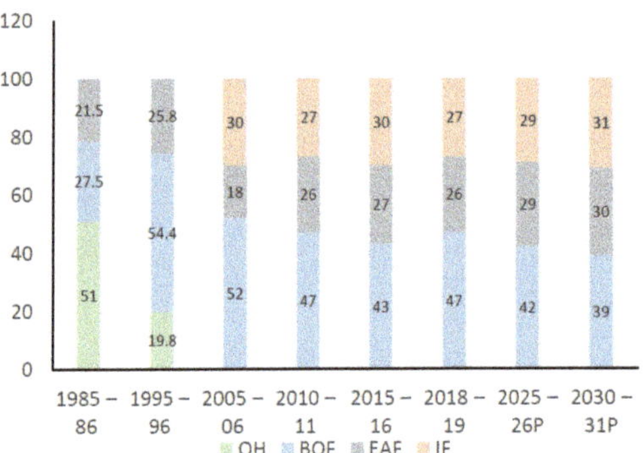

Figure 8. Process-wise crude steel production since 1985 and a forecast to 2030 (graph plotted using data from Ministry of Steel).

4.2. Processes

4.2.1. Improvements in Existing Processes

The problem of high alumina in ore can possibly be handled through appropriate modification in blast furnace design. High alumina resulting in high slag rates limits the production rates, especially because of the complex fluid dynamics in the bosh region. If the belly size is increased, the superficial gas velocity can be brought down, which can permit higher production rates.

Apart from the technologies discussed earlier (PCI, TGR-OBF), there are other methods aiming at reducing specific energy consumption and emissions such as coke dry quenching, top recovery turbine, and recycling of blast furnace slag, as well as the utilization of alternative fuels such as biomass, waste plastics, and natural gas. To implement waste heat recovery technologies such as the preheating of coal charge and coke dry quenching technology, the introduction of a waste heat recovery system is necessitated, which can bring down the requirement for fuel input in coke making from a level of 2.3 GJ/T of coal charge to 2.1 GJ/T of coal charge.

Pulverized coal injection (PCI) into blast furnaces is a common practice nowadays to improve the BF economy. For every ton of coal injected in the blast furnace, 0.85 to 0.95 tons of coke production can be avoided. To increase coal injection, one needs to adjust the parameters to achieve optimum permeability. Some ideas to achieve the same are (1) reduction of the size of coal particles—finer particles provide greater surface area and increase the rate of combustion—and (2) catalyzing the gasification reaction by treatment of coke particles with catalysts such as the slime of fine particles. Increasing the PCI is associated invariably with the pressure drop due to a decrease in coke fraction from the top. Some measures to avoid that are (1) exploring the possibilities of making coke layer thickness larger at the expense of the number of coke slits and (2) charging coarse particles of ore on the top of a coke layer [11].

Injection of plastics after proper collection, segregation and preprocessing, biomass, and even solid injectants such as flux and BOF slag into a blast furnace can be explored for improving the blast furnace production and reducing the carbon footprint. With increasing environmental awareness and the policies of the state, these efforts are expected to gain momentum.

Oxygen blast furnace with top gas recycling technique (TGR-OBF) is another process being explored for improved energy efficiency and reduced emissions to the environment [27]. This technique actually removes the carbon dioxide from top gas and the remaining gas, which is reducing in nature, is injected back into the blast furnace through tuyeres. Nogami et al. and Danloy et al. verified that heat demand will be sensibly reduced and productivity improved by top gas recycling [27,28]. N.B. Ballal reviewed this technique with the 0d model and found that top gas recycling would increase the CO partial pressure of bosh gas (from 35% to 42.7% with 20% recycle and 100 kg/ton of hot metal), resulting in faster reduction kinetics [9]. No blast furnace is running with this technique, but India can explore this possibility. In addition, dust in the blast furnace gas can be separated and can be sent back to the furnace, especially if it is low in alkali oxides. Thus, it will increase the carbon efficiency in the BF as well as reduce coke consumption.

A study on rotary kiln energy efficiency by Nishant et al. [29] identified the possible areas where energy is lost in the form of waste gas, cooling of sponge iron, high exit temperature of clean waste gas, and intrusion of air. They suggested ways to reduce these losses by process integration principles. Ideas such as these should be promoted in the coal-based rotary kiln DRI manufacturing to make the process efficient. Otherwise, with the increase in the shift toward gas-based processes because of the plethora of advantages, rotary kiln operation would see an end in the future.

The recovery of sensible heat from metallurgical slags is a challenge. An array of energy recovery means from slags such as thermal, chemical, and thermoelectric generation technologies have been investigated. Among the technologies, thermal methods are developed the most, as it does not have complex multi-step technology as well as temperature constraints. In TATA steel (Sridhar et al. [30]), in the pilot scale, waste heat from BOF slag was utilized for splitting water molecules to harvest hydrogen (H_2) that is cleaner and available at low cost. Using 15 kg of slag, they were able to get gas that contains as much as 23% hydrogen. The main challenges reported were safe handling of the large quantity of slag.

In 2020, The Energy and Resources Institute (TERI) summarized the energy-efficiency potential in the Indian steel industry [31]. By modernizing equipment and processes and adopting best available technologies (BAT), energy consumption can be reduced by 25–30% from the current level. The potentials in different process stages in the BF-BOF route are shown in Figure 9.

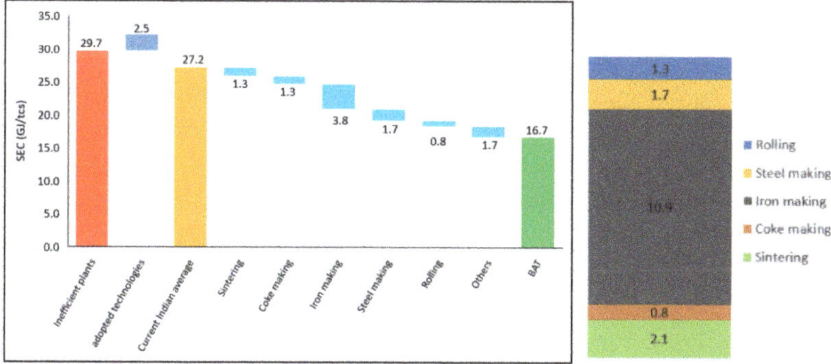

Figure 9. Specific Energy Consumption (SEC) and potentials for improved energy efficiency in BF-BOF route and breakdown of BAT (right) Reprinted with permission from TERI [31] Copyright 2020, The Energy and Resources Institute.

Due to the high phosphorus in Indian iron ore, high P in BF hot metal is inevitable. At the same time, the demand for low and ultra-low P% in high-quality steels is growing, which sets an extreme challenge to the converter operation. Dephosphorization can be done as pre-treatment of hot metal in torpedo ladle or at the basic oxygen furnace itself, depending on the phosphorous percentage in hot metal. Typically, lime or lime with fluorspar used as flux and Fe oxide with or without O_2 is used as an oxygen source for dephosphorization. It can be divided as low oxygen activity with a high basicity process or high oxygen activity with low basicity based on the slag composition. Liu et al. [32] found that the most favorable temperature for P removal is around 1300–1400 °C. The dephosphorization reaction is a slag–steel interface-controlled reaction. The distribution coefficient between slag and iron is temperature dependent, getting lower at high temperatures. On the other hand, a very low temperature under 1300 °C makes slag viscous and causes poor reaction kinetics. CO_2 injection has been applied in controlling the bath temperature instead of sending oxygen alone. In addition, it also reacts with bath elements, generating additional CO gas and thus intensifying the stirring effect. When 15% CO_2 was mixed in O_2 blown through the top lance in a 300 ton converter, the dephosphorization rate increased from 56 to 63% [31]. As a drawback, the cooling effect of CO_2 injection results in a notable reduction in scrap-melting capacity. This idea of CO_2 injection could be explored in Indian steel industries, as it helps both improve the quality of steel and the environmental point of view, provided that the required CO_2 for injection is captured through CCS technology, which is explained in a later part of this paper.

In the electric furnace route, although utilizing gas-based DRI is beneficial to the industry, the increased carbon and phosphorous contents limit the usage of it. Double slag practice can be envisaged for the induction furnace route to have better control over the phosphorus.

4.2.2. New Technologies/Processes

The regular supply of natural gas to gas-based processes such as Midrex is a real concern. Therefore, some plants are going for the in-plant generation of synthesis gas from coal. Syngas or synthesis gas is a product of the gasification of carbon-containing fuel having a mixture of hydrogen, carbon monoxide, and carbon dioxide. Some plants using the COREX route produce hot metal as well as additional gas, which can be used for gas-based reduction, e.g., via a Midrex reactor [33]. The mixing of coke oven gas with synthesis gas for reduction is also practiced by some manufacturers, and efforts are being undertaken to build coal gasifiers to produce synthesis gas. These technologies can more or less reduce CO_2 emissions compared to the current industrial practices, but they are not able to compensate for the emissions due to the planned strong growth of the steel production in India.

Recently, the use of hydrogen in ironmaking to replace coal or natural gas has received growing attention. In the first step, instead of pure hydrogen, the usage of different blends of gases, from natural gas to syngases, is necessitated. There are three major gas-based processes available in the world—namely, Midrex, HYL, and Circored. The Midrex process already has been operated with a level of 55 to 75% hydrogen concentration.

The implementation of hydrogen in the steel industry will be strongly dependent on the decarbonization of the power sector. Depending on the production process and utilization of energy source, hydrogen generation can be called green, blue, gray, or pink. Green hydrogen comes from splitting water by electrolysis, gray hydrogen is when natural gas is split into hydrogen and CO_2 through various processes. Blue hydrogen is the same as gray except the CO_2 is captured through CCUS technologies. Pink hydrogen is the same as green except it uses nuclear energy for splitting water. Since India possesses a significant capability in the generation of renewable energy sources through both solar and wind, the electrolysis process can be utilized for the production of green hydrogen. However, the process is electrically as well as water-intensive and requires high capital cost. India is expanding its natural gas grid far and wide, so the opportunities for lowering industrial emission as well as hydrogen utilization would grow stronger.

ULCOS—The Ultra-Low Carbon Steel making program is a co-operative European research and development initiative that was launched in 2004 to search for the process that could in the future when fully developed establish the potential of large cuts in CO_2 emission for steel production from iron ore. Several concepts have been investigated in parallel, using modeling and laboratory approaches to examine the potential of processes in terms of CO_2 emissions, energy consumption, cost, and sustainability. HIsarna is one such process identified through the ULCOS program; it is a combination of Isarna and HIsmelt technologies to produce liquid hot metal directly from iron ore [34]. It requires neither agglomeration of iron ore nor coke. It also has the potential to utilize the non-coking coal reserves in India. It is efficient in energy, and it has a lower carbon footprint than conventional processes. It reduces energy consumption by at least 20% and CO_2 emission by 20% [34]. One benefit of this process is the production of a very pure stream of CO_2, which can be collected cost-effectively. A pilot plant of HIsarna was constructed by Tata Steel Ijmuiden, and in November 2018, the company announced that a large-scale HIsarna will be constructed at Tata Steel, Jamshedpur [35]. It could be a path-breaking step for the steel industries in terms of energy consumption and environmental emissions. The capital costs of the HIsarna process are also 10–15% lower than the conventional BF-BOF route due to excluding sinter plant and coke ovens. It could also help in achieving a low phosphorus level, as it maintains a lower temperature in the bath.

Finex Technology—Considering the large amount of iron ore fines produced during mining as well as the abundance of non-coking coal (combined with limited resources of coking coal), Finex technology was considered to be one of the potential technologies for India [36]. In 2015, POSCO was planning to install a 12 MT steel plant using Finex technology [37]. However, because of non-technical reasons, the plan was discontinued. Considering the success of COREX technology in two plants in India, Finex technology is a viable route for the future of Indian steel sector.

Flash Ironmaking Technology (FIT) is a potential idea to utilize the huge quantity of blue dust existing in Indian mines. It is based on the reduction of fine iron ore particles to convert them directly to metallic iron with suitable reductants (such as hydrogen, natural gas, coal gas, or a combination of gases) [38]. This technology has been developed in the University of Utah as a part of American Iron and Steel Institute's CO_2 breakthrough program. Agglomeration techniques can be avoided in this technology, thus avoiding the usage of coke. Thus far, it has been tested only in bench-scale experiments. However, this idea can be explored in the future, as it helps in reducing energy consumption and CO_2 emission [39].

Biomass can be used as a renewable fossil fuel to mitigate the emission of CO_2. It can be charged at the top of a blast furnace along with coke, injected through tuyeres, or blended with coke to produce bio-coke. However, it is a challenge due to lower Coke Strength after Reaction (CSR) and higher Coke Reactivity Index (CRI) as compared to metallurgical coke. Biomass fuel shall be effective in reducing fossil CO_2 emission and more effective in the mitigation of SO_x and NO_x [40]. In Brazil, there are numerous mini blast furnaces based on charcoal [41]. Very few blast furnaces in the world tried using biomass. Stubble burning such as in Punjab and Haryana causes extreme air pollution every year. Instead, large amounts of biomass could be produced and potentially utilized by steel manufacturers. This possibility is yet to be explored for sustainable carbon footprint and its availability for the steel sector. The optimization of the biomass value chain and the efficient conversion technologies are of high importance for replacing fossil fuels in the near future.

4.2.3. Energy Efficiency in Indian Steel Industry

Significant improvements in energy efficiency can be achieved in different unit processes through improved operation of equipment via optimized integration of in-plant energy flows and by upgrading process equipment to commercially available Best Available Technology (BAT). An energy saving of around 25–30% per ton of crude steel can be achieved by improving the operational efficiency and adopting BAT for all the units of the BF-BOF production pathway, relative to the global average energy intensity for this route today (Figure 9 [31]). Using electricity to substitute for fossil fuels in the provision of process heat in equipment outside the main process units, particularly in preheaters and boilers, is another option where electrification makes a change.

4.3. Energy and Environment Conservation

Environmental emissions of the steel industries primarily relate to the air and water pollution and solid wastes [42]. The global concern over climate warming is attributed to the CO_2 emissions. An extensive decarbonization of the world steel industry until the middle of this century is a common target. India is liable to take care of its own share in this global crisis. In spite of the planned elevation in the steel production, the CO_2 emissions should peak in the next years and then turn down toward carbon neutrality in the middle of the century or shortly after. That demands the right selections and epoch-making actions. The most essential ways to solve this dilemma are discussed in the following sections.

4.3.1. Carbon Sinks, Capture, and Storage

One of the best ways to capture CO_2 is the natural sequestration by plants. India aims to create a sink of 2.5–3 billion tons of CO_2 through additional forestation and tree cover by 2030. Apparently, through large-level community involvement, India has launched the Mahatma Gandhi National Rural Employment Guarantee Act (MGNREGA), which focuses on environment and natural resource conservation. A detailed review of its potential impact on carbon sequestration through the conservation of green natural resources has been carried out by the Indian Institute of Science, Bangalore [43]. Steel plants are also actively involved in such afforestation drives [44].

Transition to processes based on low-carbon and carbon-free clean energy is a key factor to conserve the energy and environment. During the transition period, Carbon Capture and Storage (CCS) affords further means to cut CO_2 emissions. With CCS technology, up to 90% of the CO_2 emitted can be captured, compressed at high pressure, converted into a liquid, and injected to geological formation sites to be stored underground without significant leaking for hundreds of years. The implementation of this process along with TGR-OBF, HIsarna, etc., can significantly reduce the emissions levels. One major concern for CCS deployment in India is to find an accurate and suitable geological site for the installation of CCS. Another issue is that CCS significantly increases the cost of electricity while reducing net power output, which is often cited as being the biggest barrier to the acceptability of CCS in India [45].

As mentioned earlier CO_2 can be injected into geological formation sites such as saline aquifers, basalt formations, depleted oil and gas fields, and non-minable coal seams to fixate CO_2 as carbonates. An initial geological study suggests that 500–1000 Gt of CO_2 can be potentially stored around the subcontinent [46]. More specific studies pertaining to India in this direction are needed, and furthermore, India is closely studying these developments in developed countries in terms of implementation in the actual scale and subsequently planning to adopt them.

The adoption of CCS in India, especially to thermal power plants, would create major challenges, and it may be offset by the following means [45]:

- Development of IGCC (Integrated Coal Gasification Combined Cycle) technology, which gasifies the coal and uses a combined cycle (combination of gas and steam turbines) to generate electricity;
- Indigenous development of capture and compression equipment for cost efficiency;
- Improved blending and beneficiation of coal;
- Membrane-based capturing.

CCUS combined with enhanced oil recovery (EOR) can be a win–win situation for India, since it can help arrest declining output from oil and gas fields. The technology plays an increasingly important role in achieving carbon neutrality. India is now exploring its CCS potential in the power sector. A plant at the industrial port of Tamil Nadu's Tuticorin has begun capturing CO_2 from its own coal-powered boiler and using it to make baking soda [45].

Interestingly, steel slags themselves can potentially be used for CO_2 capture. Among the industrial wastes, iron and steel slags have the maximum CO_2 sequestration potential [47]. Raghavendra et al. have given an excellent review of various ways in which CO_2 can be captured such as hot route carbonation by treating the hot slag with CO_2, direct route involving gas–solid, thin film, and aqueous slurry carbonation, and an indirect route involving pH and pressure swing CO_2 absorption techniques [48].

Indian steel plants have started looking at some of these options more closely to develop in-house technologies to bring down the carbon footprint [49].

The hydrogen-based DRI route (alongside blending of electrolytic hydrogen into current blast furnaces and DRI units) and integration of CCUS in various production units shall account for substantial shares of emission reduction.

4.3.2. Prospects of Power Sector in India

India's power sector is currently dominated by coal. Toward the objective of carbon-free energy, India has set itself a target of installed capacity of 175 GW from renewable energy sources by March 2022. To control the obnoxious emissions, more efficient coal-based units are being commissioned, and inefficient units are being discarded. The total installed capacity of power generation (i.e., maximum electric power output) in India is 363 GW, and the distribution of energy generation capacity across different sources and projection for 2029–2030 is given in Table 2 [50]. The renewable energy given includes solar, wind, and biomass. The shift toward renewable energy resources can be evidenced. The emission factor of the Indian grid electricity was roughly 900 g CO_2/kWh in 2019 [51]. The Energy and Resources Institute (TERI) in Delhi launched a comprehensive study on decarbonization pathways for the power sector, including also energy-intensive industries such as steel [30]. According to their chart, the grid emission factor might decrease to 550 in 2030 and to around 100 g CO_2/kWh in 2050. India is having an ambitious figure of 52% generation through renewable sources in 2030, it seems to be an elusive task.

Table 2. Distribution of electricity generation capacity 2019–2020 and projected distribution for 2029–2030 [50].

Sector	Installed Capacity-2019–2020 (GW)	%	Estimated Capacity 2029–2030	%
Hydro	45.4	12.50	73.45	9.31
Thermal–Coal + Lignite	203.6	55.90	266.9	32.66
Thermal–Gas	24.9	6.90	25	3.07
Thermal–Diesel	0.5	0.10		-
Nuclear	6.8	1.90	16.9	2.32
Renewable energy	82.6	22.70	450.1	52.63
Total	363		831.5	

5. Emissions Mitigation in the Indian Steel Industry—Summarizing Discussion

Currently, the Indian steel sector can be broadly classified into (1) plants using oxygen route steelmaking processes and (2) plants using electric route steelmaking. Oxygen route steel makers mainly use blast furnaces, but additionally, there are two plants with four COREX units to produce hot metal for making steel. On the other hand, the electric route steel makers use DRI either produced in coal-based units or gas-based units and imported scrap for making steel. An Indian specialty is the huge number of small-size induction furnaces (the current number 1174) with an annual production of only tens of thousands of tons of steel each. The current ratio of the oxygen route versus the electric route, EAF and IF is approximately 45:30:25. Overall, the Indian steel production is expected to grow to 250 million tons per annum by 2030 and it is still growing, albeit retarding toward the year 2050 [31].

In blast furnaces, coal injection (PCI) up to 200 kg per ton of hot metal is already in use, and this is expected to only marginally increase due to constraints in operation. Additionally, steel plants are exploring possibilities of injecting gases having a significant proportion of hydrogen through tuyeres and/or in the stack to reduce the carbon footprint. The gas for the "step changes" injection may come either from imported natural gas or large gasifiers using domestic coals, and imported cheaper coals can also be envisaged. With the increasing usage of gases in these plants, DRI-based units will take a prominent role. Furthermore, if the hydrogen economy becomes a reality in the future, these plants can shift toward hydrogen-based DRI production. Further owing to the virgin ore-based production and large capacity, the oxygen route plants continue to produce high-end flat products.

As of 2020, the utilization of hydrogen in the steel industry could play a growing role in mitigating CO_2 emissions as well as improving the existing process. The cost of hydrogen is expected to fall with the rise in the utilization of gas-based processes and increase in renewable energy sources. Under a future low-carbon situation, we can foresee a tremendous growth in demand for hydrogen from 2030. Thus, hydrogen production will increase rapidly from 2030 to 2050, enabling new capacity additions utilizing hydrogen-based reduction processes.

Plants that use electrical steelmaking routes, primarily based on induction furnaces, produce construction-quality steel grades and are located close to the markets, i.e., spread across the country. Many of these units use also DRI, which is primarily produced through coal-based processes. With increasing environmental threats, these plants will be forced to shift from coal-based to gas-based processes. At the same time, a transition to larger capacity units is reasonable for economic reasons. The gas for these DRI units may be imported natural gas or from coal gasification units. These units shall supply small electrical route steel makers located near consumer markets where the scrap also will be available in the future as the consumption and recycling of steel rise in the country. At present, induction furnaces with a maximum capacity of 50 tons are being used. Depending on the demand from the consumer markets, especially near large cities, electric arc furnace-based steelmaking units with large capacities can also emerge.

Incremental advancements can be achieved through improvements in the operation of equipment and by upgrading process equipment to commercially available Best Available Technology (BAT), which reduces the energy demand required per ton of process output. An energy saving of around 20% per ton of crude steel can be achieved by improving the operational efficiency and adopting the BAT for all the units of the BF-BOF production pathway, relative to the global average energy intensity for this route today [30]. Step changes in efficiency shall be achieved by switching to alternative production pathways such as electrification or other fuel shifts. Using electricity to substitute for fossil fuels in the provision of process heat in equipment outside the main process units, particularly in preheaters and boilers, is another option where electrification can make a big change.

For meeting the global environmental goals, India is expected to shift from coal-based processes to more gas-based processes. These changes are expected in both sectors: namely oxygen route steel makers as well as electric route steel makers. It will be facilitated by the reduced prices of LNG due to the increasing number of ports, thus helping gas find a larger foothold in the Indian mix. Increased the replacement of coal-based Direct Reduced Iron (DRI) with gas-based DRI will also help in attaining reductions in final energy consumption and CO_2 emissions. This shift in processes in the iron and steel sub-sector will result in a change of the industrial ecosystem in the country. By applying the necessary improvements in the process such as moving toward gas-based DRI, increased hydrogen usage and the implementation of new technologies such as the CCUS target on specific energy consumption and specific emissions shall be achieved. Figure 10 shows a scenario of emission reduction potential for the Indian iron and steel sector. It incorporates the incremental improvements discussed previously and shown in Figure 9 and, additionally, an introduction of HIsarna and a conservative transition to hydrogen [31]. The projected production growth would mean a tremendous rise in emissions from 2020 to 2050 (Baseline). The planned actions to cut emissions would result in a 56% reduction by 2050 to a level where the total emissions would be approximately 20% higher than today! The peaking year would be around 2040. These issues proceed from the predicted radical growth of the steel production. The concurrent enterprises to raise the production and to cut the emissions are surely contradictory.

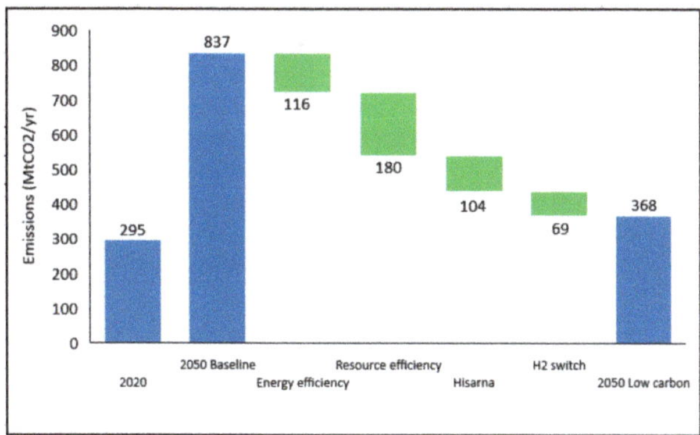

Figure 10. Emission reduction potential of the Indian iron and steel sector Reprinted with permission from TERI [31] Copyright 2020, The Energy and Resources Institute.

For fast-growing economies with a rising emissions trajectory, the need to understand the key variables that impact the choice of a peaking year is as critical as the determinants for the selection of a net-zero year. An analytical formulation [52] shows that the economic growth rate significantly impacts the 'effort gap'. For India, peaking in 2030 would be challenging given the expected economic growth rates for at least the next two decades.

Although the Indian steel industry and the government are aiming for a reduction in CO_2 emissions as well as in specific energy consumption, realization demands stronger commitment in the development of low-carbon and carbon-neutral technologies and the rapid implementation of best available technologies to improve the current processes. Major technical breakthroughs are needed as well as research promotion in the field of carbon-lean/carbon-free iron making. It can be achieved through enhanced hydrogen utilization in the process, which is linked to the production and storage for hydrogen. In addition, the growing supply of scrap is a strong trend in India, which means a growing demand on scrap-based processes, which correspondingly increases the demand for carbon-neutral electricity.

6. Conclusions

The current state of the Indian steel industry and future visions for the year 2030 and beyond were examined with the main focus on the mitigation of CO_2 emissions in relation to the global trends, goals, and agreements. The authors came to the following conclusions:

(1) As a developing country, India has set an ambitious target to more than double its steel production up to 250 MT in 2030. A significant share of the growth is based on BF ironmaking and coal-based DRI production, in other words, utilizing fossil coal as the primary energy source. This makes the CO_2 challenge extremely hard.

(2) A positive matter is that natural gas will partly substitute for coal both in BF ironmaking and direct reduction. It will generate less CO_2 emissions, but eventual methane escape should be strictly eliminated.

(3) Owing to these circumstances, it is evident that the CO_2 emissions will continue growing, and the peaking year for the steel industry's emissions is escaping far ahead. According to the plans of the government and the steel industry, the turning point is expected shortly after 2030.

(4) The current levels of energy consumption and CO_2 emissions in the Indian steel industry are much higher than the world average due to weaknesses in raw materials and energy as well as technological deficiencies. By applying the best available technologies in retrofitting plants and in new constructions, it is realistic to cut energy consumption and emission levels by 35–40% from the present levels toward the end

of the 2020s. This will greatly decelerate the emissions' rise owing to the growing production, but it will not stop it.

(5)   To get on a declining track with CO_2 emissions, stronger decarbonization means are mandatory. A considerable share of new steel plant investments should be based on hydrogen reduction and green electricity in all operations, including hydrogen production. This is the way to reach the emissions' peak and turn downward toward carbon-neutral steelmaking in the middle of the century. By strong commitment to carbon-neutral technologies in new investments, the Indian steel industry can take a forerunner position in fighting climate warming.

Author Contributions: S.P.S.—Literature survey, analysis, plotting the figures and drafting the manuscript. V.N.N.—Conceptualization of the paper, analysis and review of the manuscript. S.M.—Conceptualization of the paper, analysis and review of the manuscript. S.C.—Conceptualization of the paper, analysis and review of the manuscript. L.E.K.H.—Process development and energy inspections—critical evaluation, briefing in writing, validation. All authors contributed significantly. All authors have read and agreed to the published version of the manuscript.

Funding: This research received no external funding.

Institutional Review Board Statement: Not applicable.

Informed Consent Statement: Not applicable.

Conflicts of Interest: The authors declare no conflict of interest.

References

1.   World Steel Association. Global Crude Steel Production Statistics. Available online: https://www.worldsteel.org/steel-by-topic/statistics/annual-production-steel-data/P1_crude_steel_total_pub/CHN/IND (accessed on 22 June 2021).
2.   Steel Ministry of India. An Overview of Steel Sector Global Scenario. Available online: https://steel.gov.in/sites/default/files/An%20Overview%20of%20Steel%20Sector_mar20_0.pdf (accessed on 22 June 2021).
3.   Annual Reports, Ministry of Steel, India. Available online: https://steel.gov.in/annual-reports (accessed on 20 April 2021).
4.   Iron and Steel Industry in India. Available online: https://www.ibef.org/industry/steel.aspx (accessed on 20 April 2021).
5.   Make in India. Available online: https://en.wikipedia.org/wiki/Make_in_India (accessed on 20 April 2021).
6.   Ministry of Mines, Government of India. Indian Minerals Year Book-2014. Available online: https://ibm.gov.in/?c=pages&m=index&id=513 (accessed on 20 April 2021).
7.   Lu, L.; Holmes, R.J.; Manuel, J.R. Effects of alumina on sintering performance of Hematite Iron ores. *ISIJ Int.* **2007**, *47*, 349–358. [CrossRef]
8.   Quast, K. A review on the characterisation and processing of oolitic iron ores. *Miner. Eng.* **2018**, *126*, 89–100. [CrossRef]
9.   Government of India. Coal Statistics, Ministry of Coal. Available online: https://coal.gov.in/public-information/reports/annual-reports (accessed on 20 April 2021).
10.   Indian Steel Ministry. Draft of National Steel Policy. 2017. Available online: https://steel.gov.in/sites/default/files/draft-national-steel-policy-2017.pdf (accessed on 20 April 2021).
11.   Ballal, N.B. Some challenges and opportunities in blast furnace operations. *Trans. Indian Inst. Met.* **2013**, *66*, 483–489. [CrossRef]
12.   Ministry of Petroleum and Natural Gas. Natural Gas Scenario in India. Available online: http://petroleum.nic.in/natural-gas/about-natural-gas (accessed on 20 April 2021).
13.   Nomura, S. Use of waste plastics in coke oven: A review. *J. Sustain. Met.* **2015**, *1*, 85–93. [CrossRef]
14.   Government of India. National Steel Scrap Policy–2017, Ministry of Steel. Available online: https://pib.gov.in/newsite/PrintRelease.aspx?relid=194359 (accessed on 20 April 2021).
15.   Raw Materials. Maximising Scrap Use Helps Reduce CO2 Emissions. Available online: https://www.worldsteel.org/steel-by-topic/raw-materials.html (accessed on 26 August 2021).
16.   Mandal, A.K.; Sinha, O.P. Technological changes in Blast furnace iron making in India since last few decades. *Int. J. Sci. Res.* **2013**, *2*, 211–219.
17.   Agrawal, A.; Das, K.; Singh, B.K.; Singh, R.S.; Tripathi, V.R.; Kundu, S.; Ramna, P.R.V.; Singh, M.K. Means to cope with the higher alumina burden in the blast furnace. *Ironmak. Steelmak.* **2020**, *47*, 238–245. [CrossRef]
18.   Biswas, D.K.; Asthana, S.R.; Rau, V.G. Some studies on energy savings in sponge iron plant. *J. Energy Resour. Technol.* **2003**, *125*, 228–237. [CrossRef]
19.   Dutta, S.K.; Lele, A.B.; Pancholi, N.K. Studies on direct reduced iron melting in induction furnace. *Trans. Indian Inst. Met.* **2004**, *57*, 467–473.
20.   Bedarkar, S.S.; Dalal, N.B. Energy balance in induction furnace and arc furnace steelmaking. *Int. J. Eng. Res. Appl.* **2020**, *10*, 57–61.

21. Jones, J.A.; Bowman, B.; Lefrank, P.A. *Electric furnace steelmaking. Making Shaping and Treating of Steel*; The AISE Steel Foundation: Pittsburgh, PA, USA, 1998.
22. Krishnan, S.S.; Vunnam, V.; Sunder, P.S.; Sunil, J.V.; Ramakrishnan, M.A. *A Study of Energy Efficiency in the Iron and Steel Industry*; Centre for Science and Technology and Policy: Bangalore, India, 2013.
23. Ambasta, D.K.; Pandey, B.; Saha, N. *Utilization of Solid Waste from Steel Melting Shop*; MECON Ltd.: Ranchi, India, 2016.
24. Sharma, N.; Nurni, V.N.; Tathavadkar, V.; Basu, S. A review on the generation of solid wastes and their utilization in Indian steel industries. *Miner. Process. Extr. Met.* **2017**, *126*, 54–61. [CrossRef]
25. Nomura, T.; Yamamoto, N.; Fujii, T.; Takiguchi, Y. Beneficiation plants and pelletizing plants for utilizing low grade iron ore. *Kobelco Technol. Rev.* **2015**, *33*, 8–15.
26. Singh, B.; Kishan, K.H.; Rao, G.V.; Kumar, R. Experimental studies on reduction of coarse blue dust for production of Iron powder by Hydrogen reduction. *IJERT* **2009**, *8*, 464–470.
27. Nogami, H.; Chu, M.; Yagi, J.I. Multi-Dimensional transient mathematical simulator of blast furnace process based on multi-fluid and kinetic theories. *Comput. Chem. Eng.* **2005**, *29*, 2438–2448. [CrossRef]
28. Danloy, G.; Berthelemot, A.; Grant, M.; Borlée, J.; Sert, D.; Van der Stel, J.; Jak, H.; Dimastromatteo, M.; Hallin, N.; Eklund, N.; et al. ULCOS-Pilot experiments for low CO_2 steelmaking at Lulea experimental blast furnace. *Metall. Res. Technol.* **2009**, *106*, 1–8.
29. Dey, N.R.; Prasad, A.K.; Singh, S.K. Energy survey of the coal based sponge iron industry. *Case Stud. Therm. Eng.* **2015**, *6*, 1–15. [CrossRef]
30. Sridhar, A. *Hydrogen Harvesting Process, Tata Review*; Innovista: Oxford, UK, 2009.
31. Hall, W.; Spencer, T.; Kumar, S. *Towards a Low Carbon Steel Sector: Overview of the Changing Market, Technology and Policy Context for Indian Steel*; The Energy and Resources Institute (TERI): New Delhi, India, 2020.
32. Liu, Y.; Liu, L.; Tong, P.Q.; Zhao, J.Y.; Wu, J.P.; Li, X.M.; Wang, G.P. Process control of dephosphorization in initial smelting period of combined blowing BOF for clean steel. *Steelmaking* **2006**, *22*, 27–29. (In Chinese)
33. The MIDREX Process–Midrex Technologies. Available online: https://www.midrex.com/wp-content/uploads/Midrex-STATSbook2019Final.pdf (accessed on 25 August 2021).
34. *Hisarna: Game Changer in Steel Industry (PDF)*; Tata Steel: Mumbai, India, 2017.
35. Tata Steel Kiest Onverwacht Niet Voor IJmuiden om Nieuwe Fabriek te Vestigen, NHNieuws. Available online: https://www.nhnieuws.nl/nieuws/235611/tata-steel-kiest-onverwacht-niet-voor-IJmuiden-voor-nieuwe-fabriek (accessed on 1 December 2019).
36. Jain, U. Corex & Finex new developments in utilization of low grade raw materials. In Proceedings of the International Conference on Science and Technology of Ironmaking and Steelmaking, CSIR-NML, Jamshedpur, India, 16–18 December 2013.
37. POSCO's Finex Plant Likely to be Shifted to India by October–News Article on News 18. Available online: https://www.news18.com/news/india/poscos-finex-plant-likely-to-be-shifted-to-india-by-october-1026323.html (accessed on 28 September 2021).
38. Sohn, H.Y.; Mohassab, Y. Development of a Novel Flash Ironmaking Technology with greatly reduced energy consumption and CO_2 emissions. *J. Sustain. Met.* **2016**, *2*, 216–227. [CrossRef]
39. Sohn, H.Y. Energy consumptions and CO_2 emissions in Iron making and development of a novel flash technology. *Metals* **2020**, *10*, 54. [CrossRef]
40. Mousa, E.; Wang, C.; Riesbeck, J.; Larsson, M. Biomass application in iron and steel industry: An overview of challenges and opportunities. *Renew. Sustain. Energy Rev.* **2016**, *65*, 1247–1266. [CrossRef]
41. De Castro, J.A.; Medeiros, G.; Oliveira, E.; De Campos, M.; Nogami, H. The mini blast furnace process: An efficient reactor for green pig iron production using charcoal and Hydrogen-rich gas: A study of cases. *Metals* **2020**, *10*, 1501. [CrossRef]
42. Birat, J.-P. Society, materials, and the environment: The case of Steel. *Metals* **2020**, *10*, 331. [CrossRef]
43. Esteves, T.; Rao, K.V.; Sinha, B.; Roy, S.S. *Environmental Benefits and Vulnerability Reduction through Mahatma Gandhi NREGS: Synthesis Report*; Ministry of Rural Development and GIZL: New Delhi, India, 2013.
44. SDG Mapping of JSW Steel's Initiatives & Projects. Available online: https://www.jsw.in/groups/sustainability-framework-measuring-success-biodiversity (accessed on 28 September 2021).
45. Handa, R.; Baksi, S. CCUS: A Climate Friendly Approach to India's $5 Trillion Economy. Available online: https://www.downtoearth.org.in/blog/climate-change/ccus-a-climate-friendly-approach-to-india-s-5-trillion-economy (accessed on 28 September 2021).
46. Kalbende, K. Carbon capture and storage in geological formation; Its legal, regulatory impreatives and opportunities in India. *Int. J. Environ. Monit. Anal.* **2015**, *3*, 198. [CrossRef]
47. Pan, S.Y.; Chen, Y.H.; Fan, L.S.; Kim, H.; Gao, X.; Ling, T.C.; Chiang, P.-C.; Pei, S.-L.; Gu, G. CO_2 mineralization and utilization by alkaline solid wastes for potential carbon reduction. *Nat. Sustain.* **2020**, *3*, 399–405. [CrossRef]
48. Ragipani, R.; Bhattacharya, S.; Suresh, A.K. A review on steel slag valorisation via mineral carbonation. *React. Chem. Eng.* **2021**, *6*, 1152. [CrossRef]
49. Tata Steel and Council of Scientific & Industrial Research (CSIR) Sign MoU to Collaborate in the Area of Carbon Capture, Utilisation and Storage (CCUS). Available online: https://www.tatasteel.com/media/newsroom/press-releases/india/2020/tata-steel-and-council-of-scientific-industrial-research-csir-sign-mou-to-collaborate-in-the-area-of-carbon-capture-utilisation-storage-ccus/ (accessed on 28 September 2021).
50. Government of India; Ministry of Power; Central Electricity Authority. Report on Optimal Generation Capacity Mix for 2029–2030. Available online: https://cea.nic.in/old/reports/others/planning/irp/Optimal_generation_mix_report.pdf (accessed on 20 August 2021).

51. Takahashi, K.; Louhisuo, M. IGES List of Grid Emission Factors, Version: 10. Available online: https://www.iges.or.jp/en/pub/list-grid-emission-factor/en?_ga=2.239720963.1615283338.1630320849-201355943.1630320849 (accessed on 20 August 2021).
52. Chaturvedi, V. *Peaking and Net-Zero for India's Energy Sector CO_2 Emissions—An Analytical Exposition*; Council of Energy Environment and Water: New Delhi, India, 2021; Available online: https://www.ceew.in/publications/how-can-india-attain-net-zero-emission-by-2050 (accessed on 10 February 2021).

 *metals*

Article

The Mini Blast Furnace Process: An Efficient Reactor for Green Pig Iron Production Using Charcoal and Hydrogen-Rich Gas: A Study of Cases

Jose Adilson de Castro [1,2,*], Giulio Antunes de Medeiros [1] , Elizabeth Mendes de Oliveira [3], Marcos Flavio de Campos [1] and Hiroshi Nogami [2]

[1]   Graduate Program on Metallurgical Engineering, Federal Fluminense University, Volta Redonda, Rio de Janeiro 27255-125, Brazil; joseadilsoncastro@id.uff.br (G.A.d.M.); marcosflavio@id.uff.br (M.F.d.C.)
[2]   IMRAM-Institute of Multidisciplinary Research for Advanced Materials, Tohoku University, Sendai 980-8577, Japan; nogami@gmail.com
[3]   Federal Center for Technological Education Celso Suckow da Fonseca, Angra dos Reis, Rio de Janeiro 23953-030, Brazil; beth.mendes.oliveira@gmail.com
*   Correspondence: adilson@metal.eeimvr.uff.br; Tel.: +55-024-2107-3756

Received: 15 July 2020; Accepted: 8 November 2020; Published: 11 November 2020

Abstract: The mini blast furnace process is an efficient route to produce pig iron based on the burden with granulated charcoal. New, improved technologies have recently been introduced in the mini blast furnace process, such as pulverized charcoal and gas injections, new burden materials, and peripheral devices that improve the overall process efficiency. In this paper, we revise the new injection possibilities and discuss new aspects for further developments. The analysis is carried out with a comprehensive multiphase multicomponent mathematical model using mass, momentum, and energy conservation principles coupled with the rate equations for chemical reactions, multiphase momentum, and heat exchanges. We analyze new technological possibilities for the enhancement of this process as follows: (i) a base case of pulverized charcoal injection with industrial data comparison; (ii) a set of scenarios with raceway injections, combining pulverized charcoal with hydrogen-rich fuel gas, replacing granular charcoal in the burden; (iii) a set of scenarios with hydrogen-rich gas injection at the shaft level, replacing reducing gas in the granular zone of the reactor; and the possible combination of both methodologies. The simulated scenarios showed that a considerable decrease in granular charcoal consumption in the burden materials could be replaced by combining a pulverized charcoal injection of 150 kg/t$_{HM}$ and increasing rich gas injections and oxygen enrichment values, decreasing the specific blast injection and granular charcoal. The productivity of the mini blast furnace process was increased for all scenarios compared with the reference case. We review the aspects of these operational conditions and present an outlook for improvements on the process efficiency.

Keywords: mini blast furnace; charcoal; hydrogen; mathematical model; gas injection; kinetic models; self-reducing burden

1. Introduction

Renewable energy in the ironmaking industry is an important issue to develop the modern industry. In the steelmaking production routes, the carbon-based processes are predominant [1–7]. In this sense, the pig iron production step could significantly drive a new green steelmaking technology. The charcoal mini blast furnace process is an established technology. Although the charcoal blast furnace has been regarded as old technology, modern auxiliary and peripheral equipment have been introduced and greatly improved this technological route for hot metal production [2,4]. The mini blast furnace's efficiency based on granular charcoal has continuously been improved with the introduction of proven technological advancements formerly developed to the large blast furnaces based on coke operation and prepared burden materials [2,4]. The mini blast furnace's efficiency is strongly dependent on the fuel and reducing agent quality and the preparation of the burden materials [4]. The granular charcoal quality is variable and depends on several aspects of the biomass production and the carbonization processes. Recent improvements in the technology for biomass production have been widely reported for other sectors of applications. However, the mini blast furnace process must obtain new improvements due to its specific requirements. Moreover, the technology of pulverized materials injections in the tuyeres of large blast furnaces has achieved stable and efficient operations for quite a long period [4,8–19]. The technology enhancements for pulverized biomass injection require special attention, mainly on the grinding and pneumatic transportation systems.

The available technologies for the large coke-based blast furnaces have continuously being improved through applying novel digital paradigms [9–20]. The improvements have been mainly driven by demands to decrease the specific carbon, restrictions on the available raw materials quality, and demand for the high quality of reducing agents and renewable energy sources [7,8]. However, the total amount of fossil carbon required to produce pig iron using the coke-based blast furnace route is becoming prohibitive. The main restriction to the coke-based blast furnace process is global warming trends with a strong demand for low carbon technologies. Hence, alternative routes that minimize the fossil carbon used in the process are attractive and challenging, allowing the researchers and developers to propose several computational-based approaches. The hydrogen and the oxygen-based blast furnaces processes are recently in vogue and promise to positively impact the steel industry [2,4–7]. The developing technologies have demonstrated effectiveness in reducing CO_2 emissions. However, the effective cost of hot metal production and hydrogen-rich gas and oxygen availability are still barriers to effective implementations. The Brazilian hot metal industry has a long experience with the mini blast furnace's technology operating with high productivity, solid and liquids flow stability, and raw materials supply logistics. Recently, several small and compact granular charcoal-based blast furnaces, which used to operate only with granular charcoal produced by the carbonization process of eucalyptus wood, are incorporating and adapting new pulverized injection facilities and incorporating an off-gas treatment efficient cooling system, enhanced refractories, tapping technologies, and an efficient blast heating system [2,4]. These technologies are becoming viable due to the lowering of the investment cost and the high possibility of increasing efficiency and productivity. The mini blast furnace technology is considered net carbon-free, assuming a complete carbon cycle based on the intensive eucalyptus plantation of about seven years with high productivity in tropical countries pursuing the available land area. The technologies for addressing the carbonization of various biomasses have also been continuously improved over the years and have contributed to new carbon renewable sources. New pulverized charcoal sources can be available shortly using other plantation species with a smaller life cycle and higher productivity.

The mini blast furnace process analysis has been performed by several researchers using a variety of methodologies [12–26]. Nonetheless, the empirical and theoretical approaches aiming to analyze appropriate operational practices using variable sources of raw materials require a comprehensive approach using detailed models.

The use of sophisticated and comprehensive mathematical models based on transport phenomena principles that consider detailed chemical reaction mechanisms and rate equations demands time-consuming computations [18–27]. This approach is particularly useful for studying the impacts of multifactorial aspects of testing a new technological concept. For example, the operational practices have been dramatically improved by using operational diagrams based on mass and energy zone balances applied to the process analysis or models based on statistical and data analysis to implement control tools and automatization after the new modern industry concept. Nevertheless, further improvements and detailed analysis are only possible by applying comprehensive mathematical modeling, which uses fundamental equations based on conservation principles of mass, momentum, energy, and detailed chemical species. The present study uses this methodology. A considerable number of mathematical models to predict the large coke-based blast furnace operations and pulverized coal injection have been developed and shown their ability to predict the actual operation. These are currently used to recommend new technologies that improve the efficiency of the processes [12–27]. These continuous developments have maintained the coke-based blast furnace process as the leading technological route to produce pig iron for a long period. The tendency is that this route will still be dominant in the near future. However, other competitive reactors have been developed and proved commercially profitable, depending on local conditions [7,27–29]. Comprehensive mathematical models to predict the mini blast furnace process based on charcoal operation are scarce in the current literature. Some of the recent attempts were addressed by Matos et al. [30] and Castro et al. [4,21].

The focus was to analyze the granular charcoal-based mini blast furnace process's actual operating conditions in their works. In the current study, we established new mathematical modeling, which considers six phases and their mutual interactions of momentum, mass, and energy. New rate equations are included in the model with their respective individual chemical species. The model employed the representative elementary control volume concept, which is especially valuable to deal with the simultaneous coexistence of the phases in the lower part of the furnace under pulverized charcoal injection conditions. Figure 1 schematically shows the theoretical approach for dealing with the complex and simultaneous motion of gas, solid, pulverized charcoal, hot metal, and slag coexisting in the same representative control volume and the corresponding interphase interactions of momentum, mass (chemical species), and energy. This model differs from the previous attempts due to the new phase implemented to account for the pulverized charcoal injection and the new chemical species and the implementations of the chemical reactions involving the new phase and chemical species.

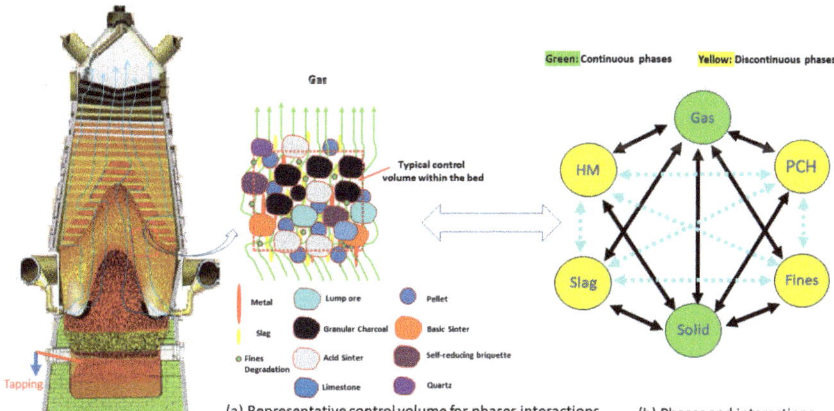

Figure 1. The multiphase and multicomponent approach of the internal state variables of the charcoal mini blast furnace modeling. (**a**) A representative control volume of the inner furnace. (**b**) The corresponding local multiphase interactions of mass, momentum, and enthalpy. In (**a**), the color particles represent the type of the granular materials inside the reactor, and (**b**) color represents the phases and their interactions: green is continuous phases, and yellow is discontinuous phases. Continuous lines indicate momentum, chemical reactions, and energy exchange. Dotted lines are for only chemical reactions and energy exchanges among the phases with single interactions for momentum.

Additionally, new operational conditions with high pulverized charcoal injection are analyzed. In this study, the hot hydrogen-rich gas injection is addressed for analyzing the mini blast furnace process. The impacts on the inner reactor variable's distributions are discussed in the view of the carbon intensity parameter, which is defined as the total amount of carbon used to produce a unit of pig iron, as a parameter for low-carbon technological analysis. The operational practices and scenarios analyzed in this study are fully based on a renewable source of carbon and hydrogen-rich gas. Thus, the mini blast furnace process product could be referred to as the green pig iron.

2. Materials and Methods

Methods for analyzing the mini blast furnace process based on simulation tools have been continuously developed since the last century. This paper presents a comprehensive multiphase, multicomponent model based on the full physical and chemical interactions with the consideration of detailed rate equations accounting for the kinetics of the chemical reactions and phase transformations dynamically coupled with the internal state variables of the mini blast furnace reactor. The model is coupled with the thermophysical properties of the phases and chemical species being locally considered. Figure 1 presents the concept of the multiphase and multicomponent approach. In this concept, the process's inner state variables are associated with a representative volume of the bed so that the phases can coexist. The phases present their own chemical species and thermophysical properties, which are dynamically changed during their motion inside the reactor. The phases interact with one another, exchanging momentum, mass, and enthalpy through interfaces phenomena, chemical reactions, and phase transformations. The phases and interactions considered in this model are illustrated in Figure 1b.

The solid lines connecting two phases represent full interactions of mass, momentum, and energy between the phases. The dotted lines indicated that a simplified model for momentum interactions are assumed besides the full interactions for mass and chemical reactions. The gas-phase includes the blast chemical species and those generated with the reactor due to the chemical reactions. The solid phase includes all the granular materials charged in the furnace through the burden materials charging system (granular charcoal, lump ore, sinter, small sinter, pellets, and slag agents). The pulverized charcoal (PCH) corresponds to the pulverized material injected through the tuyeres' injection lances. The hot metal (HM) comprises the liquid reduced materials that form the liquid pig iron. The slag is formed by the liquid oxides incorporated during the melting down phenomenon, silicon transfer, phosphorus, manganese, and other chemical reactions during alloy elements incorporated in the pig iron. The fines are regarded as the fine particles produced during the burden materials degradation and carried by the upward gas flow. Thus, fines can be formed by all the chemical species present in the burden materials and those formed during the solid transformations.

The phases and their chemical species are presented in Table 1. The chemical reactions involve the multiphase and multicomponent mass transfer, accompanied by momentum and heat, depending on the local rate equations and specific mechanisms. The chemical reactions and phase transformations considered in this model are listed in Table 2. The rate equations considered for these reactions were implemented using available models and data previously tested and adapted by the authors from the literature [2,4,5,8–45]. The momentum and energy interphases exchanges are modeled based on fundamental relations using local flowing conditions [8,31,34–36]. These relations were adapted to the mini blast furnace conditions [2,4]. The boundary conditions and analysis cases are specified, obeying the mini blast furnace's actual operation practice using a pulverized charcoal injection system and peripheral facilities.

Table 1. The phases and chemical species assumed in the charcoal mini blast furnace model.

Phases	Chemical Species
Gas	N_2, O_2, CO, CO_2, H_2, H_2O, NO_x, SO_x, H_2S, SiO, fuel and reducing gas (H_2, CO, CO_2, H_2O, CH_4, C_2H_6, C_3H_8, C_4H_{10}, C_nH_m)
Solid	Granular charcoal: C, volatile, H_2O, Ash (SiO_2, CaO, Al_2O_3, MgO, MnO, P_2O_5, S) Sinter: Fe_2O_3, Fe_3O_4, FeO, Fe, H_2O, SiO_2, CaO, Al_2O_3, MgO, MnO, P_2O_5, FeS Pellet: Fe_2O_3, Fe_3O_4, FeO, Fe, H_2O, SiO_2, CaO, Al_2O_3, MgO, MnO, P_2O_5, FeS Lump ore: Fe_2O_3, Fe_3O_4, FeO, Fe, H_2O, SiO_2, CaO, Al_2O_3, MgO, MnO, P_2O_5, FeS Briquette: C, volatile, Fe_2O_3, Fe_3O_4, FeO, Fe, H_2O, SiO_2, CaO, Al_2O_3, MgO, MnO, P_2O_5, FeS Quartz: SiO_2, H_2O, gangue (Al_2O_3, MgO, MnO, CaO, P_2O_5, Fe_2O_3, Fe_3O_4, FeO) Limestone: $CaCO_3$, $MgCO_3$, H_2O, gangue (SiO_2, Al_2O_3, MgO, MnO, CaO, P_2O_5, Fe_2O_3, Fe_3O_4, FeO)
PCH [1]	C, volatile, H_2O, ash (SiO_2, CaO, Al_2O_3, MgO, MnO, P_2O_5, S)
Fines [2]	C, volatile, H_2O, Fe_2O_3, Fe_3O_4, FeO, Fe, SiO_2, CaO, Al_2O_3, MgO, MnO, P_2O_5, FeS
Hot metal	Fe, C, Si, Mn, P, S, Ti. TiC
Slag	SiO_2, CaO, Al_2O_3, MgO, MnO, TiO_2, MnS, P_2O_5

[1] PCH: Pulverized charcoal injection; [2] Fine materials generated during burden degradation.

Table 2. The assumed chemical reactions and phase transformations in the biomass-based mini blast furnace process.

Reactions	Reactions
Combustion and Gasification	
$C\,(i)\,+O_2(g)\rightarrow CO_2(g)$	$C\,(i)\,+\tfrac{1}{2}O_2(g)\rightarrow CO(g)$
$C\,(i)\,+CO_2(g)\rightarrow 2CO(g)$	$C\,(i)\,+H_2O(g)\rightarrow CO(g)\,+H_2(g)$
$S_2\,(i)\,+2\,O_2(g)\rightarrow 2\,SO_2(g)$	$S_2\,(i)\,+O_2(g)\rightarrow 2\,SO(g)$
$SiO_2\,(i)\,+C(i)\leftrightarrow SiO(g)+CO(g)$	$SO_2\,(g)\,+0.5\,S_2(i)\rightarrow 2\,SO(g)$
$SiO_2\,(i)\,+3C(i)\leftrightarrow SiC(i)+2CO(g)$	$SiC\,(i)\,+CO(g)\rightarrow SiO(g)+2C(i)$
$H_2O\,(vapor)\,+C(i)\rightarrow H_2(g)+CO(g)$	$CH_4(g)\,+\tfrac{1}{2}O_2(g)\rightarrow CO(g)+2H_2(g)$
volatiles $(i)\,+\alpha_1O_2\rightarrow\alpha_2CO_2(g)\,+\alpha_3H_2O(g)\,+\alpha_4N_2(g)$	$C_2H_6(g)\,+O_2(g)\rightarrow 2CO(g)+3H_2(g)$
volatiles $(i)\,+\alpha_5CO_2(g)\rightarrow\alpha_6CO(g)\,+\alpha_7H_2(g)\,+\alpha_8N_2(g)$	$C_3H_8(g)\,+\tfrac{3}{2}O_2(g)\rightarrow 3CO(g)+4H_2(g)$
$C_4H_{10}(g)\,+2O_2(g)\rightarrow 4CO(g)+5H_2(g)$	$CO(g)\,+\tfrac{1}{2}O_2(g)\rightarrow CO_2(g)$
$H_2(g)\,+\tfrac{1}{2}O_2(g)\rightarrow H_2O(g)$	$CH_4(g)\,+CO_2(g)\rightarrow 2CO(g)+2H_2(g)$
$C_2H_6(g)\,+2CO_2(g)\rightarrow 4CO(g)+3H_2(g)$	$C_3H_8(g)\,+3CO_2(g)\rightarrow 6CO(g)+4H_2(g)$
$CH_4(g)\,+2O_2(g)\rightarrow CO_2(g)+2H_2O(g)$	$C_2H_6(g)\,+5O_2(g)\rightarrow 2CO_2(g)+3H_2O(g)$
$C_3H_8(g)\,+5O_2(g)\rightarrow 3CO_2(g)+4H_2O(g)$	$C_4H_{10}(g)\,+4CO_2(g)\rightarrow 8CO(g)+5H_2(g)$
$C_4H_{10}(g)\,+\tfrac{13}{2}O_2(g)\rightarrow 4CO_2(g)+5H_2O(g)$	$2SO_2(g)\,+3H_2(g)\leftrightarrow 2H_2O(g)+H_2S(g)$
$CO_2\,(g)\,+H_2(g)\leftrightarrow CO(g)\,+H_2O(g)$	$C_nH_m(g)\,+\left(n+\tfrac{m}{2}\right)O_2(g)\rightarrow nCO_2(g)+mH_2O(g)$
$C_nH_m(g)\,+\left(\tfrac{n}{2}\right)O_2(g)\rightarrow nCO(g)+mH_2(g)$	$N_2(g)+O_2(g)\rightarrow 2NO(g)$
$H_2O\,(i)\;\;\leftrightarrow\;\;H_2O(g)$	$N_2(g)+2O_2(g)\rightarrow 2NO_2(g)$
i = granular charcoal, pulverized charcoal, briquettes, biomasses, and fines	
Gas-Solid Gaseous Reduction	
$3\,Fe_2O_3\,(i)\,+CO(g)\rightarrow CO_2(g)+2\,Fe_3O_4(i)$	$3\,Fe_2O_3\,(i)\,+H_2(g)\rightarrow H_2O(g)+2\,Fe_3O_4(i)$
$\tfrac{w}{4w-3}\,Fe_3O_4\,(i)\,+CO(g)\rightarrow CO_2(g)+Fe_wO(i)$	$\tfrac{w}{4w-3}\,Fe_3O_4\,(i)\,+H_2(g)\rightarrow H_2O(g)+Fe_wO(i)$
$Fe_wO\,(i)\,+CO(g)\rightarrow CO_2(g)+w\,Fe(i)$	$Fe_wO\,(i)\,+H_2(g)\rightarrow H_2O(g)+w\,Fe(i)$
$FeS(i)+O_2(g)\rightarrow FeO(i)+SO(g)$	$FeS(i)+1.5\,O_2(g)\rightarrow FeO(i)+SO_2(g)$
$CaCO_3(i)\rightarrow CaO(i)+CO_2(g)$	$MgCO_3(i)\rightarrow MgO(i)+CO_2(g)$
i = lump ore, sinter, pellets, briquette, and fines	
Melting/Slagging	
$Fe_2O_3(i)\rightarrow Fe_2O_3\,(slg)$	$Fe_3O_4(i)\rightarrow Fe_3O_4\,(slg)$
$FeO(i)\rightarrow FeO\,(slg)$	$Fe(i)\rightarrow Fe\,(hm)$
$CaO(i)\rightarrow CaO\,(slg)$	$Al_2O_3(i)\rightarrow Al_2O_3\,(slg)$
$SiO_2(i)\rightarrow SiO_2\,(slg)$	$MnO(i)\rightarrow MnO\,(slg)$
$SiO_2(slg)+C(i)\rightarrow SiO\,(g)+CO(g)$	$TiO_2(i)\rightarrow TiO_2\,(slg)$
$MgO(i)\rightarrow MgO\,(slg)$	$P_2O_5(i)\rightarrow P_2O_5\,(slg)$
$SO(g)+Si(hm)\rightarrow S\,(hm)+SiO(g)$	
i = lump ore, sinter, pellets, briquette, granular charcoal, pulverized charcoal, biomasses, and fines	
Hot Metal Incorporation Elements	
$C(i)\rightarrow C\,(hm)$	$SiO(g)+C(i)\rightarrow Si\,(hm)+CO(g)$
$SiO(g)+C(hm)\rightarrow Si\,(hm)+CO(g)$	$SO(g)+C(i)\rightarrow S\,(hm)+CO(g)$
$SO(g)+C(hm)\rightarrow S\,(hm)+CO(g)$	$CaO(slg)+C(hm)+S(hm)\leftrightarrow CaS\,(slg)+CO(g)$
$MnO(slg)+C(hm)\leftrightarrow Mn\,(hm)+CO(g)$	$2\,MnO(slg)+Si(hm)\leftrightarrow 2\,Mn\,(hm)+SiO_2(slg)$
$P_2O_5\,(slg)+5C(hm)\rightarrow 2P(hm)+5CO(g)$	$TiO_2(slg)+Si(hm)\leftrightarrow Ti\,(hm)+SiO_2(slg)$
$TiO_2(slg)+2C(hm)\leftrightarrow Ti\,(hm)+2CO(g)$	
i = granular charcoal, pulverized charcoal, briquettes, biomasses, and fines	
Liquid-Solid Direct Reduction	
$Fe_2O_3\,(slg)+3C(i)\rightarrow 2Fe\,(hm)+3CO(g)$	$Fe_3O_4\,(slg)+4C(i)\rightarrow 3Fe\,(hm)+4CO(g)$
$FeO\,(slg)+C(i)\rightarrow Fe\,(hm)+CO(g)$	$P_2O_5\,(slg)+5C(i)\rightarrow 2P(hm)+5CO(g)$
$TiO_2\,(slg)+2C(i)\rightarrow Ti(hm)+2CO(g)$	$MnO(slg)+C(i)\leftrightarrow Mn\,(hm)+CO(g)$
i = Granular charcoal, briquette, pulverized charcoal, and fines	

$\alpha_1,\alpha_2,\alpha_3,\alpha_4,\alpha_5,\alpha_6,\alpha_7,\alpha_8$: coefficients depend on the biomass materials volatilization characteristics, m and n are stoichiometric coefficients, g: gas, hm: hot metal/pig iron, slg: slag species.

Mass and energy transfer mechanisms are locally considered in the rate equations, resulting in a complex and coupled kinematic model, allowing the model equations to compute variable raw materials and operational conditions. A complete kinetic database for a range of raw materials used has been constructed in this sense. As the model implementation is done in an open-source computational code, new kinetic data have been continuously developed and updated in the model based on both experimental work and industrial trials [2,4]. Therefore, it is worthy to emphasize that the capability of the model prediction and its accuracy have been continuously enhanced, with new implementations and consistent experimental data incorporation to new raw materials.

The kinetic models' fundamentals were implemented using the same model structure but with a new constant rate. The parameters were adjusted for the new materials' experimental data obtained under controlled dimensionless parameters. Therefore, the computational code and simulation tools are open and able to consider the raw materials availability's continuous changes. Thus, a useful computational code for the analysis of the process is welcome. Using this model, we can predict the value of using a new raw material in the green pig iron production route based on its performance in the process. Moreover, it is possible to determine the operational conditions that maximize the new raw materials' competitive characteristics' efficiency. Some new features are included in this paper as reacting to new species and their corresponding new rate equations. It is necessary to obtain suitable rate equations for the reactions taken under the mini blast furnace environment conditions regarding the biomass utilization. These rate equations were based on the previous studies and adapted for the mini blast furnace process conditions under hydrogen-rich gas and pulverized charcoal injections [2,4,8–45].

2.1. The Multiphase and Multicomponent Model Equations Applied to the Charcoal Mini Blast Furnace Reactor

The model formulation assumes that the conservation principle holds in the calculation domain. The phases are denominated as continuous and discontinuous for accounting for the phases interactions. The model is based on a set of conservation equations for momentum, energy, and chemical species of the coexisting phases, summarized in Equations (1)–(7).

Momentum and continuity of continuous phases:

$$\frac{\partial\left(\rho_i\varepsilon_i u_{i,j}\right)}{\partial t} + \frac{\partial\left(\rho_i\varepsilon_i u_{i,k}u_{i,j}\right)}{\partial x_k} = \frac{\partial}{\partial x_k}\left(\mu_i\frac{\partial u_{i,j}}{\partial x_k}\right) - \frac{\partial P_i}{\partial x_j} - F_j^{i-l} \tag{1}$$

$$\frac{\partial(\rho_i\xi_i)}{\partial t} + \frac{\partial\left(\rho_i\xi_i u_{i,k}\right)}{\partial x_k} = \sum_{m=1}^{Nreacts} M_n r_m \tag{2}$$

Momentum and mass conservations of discontinuous phases:

$$\frac{\partial\left(\rho_i\varepsilon_i u_{i,j}\right)}{\partial t} + \frac{\partial\left(\rho_i\varepsilon_i u_{i,k}u_{i,j}\right)}{\partial x_k} = \rho_i\varepsilon_i g_j - F_j^{i-l} \tag{3}$$

$$\frac{\partial(\rho_i\xi_i)}{\partial t} + \frac{\partial\left(\rho_i\xi_i u_{i,k}\right)}{\partial x_k} = \sum_{m=1}^{Nreacts} M_n r_m \tag{4}$$

Conservation of the phase's chemical species:

$$\frac{\partial(\rho_i\varepsilon_i\varphi_n)}{\partial t} + \frac{\partial\left(\rho_i\varepsilon_i u_{i,k}\varphi_n\right)}{\partial x_k} = \frac{\partial}{\partial x_k}\left(D_n^{eff}\frac{\partial\varphi_n}{\partial x_k}\right) + \sum_{m=1}^{Nreacts} M_n\, r_m \tag{5}$$

Conservation of the phase's enthalpies:

$$\frac{\partial(\rho_i\varepsilon_i h_i)}{\partial t} + \frac{\partial\left(\rho_i\varepsilon_i u_{i,k}h_i\right)}{\partial x_k} = \frac{\partial}{\partial x_k}\left(\frac{k_i}{Cp_i}\frac{\partial h_i}{\partial x_k}\right) + E^{i-l} + \sum_{m=1}^{Nreacts} \Delta h_m r_m \tag{6}$$

Restriction for consistency of coexisting phases' volumes fractions:

$$\sum_{i=1}^{Nphases} \varepsilon_i = 1 \tag{7}$$

The x variable represents the physical coordinates with the indexes i, j, and k representing the coordinate directions. The phase properties ρ, ε, μ are the effective density, volume fraction, and viscosity with the indexes i and l accounting for the phases. The variables u, P, and F stand for the velocity component, pressure, and interphase interaction forces with the indexes i and l indicating the phases and j and k indicating the coordinate directions. M and r are the chemical species molecular weight and the chemical reactions rates, with the indexes n and m representing the chemical species and their respective chemical reactions. The variables φ, D_n^{eff} are mass fractions and effective species diffusion coefficient of the chemical species within the phases. E, h, k, C_p and Δh are the heat interphase exchanges, phase enthalpy, thermal conductivity, specific heat capacity, and enthalpy of reactions, respectively.

Table 1 lists the phases and their chemical species considered in this study. The solid phase is divided into burden components with their own compositions and chemical species, undergoing particular chemical reactions kinetics. The rate parameters obtained from experimental work are carried out under similarity, enabling the modeling implementations' scaleup. Thus, the fundamental theoretical principles are compatible with the reactor conditions and can be applied to model the whole process features.

The set of partial differential equations is resolved, assuming the boundary conditions representing the mini blast furnace conditions with the process monitoring's operational data. Therefore, the boundary conditions are applied to the computational domain boundary delimitated at the bottom by the slag surface, at the top by the burden surface profile, and by lateral walls. At the top, the gas phase is assumed as a fully developed flow, and solid inflow is modeled based on the inflow rate given by local solid mass consumption due to chemical reactions and melting or gravity-driven flows. At the tuyere injection, the blast's inlet, additional oxygen, hydrogen, or fuel gas, and pulverized charcoal are given by their inflow rates.

The blast flowrate and gaseous fuel injection are fixed. In contrast, the pulverized charcoal injection is iteratively calculated to reach the aimed injection rate, which is specified at the beginning of the iterative calculation [2,4]. At the sidewall, mass fluxes across the wall are assumed null. At the same time, heat loss is allowed by setting an overall cooling heat transfer coefficient based on the cooling water system's temperature variation and flow rate using the operation's acquisition data.

The source terms appearing in the momentum equations are calculated using sub-models based on semi-empirical correlations obtained under blast furnace conditions. The gas-solid momentum interactions consider the local layer structure evolution inside the reactors. The evolution of their volume fractions traces the solid component's coexistence within each representative control volume. The solid phase properties are calculated by taking into account the mass fractions of the chemical species within the solid component pondered by their volume fractions in the solid phase components. In this sense, the layer structured descending properties in the burden is locally considered by tracing the solid components burden layer and their volume fractions, accounting for the solid layer local motion dynamics.

$$\sum_{i=1}^{solid} \varepsilon_i = \varepsilon_{fo} + \varepsilon_{fs} + \varepsilon_{fp} + \varepsilon_{fb} + \varepsilon_{fc} = \varepsilon_s \tag{8}$$

The variables ε_{fo}, ε_{fs}, ε_{fp}, ε_{fb}, ε_{fc}, ε_s represent the volume fractions of the burden materials in the solid phase, lump ore, sinter, pellets, briquets, granular charcoal, and the total amount of the solid phase using the additive rule, respectively. The volume fractions of the burden materials are calculated using the burden layers' evolution within the furnace using as inlet boundary conditions the burden structure calculated depending on the charging practice [4,8].

2.2. Numerical Solution of the Conservation Equations

The numerical method selected to resolve the transport equations is based on the finite volume method (FVM). The formulation is developed in a general non-orthogonal coordinate system to account for the accurate geometry of the physical space domain [46–48]. The numerical mesh is constructed using the body-fitted coordinate system to accurately describe the blast furnace wall shape [48].

To resolve the coupling of the governing equations of mass and momentum of the continuous phases, the SIMPLE (Semi-Implicit Method for Pressure-Linked Equations) algorithm is applied on a staggered grid frame using covariant projections of the velocities. The numerical calculation of the discretized equations' coefficients is the widely used power-law scheme [47,48]. The discretized algebraic equations are solved using the line-by-line method based on the iterative procedure constructed using the tridiagonal matrix solver. This procedure allows the tridimensional calculations with strongly no-linear differential equations [46–48].

The iterative procedure used to couple the discontinuous phase motions use the momentum equations calculations with the phase volume fractions iteratively calculated by the phase's total mass conservation. In the iterative procedure of the momentum and volume fraction momentum equations, the numerical coefficients are calculated combining the power-law scheme for the momentum equations and the upwind scheme for the volume fraction based on the mass balance of the corresponding phase. For all chemical species in the phase calculations, the numerical procedure assumes the power-law scheme using iterative calculations [8,46–48].

Each phase's enthalpy conservation equations are calculated, coupling the mass and enthalpy exchanges using the rate equations for the chemical reactions. A polynomial function relating to the phase enthalpy and its temperature are constructed based on the local composition. The specific heat capacities of the chemical species are locally used, assuming the additive rule. The secant method is used to obtain the polynomial equations' real roots to obtain the temperatures of the phase [8].

2.3. A New Concept for Hydrogen-Rich Gas Injections in Two Tuyeres Levels of the Charcoal Mini Blast Furnace

The rich gas hydrogen injection can be used to replace the reducing gas at the shaft of the furnace or as fuel gas at the raceway. Figure 2 shows these two concepts schematically. Figure 2a presents the possible injection simultaneously at the raceway level and the second injection level at the beginning of the shaft. Figure 2b shows the concept of using hydrogen-rich fuel gas injection in a separate lance at the tuyeres in the raceway region. These practices are substantially different and have strong effects on the whole blast furnace operation. Although some attempts to analyze these possibilities have been proposed, the analysis has been carried out simplified models and experiments that did not consider the complex changes that the whole blast furnace undergoes. Thus, using a comprehensive total blast furnace model using the multiphase and multispecies interchange principle is adequate.

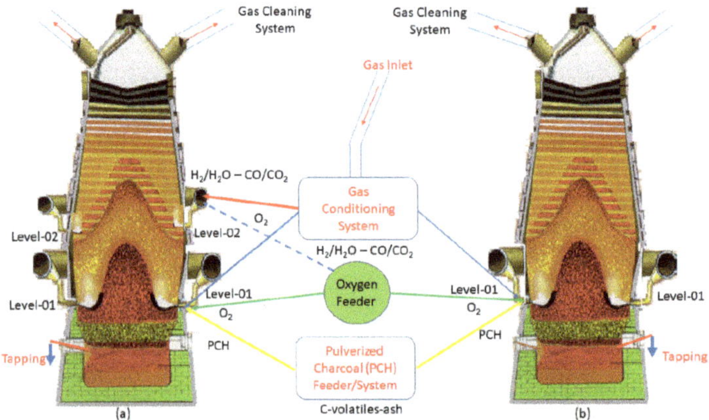

Figure 2. Schematic of the concept for the multilevel gas injection proposed for enhancing the mini blast furnace efficiency. (**a**) Two levels of hydrogen-rich gas injection with oxygen enrichment and pulverized charcoal. (**b**) One level of fuel gas injection with pulverized charcoal and oxygen and steam enrichments.

In our previous model, we implemented the new features and chemical reaction to consider the coupled phenomena involving all the phases and chemical species listed in Tables 1 and 2. The rate equations for the chemical reactions were obtained using experimental data and available rate equations previously published [2,4,8–45]. To investigate these new conditions, we examined four cases of increasing injection at the second level, four injection cases at the raceway, and combined cases of both. These cases were compared with a reference case of actual operation in a mini blast furnace with 390 m^3 of inner volume with a long experience of operation with high rates of pulverized charcoal injection, and granular charcoal charged using a double-bell system with moving armor to control the thickness of the layers.

3. Results and Discussion

In this study, we analyze these possibilities of injection organizing into three sets of calculations. Firstly, a reference case using the actual operation is adjusted, and the measured data are used to compare the calculated results. A good accurate reference case is an important step to guarantee the model's reliability and sensibility against controllable operational conditions such as blast, burden materials, and suitable boundary conditions. Two other sets of analysis cases were considered as well as the combination of both. These cases cover the increasing amount of injections at the shaft level and raceway and necessary adjustments on the oxygen and vapor supply.

The composition of the injected gases at shaft and tuyeres levels are specified, as shown in Table 3. The mixed gases are obtained from the gasification of torrefied biomass of eucalyptus under water vapor to obtain a high H_2 content. The experimental trials used pressurized gas to inject in the shaft and tuyere levels. The effects of these injection possibilities are discussed in the view of the fully operational status. The chemical composition of the burden materials used is listed in Table 4.

Table 3. Composition of the shaft and tuyere gas injections considered in this study.

Gas (%vol)	CO	CO_2	H_2	H_2O	CH_4	C_mH_n	N_2	SO_2
Shaft	16.1	3.9	76.4	0.1	1.2	0.4	0.8	0.1
Tuyere	22.0	3.2	51	0.5	14.8	0.9	7.6	0.0

Table 4. The phases and chemical species assumed in the charcoal mini blast furnace model.

Species	Granular Charcoal	Pulverized Charcoal (PCH)	Sinter	Pellets	Briquette	Lump Ore
C (fix)	75.5	73.4	0.0	0.0	27.1	0.0
Volatiles	22.4	25.3	0.0	0.0	1.2	0.0
H_2O	1.5	0.3	1.3	0.4	0.2	2.5
Fe_2O_3	0.0	0.0	71.3	87.3	36.5	92.6
Fe_3O_4	0.0	0.0	15.2	2.0	21.5	0.0
$Fe_{0.95}O$	0.0	0.0	0.2	1.1	0.0	0.0
FeS	0.3	0.1	0.0	0.0	0.1	0.1
S_2	0.05	0.02	0.0	0.0	0.0	0.0
P_2O_5	0.05	0.08	0.01	0.06	0.04	0.05
CaO	0.1	0.2	4.1	1.1	2.7	0.0
SiO_2	0.6	0.4	5.6	3.7	4.3	3.1
MnO	0.01	0.01	0.11	0.12	0.10	0.05
MgO	0.01	0.02	0.89	0.04	0.03	0.01
TiO_2	0.00	0.00	0.06	0.02	0.02	0.01

The shaft gas presents high H_2 and CO concentration and other hydrocarbons with small quantities of impurities. The injection temperatures were maintained at 1050 °C for all calculations. The hydrogen-rich gas injection at the tuyere has calorific values compatible with a fuel of medium value of traditional pulverized coal but negligible impurities and controlled humidity.

The burden materials are composed of granular charcoal, sinter, pellets, briquette, and lump ore. The charging practice and materials proportions to form the layer structure depend on the operational practice and the materials' availability. In this study, all the calculations assumed the same proportion used in the reference case (sinter: 65%, pellet: 20%, briquette: 10%, and lump ore: 5%). As shown in Table 4, the volatile matter in the granular charcoal and pulverized charcoal are present with significant amounts and important parameters to determine the solid carbon source's heat input.

However, it is worth mentioning that all the carbon source used in this mini blast furnace process is renewable, as it was obtained from the planted forest's carbonization process. Nevertheless, the partial replacement of the granular charcoal is attractive to widen the fuel and reducing agent available to the process with the possibility of combining gasification processes of a variable source of biomasses and allow this technology also to include other sources of hydrogen in the ironmaking processes. The pulverized charcoal system comprises comminution facilities and pneumatic transportation in a closed circuit to attain particle size and composition homogeneities. To attain the mini blast furnace process's slag volume and basicity characteristics, the acid sinter and pellet are charged. The phosphorous and sulfur impurities are present due to the trace elements in the iron ore and charcoal.

It is important to emphasize the noticeable low amount of the carbonaceous raw materials' contribution regarding the sulfur content. The briquette charged is regarded as self-reducing materials and furnishes a considerable amount of carbon, depending on its contribution to the burden fraction. In this study, a fixed fraction of 10% of the metallic burden was used.

3.1. The Actual Mini Blast Furnace Operation and Model Comparison

The reference of the operational practice used to analyze the mini blast furnace process was selected from a long period of stable operation with the metallic burden materials in the proportion of 65% sinter, 20% pellets, 10% briquette, and 5% lump ore.

The reference case's oxygen enrichment was 9.5%, with the blast temperature maintained at 1100 °C and fixed blast flowrate. The iterative calculation of the reference case is an important task to ensure its reliability and adherence to the actual industrial practice. The reference base case's initial adjustment is crucial for accurate overall profile determination of the burden descending materials' metallic to granular charcoal ratio.

This refinement procedure is iteratively done using the initial top radial burden probe profiler and the gas composition and temperature data measurements. We apply a simple optimization algorithm to iteratively correct the layer thickness until simultaneously achieving the measured gas composition and temperature predicted by the model reasonably agree. The overall temperature field calculation using this procedure is shown in Figure 3 for the gas, solid, pig iron, and slag phases, respectively. Figures 4 and 5 are obtained for the normalized burden layers thickness radial distribution and the comparison with the measured and monitored gas temperature and composition from the analysis probe. Observing Figures 4 and 5, the central flow of the gas phase is evident.

The resulting radial distribution of the average burden layer thickness and the size distributions of the burden materials are compatible with the whole temperature distribution and presented in Figure 5a,b. These figures are obtained after the continuous iterative search of the simultaneous attainment of the gas phase's temperature and composition with the average measurements. As shown in Figure 5, a good agreement of the calculated results with the measured data is achieved. The measured data were averaged for a dataset of 10 continuous measurements of a stable operation.

This agreement of the calculated results with the outlet gas's monitored data was confirmed comparing the parameters related to the gas composition and temperature listed in Table 5, representing the major overall operational parameters of the process.

Table 5 compares the calculated and measured values of the mini blast furnace's operational data, which was carried out. The calculated results agree with the averaged measured data with a maximum error of less than 5%.

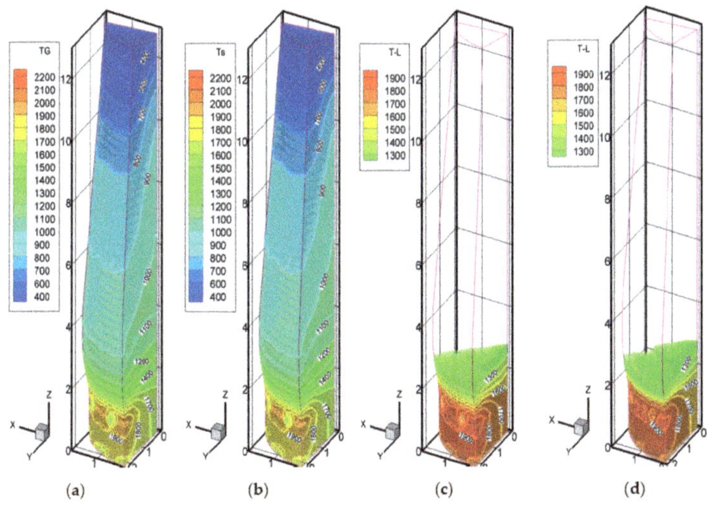

Figure 3. 3D temperature distributions for the reference case. (**a**) Gas; (**b**) Solid; (**c**) Pig iron; (**d**) Slag.

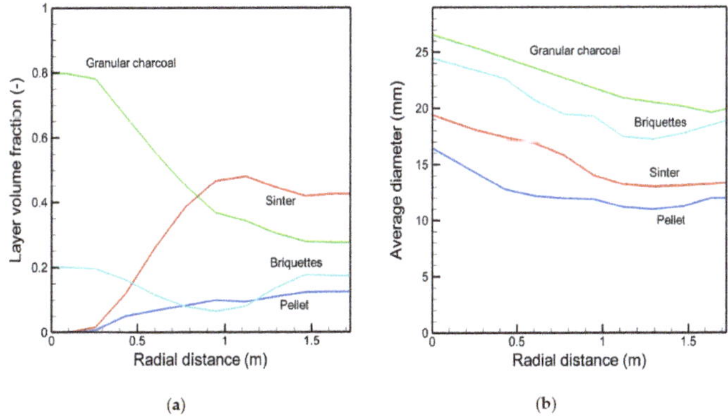

Figure 4. The radial distributions of the burden materials for the reference case. (**a**) Layer thickness volume fractions; (**b**) Average diameters of the burden materials.

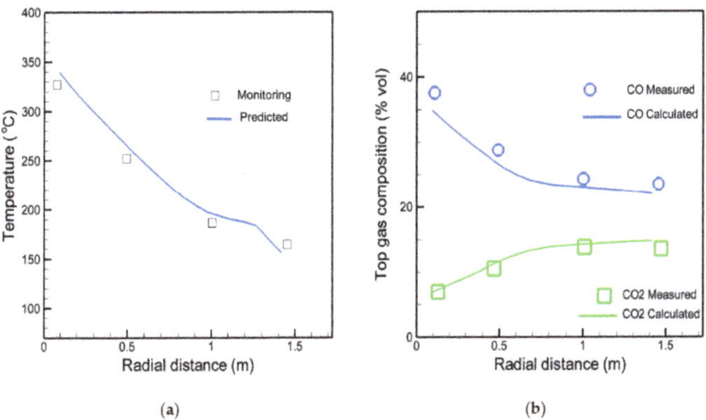

Figure 5. Comparison of the calculated and measured values of the probes for the reference case. (**a**) Gas temperatures; (**b**) Gas composition.

Table 5. Comparison of the measured and calculated operational parameters for the reference case.

Operational Parameters	Measured	Calculated	Error (%) [1]
Productivity ($t_{HM/day/m^3}$)	2.80	2.82	0.71
Granular charcoal (kg/t_{HM})	398.5	395.4	0.77
Blast rate (Nm^3/t_{HM})	1042	1037	0.48
O_2 rate (Nm^3/t_{HM})	334.5	331.4	0.93
O_2 enrichment (%)	9.4	9.5	1.06
Outlet gas (Nm^3/t_{HM})	1548	1531	1.09
CO (kg/t_{HM})	632.1	653.7	2.18
CO_2 (kg/t_{HM})	722.3	701.9	3.59
H_2 (kg/t_{HM})	1.8	1.9	5.55
H_2O (kg/t_{HM})	64.8	62.7	3.32
Temperature (°C)	202.5	200.4	1.04
Slag rate (kg/t_{HM})	132.0	135.2	2.42
Slag basicity ($B_2 = CaO/SiO_2$)	1.05	1.10	4.76
Silicon pig iron (%)	1.73	1.65	4.62
Carbon intensity ($kg\ C/t_{HM}$)	499.5	495.8	0.74

[1] $Error(\%) = |Measured - Calculated| \times 100/Measured$.

The computational effort on this step is the most time consuming due to the iterative search of the good correspondence of the burden distribution adjustments using the initial distribution of the charging system estimation positions corresponding to the burden layer thickness and the settled materials obtained due to the dynamics of descending burden and upward gas flow. Using this procedure, we can take into account the effects of multifactorial parameters such as burden materials rolling characteristics, shape factors and granulometric distributions, and the gas flowing pattern and variable gas thermophysical properties, which is only possible by using a multiphase multicomponent comprehensive approach. Thus, although this computational time-consuming at the beginning of the calculation, the overall prediction can be greatly improved. A detailed 3D visualization of the gas flow with the trajectories and the residence time are shown in Supplementary video S1.

Therefore, the model prediction is consistent and presents close adherence with the measured data obtained during an actual process operation's regular operational practice. Thus, the model was used to discuss new scenarios with gas injection and verify their feasibility and potential to switch from a carbon to a hydrogen-based ironmaking industry.

3.2. Scenarios with High Pulverized Charcoal Combined with Hydrogen-Rich Gas Injections in the Raceway and Oxygen Enrichment

The pulverized charcoal (PCH) injection through the mini blast furnace raceway is a recent technology compared with the large blast furnace operation. The pulverized charcoal presents attractive properties related to the high reactivity in carbon and hydrogen gas environments. The high reactivity of the PCH allows a large amount of injection and, at the same time, keeps the small volume of the slag, which is characteristic of the mini blast furnace operation. To demonstrate the feasibility of the proposed injection practices, simulated blast furnace operations with increasing hydrogen-rich fuel gas injection at the tuyere level are considered, as shown in Figure 6. Cases 1 and 2 shows increasing gas injections only. Cases 3 and 4 present increasing gas injections with oxygen additions to compensate for the raceway temperature distributions.

As can be observed comparing cases 1 and 2, the lower part of the furnace's temperature distribution slightly decreases but with enlargement of the high-temperature zone and decreasing the raceway region temperature. Further gas injections are promoted, increasing the injection rate accompanied by oxygen enrichment (cases 3 and 4).

The global operational parameters are shown in Table 6. As can be observed, the mini blast furnace reactor's performance could be improved in various combined scenarios. These results allow the decision-makers to select the best practice depending on the market requirement and conditions, depending on the operational practice goals and raw materials availability.

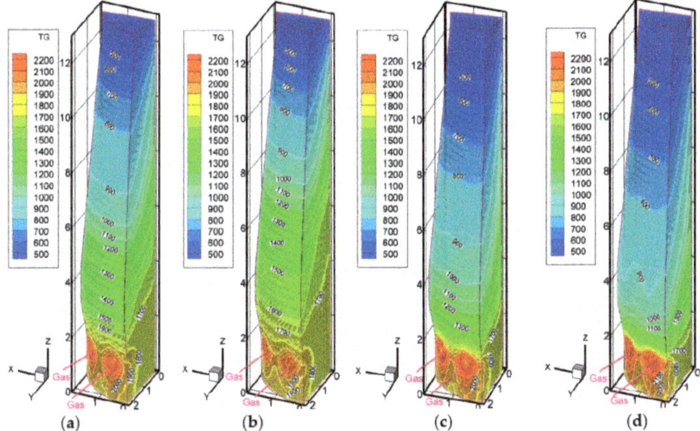

Figure 6. Three-dimensional (3D) temperature distributions for the shaft injection at the second level within a bustle ring. (**a**) Case 1; (**b**) Case 2; (**c**) Case 3; (**d**) Case 4.

Table 6. The selected operational parameters for the scenarios discussed in this review paper.

	Scenarios Analyzed									
Parameters	**Case 1**	**Case 2**	**Case 3**	**Case 4**	**Case 5**	**Case 6**	**Case 7**	**Case 8**	**Case 9**	**Case 10**
Productivity ($t_{HM}/day/m^3$)	2.8	2.9	3.3	3.6	3.0	3.2	3.7	3.2	2.9	3.1
Granular charcoal (kg/t_{HM})	395.4	400.6	392.1	401.7	283.1	283.3	273.1	283.2	314.3	278.1
Blast (Nm^3/t_{HM})	1037	1018	903	832	825	822	792	822	633	536.3
Outlet gas (Nm^3/t_{HM})	1663	1679	1676	1593	2135	2066	1950	2086	1692	1750
Outlet gas temperature (°C)	200	205	167	152	391	368	339	371	291	312
CO (kg/tHM)	654	717	1031	926	888	855	791	866	779	788
CO_2 (kg/t_{HM})	702	655	512	379	330	356	389	349	481	406
H_2 (kg/t_{HM})	1.9	2.4	4.4	3.8	35.6	32.7	29.3	33.7	17.2	25.3
H_2O (kg/t_{HM})	62.7	63.1	86.5	60.6	244.1	226.1	207.5	230.8	164.4	215.2
Slag (kg/t_{HM})	135	118	108	143	137	132	135	133	124	126
Basicity (-)	1.15	1.18	1.10	1.06	1.21	1.17	1.19	1.20	1.35	1.26
Carbon intensity (kg C/t_{HM})	472	486	582	500	471	464	446	466	465	448
Parameters	**Case 11**	**Case 12**	**Case 13**	**Case 14**	**Case 15**	**Case 16**	**Case 17**	**Case 18**	**Case 19**	**Case 20**
Productivity ($t_{HM}/day/m^3$)	4.6	4.9	2.8	2.8	3.4	3.6	3.8	4.4	6.1	6.5
Granular charcoal (kg/t_{HM})	299	278	315	300	299	279	278	273	263	263
Blast (Nm^3/t_{HM})	329	313	845	858	901	814	753	679	550	515
Outlet gas (Nm^3/t_{HM})	1442	1391	1827	2050	2218	2080	1928	1793	1840	1930
Outlet gas temperature (°C)	233	217	299	345	374	337	329	315	327	346
CO (kg/t_{HM})	717	664	791	843	836	793	771	725	793	850
CO_2 (kg/t_{HM})	501	529	469	403	415	425	425	453	380	321
H_2 (kg/t_{HM})	17.8	17.2	18.4	31.6	38.9	35.8	32.7	26.6	33.9	41.8
H_2O (kg/t_{HM})	167.9	162.1	150.7	228.6	299.7	265.0	255.5	251.3	277.6	315.8
Slag (kg/t_{HM})	143	134	130	131	130	132	135	124	129	121
Basicity (-)	1.12	1.05	1.31	1.25	1.23	1.08	1.14	1.26	1.18	1.32
Carbon intensity [1] (IC= kg C/t_{HM})	444	429	467	449	471	456	446	434	443	452

[1] IC: carbon intensity of the scenario or specific carbon consumption.

3.3. Scenarios with High Rates of Hydrogen-Rich Gas Injections in the Shaft Zone

The reducing gas injection at the shaft zone is explored in cases 5–8. The strong effect on the furnace temperature is evidenced in Figure 7. Cases 5–8 were obtained by incrementally substituting the granular charcoal contribution on the reduction steps by hydrogen-rich gas at the shaft zone. These cases were designed to improve the productivity of the process allowed by enhancing the kinetics of the reduction steps by hydrogen. It is a consensus that the hydrogen reduction in the gas-solid is enhanced at a higher temperature.

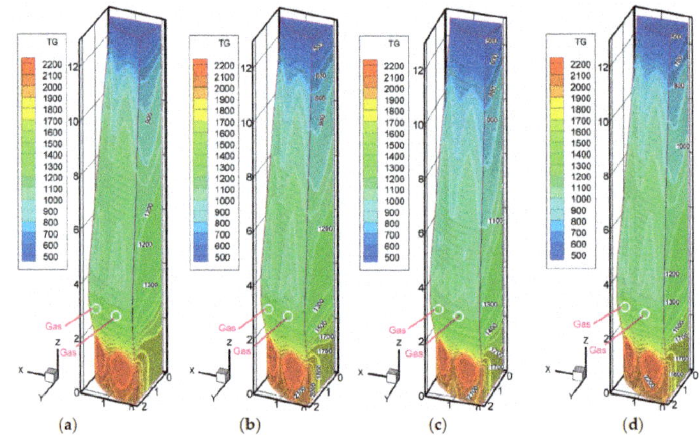

Figure 7. Three-dimensional (3D) temperature distributions for the reference case. (**a**) Case 5; (**b**) Case 6; (**c**) Case 7; and (**d**) Case 8.

Conversely, these features are concurrent with the energy supply for the endothermic reactions ruling in the local fluid dynamics conditions, which are strongly affected when massive gas injections are taking place. In this sense, it is also important to mention that both methods analyzed here are complementary and offer the possibility of symbiosis effects, leading to improving the reactor's general performance under such conditions. These aspects are only properly computed by using a comprehensive model, as proposed in this analysis. As shown in Figure 7, good temperature distribution on the whole furnace is attained for all cases, confirming the feasibility of the proposed operational practices.

Figure 8 shows the temperature distributions for cases 9–12. In these cases, we explore the possibility of the enhancement of the reactions rates with hydrogen gas within the shaft of the furnace by taking the advantages of self-reforming gas and locally increasing the temperature by using the self-catalytic effects of partially reduced iron by promoting the gas reforming with the energy supply obtained by a small amount of oxygen furnished during the gas distribution on the bustle ring, which is similar to the technology used in the shaft direct reduction processes such as the variants of the MIDREX reactor [27,28]. In this sense, the proposed scenarios of operation with the mini blast furnace process using mixed gas are flexible and attractive. The operation of the mini blast furnace process is competitive, depending on the hydrogen-rich gas availability. The process operation's symbiosis is straightforward with the processes for producing direct reduction iron (DRI). Moreover, the mini blast furnace process can be complementary to the self-reducing biomass version-based Tecnored process with the mutual supply of raw materials and by-products [49–52]. The inner pattern of the gas dynamics with trajectories and residence time for case 10, representing the second level hydrogen injection, is shown in Supplementary video S2.

This idea is explored in cases 9–12, as presented in Figure 8a–d. In Figure 8, the amounts of the oxygen supply are continuously increased. As can be observed, as the oxygen supply is increased, the increases in the shaft region's temperatures are increased and enlarged. The general effects on the reactor's performance are shown in Table 6, which shows the comparative results for all cases considered in this study. Figure 9 shows the combined cases where the injection of reducing gas at the shaft is combined with fuel and water vapor injection at the tuyeres.

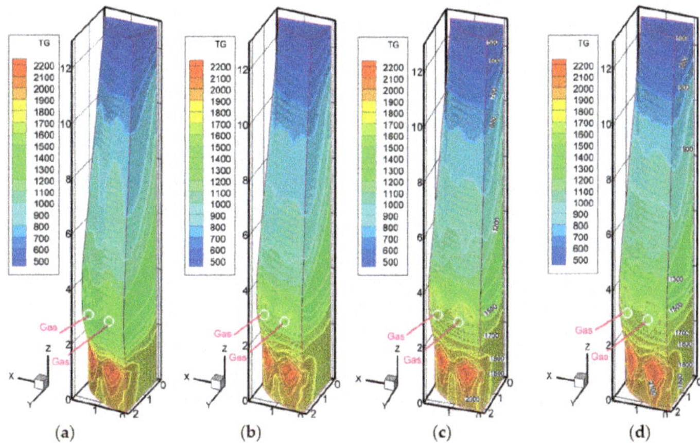

Figure 8. Three-dimensional (3D) temperature distributions for high hydrogen gas injections at the shaft zone with controlled oxygen to enhance self-catalytic reactions. (**a**) Case 9; (**b**) Case 10; (**c**) Case 11; (**d**) Case 12.

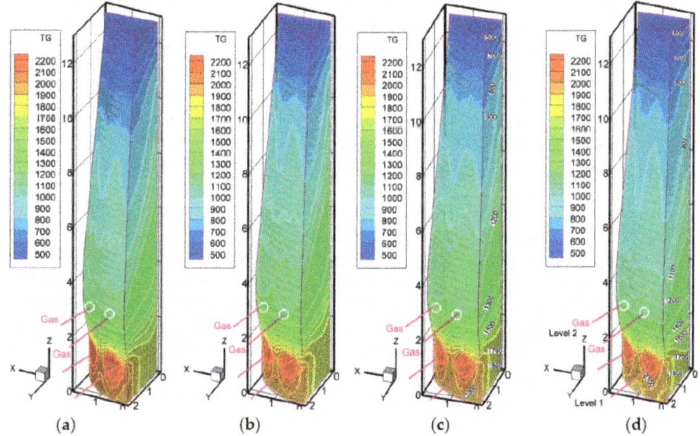

Figure 9. Three-dimensional (3D) temperature distributions for the analysis cases of combined injections of high hydrogen gas at the shaft and fuel gas at the tuyeres. (**a**) Case 13; (**b**) Case 14; (**c**) Case 15; (**d**) Case 16.

Figure 9 (cases 13–16) shows that when the fuel gas is injected and the correspondent increase of the oxygen in the raceway is used, the furnace's performance is improved. The general aspect of the whole blast furnace's temperature distribution is well behaved and indicated stable operation with the shaft zone in good shape.

In the sequence, we explore the possibility of further improvement of the performance of the mini blast furnace process by using the combined effect of high hydrogen gas injection on the shaft combining self-catalytic reactions in the reducing zone and the temperature control of the raceway by combining fuel gas injections, oxygen enrichment, and water vapor injection. The temperature distributions for the stable operational predictions simulating these scenarios are shown in Figure 10. In Figure 10 (cases 17–20), the amount of fuel gas injections and bustle gas are balanced by decreasing the blast rate, simultaneously controlling the lower part of the blast furnace temperature, which is important to keep the dropping liquids flowing conditions and the cohesive zone permeability. The detailed flow trajectories showing the gas phase's residence time for case 20 is presented in Supplementary video S3. The Supplementary videos S4 and S5 show the flow patterns with the temperature and H2 concentrations distributions for the two-level representative case 20.

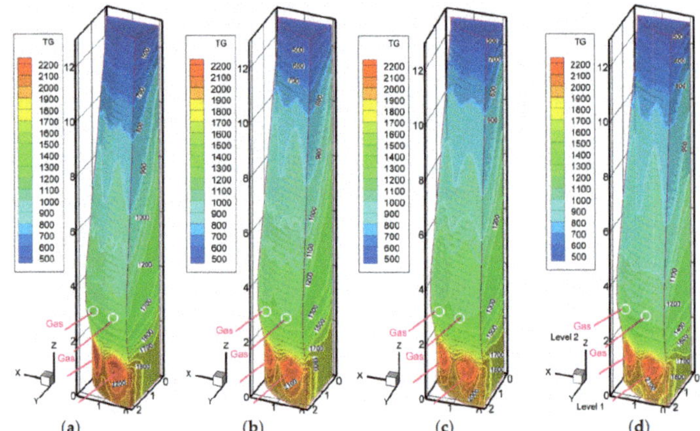

Figure 10. Three-dimensional (3D) temperature distributions for the combined scenarios of gas injections at two levels in the mini blast furnace process. (**a**) Case 17; (**b**) Case 18; (**c**) Case 19; (**d**) Case 20.

One important parameter to control such severe flowing conditions in a small reactor as the charcoal mini blast furnace is the pressure drop. Thus, the blast flowrates and production march parameters are iteratively searched for all scenarios using a simple blast correction algorithm that uses, as a tumble rule, the maximum pressure drop allowed value. The tumble rule uses a proportional correction of the blast rate until the stable temperature distribution is achieved for the specified admitted pressure drop range for the whole furnace.

Nevertheless, we must emphasize that the stable operational conditions obtained during this iterative procedure. However, it could not be a unique solution. They are compatible with the whole operational aspects involved in such strong internal changes of the multiphase and multicomponent dynamics that undergo the process. Thus, these scenarios truly represent the complex equilibrium and coupled phenomena achieved for the new operating conditions and could guide new industrial trials. Therefore, we hope that this research could drive new developments and improvements toward a clean and environmentally friendly green pig iron production process with low carbon intensity and 100% of the renewable route's carbon resources.

The scenarios analyzed comprise several combined possibilities to cover massive injections of hydrogen-rich gas injections at the second level using a bustle system with the second level of tuyeres and gas injection lances at the raceway level. As shown in Table 6, the process's productivity could be significantly increased (about 130% for the best case).

The granular charcoal consumption can decrease (about 30% for the best scenario). The carbon intensity parameter, which relates the whole process's specific carbon, can be decreased by about 9% for the best combination. The lower carbon intensity feature is a clear advantage of these scenarios that decreases costs and allows the plantation land's efficient usage. Thus, based on the scenario's predictions, a selection of operational setup can be chosen that meets the optimization goals.

3.4. Discussion on the Enhancement of the Reduction Kinetics under the Conditions of the Injection

The hydrogen-rich gas injection in the blast furnace strongly depends on the flow gas's dynamics, the gas phase's local concentration, and the burden materials reduction degree. Several studies have separately discussed the hydrogen and mix gas effectivity of the burden materials under blast furnace shaft gas conditions. However, the concurrent phenomena related to the convective and diffusive mass transfer in the bulk gas and in the inner particles and the heat transfer mechanisms to supply the endothermic reaction steps limit the shaft zone furnace's efficiency.

Nevertheless, the water gas (H_2O ($vapor$) $+ C(i) \rightarrow H_2(g) + CO(g)$) and the water shift reactions (CO_2 $(g) + H_2(g) \leftrightarrow CO(g) + H_2O(g)$) can play an important role in enhancing the reducing conditions. Therefore, these discussions must consider all the concurrent reactions strongly affected by the new hydrogen-rich gas injections. The contributions of each chemical reaction step are presented in Table 7, comparing the cases analyzed. Table 7 summarizes the consumption of gas species amount involved in each reduction step. As can be observed, the amount of hydrogen and carbon monoxide involved in the scenarios proposed strongly depends on the injection practice. The model predictions indicated that hydrogen plays the most relevant role in the final reduction step. At this moment, we must emphasize that this research proposed the injection temperature at the bustle level higher than that of the thermal reserve zone, which is expected to improve the hydrogen reduction kinetics at the bottom position of the shaft zone. As expected, the results indicated that the gas temperature distributions allowed the model to predict smooth operation with stable cohesive zone locations, as evidenced in Figures 6–10. The calculations of the global parameters shown in Table 6 indicated that cases with a higher contribution of the hydrogen reduction allow the decrease of blast volume and granular charcoal, which are attractive possibilities to enhance the operation with higher productivity. These aspects are confirmed by comparing the calculated results of Tables 6 and 7.

Table 7. Comparison of the carbon-based and hydrogen gas-solid reduction steps for all scenarios discussed in this research.

Reactions [1]

(R1C) $3Fe_2O_3(i) + CO(g) \rightarrow CO_2(g) + 2Fe_3O_4(i)$

(R2C) $\dfrac{0.95}{4(0.95)-3}Fe_3O_4(i) + CO(g) \rightarrow CO_2(g) + Fe_{0.95}O(i)$

(R3C) $Fe_{0.95}O(i) + CO(g) \rightarrow CO_2(g) + 0.95Fe(i)$

(R1H) $3Fe_2O_3(i) + H_2(g) \rightarrow H_2O(g) + 2Fe_3O_4(i)$

(R2H) $\dfrac{0.95}{4(0.95)-3}Fe_3O_4(i) + H_2(g) \rightarrow H_2O(g) + Fe_{0.95}O(i)$

(R3H) $Fe_{0.95}O(i) + H_2(g) \rightarrow H_2O(g) + 0.95Fe(i)$

Reactions	Reference	Case 1	Case 2	Case 3	Case 4	Case 5	Case 6	Case 7	Case 8	Case 9	Case 10
(R1C) [1]	61.5	62.3	57.2	60.7	71.1	63.2	63.2	63.1	62.7	63.2	62.9
(R1H) [2]	0.017	0.019	0.050	0.046	0.046	0.027	0.032	0.036	0.028	0.029	0.039
(R2C) [1]	124.9	125.6	125.8	162.5	173.1	51.6	57.2	63.8	236.24	57.3	43.1
(R2H) [2]	0.015	0.017	0.049	0.013	0.060	4.953	5.130	4.659	5.207	6.14	6.15
(R3C) [1]	393.6	388.2	391.7	401.2	428.8	272.9	91.3	302.1	85.6	372.9	247.2
(R3H) [2]	0.000	0.001	0.002	0.000	0.000	14.050	12.458	10.816	12.859	7.645	16.540

Reactions	Reference	Case 11	Case 12	Case 13	Case 14	Case 15	Case 16	Case 17	Case 18	Case 19	Case 20
(R1C) [1]	61.5	62.9	62.8	63.2	63.3	63.2	63.2	63.0	63.1	62.9	64.9
(R1H) [2]	0.017	0.042	0.045	0.036	0.038	0.038	0.044	0.049	0.048	0.046	0.044
(R2C) [1]	124.9	57.9	59.4	73.6	54.5	52.2	58.8	59.2	58.6	43.6	40.3
(R2H) [2]	0.015	6.06	4.96	3.97	5.35	5.53	5.05	4.97	5.03	6.13	7.93
(R3C) [1]	393.6	359.9	363.0	346.5	301.7	315.1	301.1	305.6	312.6	260.8	227.8
(R3H) [2]	0.000	7.15	5.69	7.70	11.53	10.07	10.58	10.21	9.63	13.65	15.85

[1] The values refer to kg CO/t_{HM}; [2] The values refer to kg H_2/t_{HM}; (i) = lump ore, sinter, pellets, briquette, fines.

4. Summary, Outlook, and Future Trends

A review of the mini blast furnace process operation using renewable sources of fuel and hydrogen-rich gas was presented and discussed. These compilation and new calculation results clearly indicated that the mini blast furnace process efficiency has continuously improved by introducing peripheral devices, available technologies for pulverized charcoal injection, and the combined fuel and reducing gas injections at different reactor levels. The full operation with different sources of biomass and biogas became feasible. New prepared burden materials have been introduced in the current practice, allowing to recycle the iron-bearing fine materials and carbonaceous waste from the steelmaking facilities. We developed a computational code implementing a comprehensive multiphase mathematical modeling, which considers the specific rate equations suitable for the mini blast furnace process operation with flexible raw materials. The model was successfully compared with the operational data using a reference case for validation purposes. The model was applied to simulate 20 cases, representing feasible fuel gas scenarios and high hydrogen-reducing gas injections, maintaining a high pulverized charcoal injection (150 kg/t$_{HM}$). The results suggested that a considerable increase in the productivity and decrease of the carbon intensity could be achieved using these operational scenarios, leading to important impacts on the sustainability of the ironmaking route. The model predictions indicated that for the considered conditions of simultaneous injections of 150 kg/t$_{HM}$ of pulverized charcoal in the tuyeres and hot hydrogen-rich gas in the shaft, the productivity could increase 130% (for the best combination).

In contrast, the consumption of granular charcoal in the burden would decrease by 30%. The results were iteratively obtained by searching stable operational conditions and adjusting the blast volume and oxygen enrichment, which are compatible with the burden materials flowrate. The comprehensive multiphase and multicomponent model approach allows the decision-makers to accurately determine the equilibrium of the energy and materials requirements to keep the operation stable. The selected scenarios indicated that the specific carbon consumption could decrease about 9% for these operational conditions, depending on the overall iterative adjustments, which allows the decision-makers to select the suitable conditions for the available raw materials and productivity targets. Finally, we furnish some video materials animations to the reader to visualize the inner variables of some key calculated results for the reference cases and compare them with the selected analysis cases.

Supplementary Materials: The following are available online at http://www.mdpi.com/2075-4701/10/11/1501/s1, Video S1: Gas flow simulation for the reference case, Video S2: Gas flow simulation for case 10, Video S3: Gas flow simulation for case 20, Video S4: 3D visualization of the temperature distributions (case 20), Video S5: 3D visualization of the hydrogen reducing gas in the shaft zone for the case analysis (case 20).

Author Contributions: Conceptualization, J.A.d.C. and E.M.d.O.; methodology, J.A.d.C. and H.N.; software, J.A.d.C.; validation, J.A.d.C., E.M.d.O. and G.A.d.M.; investigation, J.A.d.C. and G.A.d.M.; writing—original draft preparation, J.A.d.C.; writing—review and editing, M.F.d.C. All authors have read and agreed to the published version of the manuscript.

Funding: The APC was funded by "Fundação Carlos Chagas Filho de Amparo a Pesquisa do Estado do Rio de Janeiro (FAPERJ E-26/202.888/2018 grant No 239505)".

Acknowledgments: The authors thank the Federal University's multiphase system laboratory and the graduate program on Metallurgical Engineering and their students and researchers for supporting the theoretical analysis and experimentations. Their efforts in the experimental research and analysis are acknowledged.

Conflicts of Interest: The authors declare no conflict of interest.

References

1. Tan, X.; Li, H.; Guo, J.; Gu, B.; Zeng, Y. Energy-saving and emission-reduction technology selection and CO_2 emission reduction potential of China's iron steel industry under energy substitution policy. *J. Clean. Prod.* **2019**, *222*, 823–834. [CrossRef]
2. Rocha, E.P.; Guilherme, V.S.; Castro, J.A.; Sasaki, Y.; Yagi, J. Analysis of synthetic natural gas injection into charcoal blast furnace. *J. Mater. Res. Technol.* **2013**, *2*, 255–262. [CrossRef]
3. Patisson, F.; Mirgaux, O. Hydrogen Ironmaking: How it Works. *Metals* **2020**, *10*, 922. [CrossRef]
4. Castro, J.A.; Medeiros, G.A.; Oliveira, E.M.; Silva, L.M.; Nogami, H. A Comprehensive Modeling as a Tool for Developing New Mini Blast Furnace Technologies Based on Biomass and Hydrogen Operation. *J. Sustain. Metall.* **2020**, *6*, 281–293. [CrossRef]
5. Yilmaz, C.; Jens Wendelstorf, J.; Turek, T. Modeling and simulation of hydrogen injection into a blast furnace to reduce carbon dioxide emissions. *J. Clean. Prod.* **2017**, *154*, 488–501. [CrossRef]
6. Ni, M.; Leung, D.Y.C.; Leung, M.K.H.; Sumathy, K. An overview of hydrogen production from biomass. *Fuel Process. Technol.* **2006**, *87*, 461–472. [CrossRef]
7. Abdul Quader, M.; Ahmed, S.; Dawal, S.Z.; Nukman, Y. Present needs, recent progress and future trends of energy-efficient Ultra-Low Carbon Dioxide (CO_2) Steelmaking (ULCOS) program. *Renew. Sustain. Energy Rev.* **2016**, *55*, 537–549. [CrossRef]

8.   Castro, J.A. A Multi-Dimensional Transient Mathematical Model of Blast Furnace Based on Multi-Fluid Model. Ph.D. Thesis, Tohoku University, Sendai, Japan, 15 March 2001.

9.   Omori, Y. *The Blast Furnace Phenomena and Modelling*, 1st ed.; Elsevier Applied Sciences: London, UK, 1987; pp. 20–480.

10. Castro, J.A.; Araujo, G.M.; Mota, I.O.; Sasaki, Y.; Yagi, J. Analysis of the combined injection of pulverized coal and charcoal into large blast furnaces. *J. Mater. Res. Technol.* **2013**, *2*, 308–314. [CrossRef]

11. Castro, J.A.; Nogami, H.; Yagi, J. Three-dimensional multiphase mathematical modeling of the blast furnace based on the multifluid model. *ISIJ Int.* **2002**, *42*, 44–52. [CrossRef]

12. Li, H.; Saxén, H.; Liu, W.; Zou, Z.; Shao, L. Model-Based Analysis of Factors Affecting the Burden Layer Structure in the Blast Furnace Shaft. *Metals* **2019**, *9*, 1003. [CrossRef]

13. Castro, J.A.; Silva, A.J.; Sasaki, Y.; Yagi, J. A six-phases 3-D model to study simultaneous injection of high rates of pulverized coal and charcoal into the blast furnace with oxygen enrichment. *ISIJ Int.* **2011**, *51*, 748–758. [CrossRef]

14. Austin, P.R.; Nogami, H.; Yagi, J. A mathematical model for blast furnace reactions analysis based on the four-fluid model. *ISIJ Int.* **1997**, *37*, 748–755. [CrossRef]

15. Castro, J.A.; Nogami, H.; Yagi, J. Numerical investigation of simultaneous injection of pulverized coal and natural gas with oxygen enrichment to the blast furnace. *ISIJ Int.* **2002**, *42*, 1203–1211. [CrossRef]

16. Castro, J.A.; Takano, C.; Yagi, J. A theoretical study using the multiphase numerical simulation technique for effective use of H_2 as blast furnaces fuel. *J. Mater. Res. Technol.* **2017**, *6*, 258–270. [CrossRef]

17. Hamadeh, H.; Mirgaux, O.; Patisson, F. Detailed Modeling of the Direct Reduction of Iron Ore in a Shaft Furnace. *Materials* **2018**, *11*, 1865. [CrossRef]

18. Castro, J.A.; Nogami, H.; Yagi, J. Numerical analysis of multiple injections of pulverized coal, pre-reduced iron ore and flux with oxygen enrichment to the blast furnace. *ISIJ Int.* **2001**, *41*, 18–24. [CrossRef]

19. Ueda, S.; Natsui, S.; Nogami, H.; Yagi, J.; Ariyama, T. Recent progress and future perspective on mathematical modeling of blast furnace. *ISIJ Int.* **2010**, *50*, 914–923. [CrossRef]

20. Ariyama, T.; Natsui, S.; Kon, T.; Ueda, S.; Kikuchi, S.; Nogami, H. Recent Progress on Advanced Blast Furnace Mathematical Models Based on Discrete Method. *ISIJ Int.* **2014**, *54*, 1457–1471. [CrossRef]

21. Castro, J.A.; Silva, A.J.; Nogami, H.; Yagi, J. Simulação computacional da injeção de carvão pulverizado nas ventaneiras de mini altos-fornos. *TMMM Tecnol. Metal. Mater. Min.* **2004**, *1*, 59–62. [CrossRef]

22. Li, M.; Wei, H.; Ge, Y.; Xiao, G.; Yu, Y. A Mathematical Model Combined with Radar Data for Bell-Less Charging of a Blast Furnace. *Processes* **2020**, *8*, 239. [CrossRef]

23. Hashimoto, Y.; Okamoto, Y.; Kaise, T.; Sawa, Y.; Kano, M. Practical Operation Guidance on Thermal Control of Blast Furnace. *ISIJ Int.* **2019**, *59*, 1573–1581. [CrossRef]

24. Castro, J.A.; Nogami, H.; Yagi, J. Transient mathematical model of blast furnace based on multi-fluid concept, with application to high PCI operation. *ISIJ Int.* **2001**, *40*, 637–646. [CrossRef]

25. Kou, M.; Zhou, H.; Hong, Z.; Shun Yao, S.; Wu, S.; Xu, H.; Xu, J. Numerical Analysis of Effects of Different Blast Parameters on the Gas and Burden Distribution Characteristics Inside Blast Furnace. *ISIJ Int.* **2020**, *60*, 905–914. [CrossRef]

26. Castro, J.A.; Silva, L.M.; Medeiros, G.A.; Oliveira, E.M.; Nogami, H. Analysis of a compact iron ore sintering process based on agglomerated biochar and gaseous fuels using a 3D multiphase multicomponent mathematical model. *J. Mater. Res. Technol.* **2020**, *9*, 6001–6013. [CrossRef]

27. Da Rocha, E.P.; de Castro, J.A.; Silva, L.; de Souza Caldas, R. Computational Analysis of The Performance of Shaft Furnaces with Partial Replacement of The Burden with Self-Reducing Pellets Containing Biomass. *Mater. Res.* **2019**, *22*, e20190533. [CrossRef]

28. Castro, J.A.; Rocha, E.P.; Oliveira, E.M.; Campos, M.F.; Francisco, A.S. Mathematical modeling of the shaft furnace process for producing DRI based on the multiphase theory. *REM Int. Eng. J.* **2018**, *71*, 81–87. [CrossRef]

29. Wang, X.; Fu, G.; Li, W.; Zhu, M. Numerical Analysis of the Effect of Water Gas Shift Reaction on Flash Reduction Behavior of Hematite with Syngas. *ISIJ Int.* **2019**, *59*, 2193–2204. [CrossRef]

30. Matos, U.F.; Castro, J.A. Modeling of self-reducing agglomerates charging in the mini blast furnace with top gas recycling. *REM Int. Eng. J.* **2012**, *65*, 65–71.

31. Yagi, J. Mathematical modeling of the flow of four fluids in a packed bed. *ISIJ Int.* **1993**, *33*, 619–639. [CrossRef]

32. Zare Ghadi, A.; Valipour, M.S.; Vahedi, S.M.; Sohn, H.S. A Review on the Modeling of Gaseous Reduction of Iron Oxide Pellets. *Steel Res. Int.* **2020**, *91*, 1900270. [CrossRef]

33. Bonalde, A.; Henriquez, A.; Manrique, M. Kinetic analysis of the iron oxide reduction using hydrogencarbon monoxide mixtures as reducing agent. *ISIJ Int.* **2005**, *45*, 155–1260. [CrossRef]

34. Perry, R.H.; Green, D. *Perry's Chemical Engineers Handbook*, 6th ed.; McGraw-Hill: New York, NY, USA, 1984.

35. Ergun, S. Fluid Flow through Packed Columns. *Chem. Eng. Prog.* **1952**, *48*, 89–94.

36. Ranz, W.E.; Marshall, W.R. Evaporation from drops. *Chem. Eng. Prog.* **1952**, *48*, 141–146.

37. El-Tawil, A.A.; Ahmed, H.M.; Ökvist, L.S.; Björkman, B. Self-Reduction Behavior of Bio-Coal Containing Iron Ore Composites. *Metals* **2020**, *10*, 133. [CrossRef]

38. Lemos, L.R.; Rocha, S.H.S.F.; Castro, L.F.A.; Assunção, G.B.M.; Silva, G.L.R. Mechanical strength of briquettes for use in blast furnaces. *REM Int. Eng. J.* **2019**, *72*, 63–69. [CrossRef]

39. Paknahad, P.; Askari, M.; Shahahmadi, S.A. Cold-Briquetted Iron and Carbon (CBIC): Investigation of the Influence of Environmental Condition on Its Chemical and Physical Properties. *J. Sustain. Metall.* **2019**, *5*, 497–509. [CrossRef]

40. Castro, J.A.; Baltazar, A.W.S.; Silva, A.J.; DAbreu, J.C.; Yagi, J. Evaluation of the performance of the blast furnace operating with self-reducing pellet using the computational simulation. *TMMM Tecnol. Metal. Mater. Min.* **2005**, *2*, 45–50.

41. Tang, H.; Sun, Y.; Rong, T. Experimental and Numerical Investigation of Reaction Behavior of Carbon Composite Briquette in Blast Furnace. *Metals* **2020**, *10*, 49. [CrossRef]

42. Chu, M.; Nogami, H.; Yagi, J. Numerical Analysis on Blast Furnace Performance under Operation with Top Gas Recycling and Carbon Composite Agglomerates Charging. *ISIJ Int.* **2004**, *44*, 2159–2167. [CrossRef]

43. Mousa, E.; Lundgren, M.; Ökvist, L.S.; From, L.E.; Robles, A.; Hällsten, S.; Sundelin, B.; Friberg, H.; El-Tawil, A. Reduced Carbon Consumption and CO_2 Emission at the Blast Furnace by Use of Briquettes Containing Torrefied Sawdust. *J. Sustain. Metall.* **2019**, *5*, 391–401. [CrossRef]

44. Tang, H.; Rong, T.; Fan, K. Numerical investigation of applying high-carbon metallic briquette in blast furnace ironmaking. *ISIJ Int.* **2019**, *59*, 810–819. [CrossRef]

45. Natsui, S.; Kikuchi, T.; Suzuki, R.O. Numerical analysis of carbon monoxide–hydrogen gas reduction of iron ore in a packed bed by an Euler–Lagrange approach. *Metall. Mater. Trans. B* **2014**, *45*, 2395–2413. [CrossRef]

46. Patankar, S.V.; Spalding, D.B. A calculation procedure for heat, mass and momentum transfer in three-dimensional parabolic flows. *Int. J. Heat Mass Transf.* **1972**, *15*, 1787–1806. [CrossRef]

47. Karki, K.C.; Patankar, S.V. Calculation procedure for viscous incompressible flows in complex geometries. *Numer. Heat Transf. B* **1988**, *14*, 295–307. [CrossRef]

48. Melaen, M.C. Calculation of fluid flows with staggered and nonstaggered curvilinear non-orthogonal grids-the theory. *Numer. Heat Transf. B* **1992**, *21*, 1–19. [CrossRef]

49. Castro, J.A.; Dabreu, J.C.; Silva, A.J. Two-phase model and computational simulation of self-reduction in shaft furnaces. *TMMM Tecnol. Metal. Mater. Min.* **2006**, *2*, 45–50.

50. Noldin, J.H., Jr.; Cox, I.J.; DAbreu, J.C. The Tecnored ironmaking process part 1-competitiveness and pilot development work. *Ironmak. Steelmak.* **2008**, *37*, 334–340.

51. Santos, D.M.; Mourão, M.B.; Takano, C. Reaction rate and product morphology in carbon composite iron ore pellets with and without portland cement. *Ironmak. Steelmak.* **2010**, *35*, 245–250. [CrossRef]

52. Dabreu, J.C.; Kohler, H.M.; Noldin, J.H., Jr. Mathematical model for descending self-reducing agglomerates in lumpy zone of Tecnored furnace. *Ironmak. Steelmak.* **2013**, *40*, 484–490. [CrossRef]

Publisher's Note: MDPI stays neutral with regard to jurisdictional claims in published maps and institutional affiliations.

 *metals*

Article

Energy Consumption and CO$_2$ Emissions in Ironmaking and Development of a Novel Flash Technology

Hong Yong Sohn 

Department of Materials Science & Engineering, University of Utah, 135 South 1460 East, Salt Lake City, UT 84112, USA; h.y.sohn@utah.edu; Tel.: +1-801-581-5491

Received: 22 November 2019; Accepted: 20 December 2019; Published: 27 December 2019

Abstract: The issues of energy consumption and CO$_2$ emissions of major ironmaking processes, including several new technologies, are assessed. These two issues are interconnected in that the production and use of fuels to generate energy add to the total amount of CO$_2$ emissions and the efforts to sequester or convert CO$_2$ require energy. The amounts of emissions and energy consumption in alternate ironmaking processes are compared with those for the blast furnace, currently the dominant ironmaking process. Although more than 90% of iron production is currently through the blast furnace, intense efforts are devoted to developing alternative technologies. Recent developments in alternate ironmaking processes, which are largely driven by the needs to decrease CO$_2$ emissions and energy consumption, are discussed in this article. This discussion will include the description of the recently developed novel flash ironmaking technology. This technology bypasses the cokemaking and pelletization/sintering steps, which are pollution prone and energy intensive, by using iron ore concentrate. This transformational technology renders large energy saving and decreased CO$_2$ emissions compared with the blast furnace process. Economic analysis indicated that this new technology, when operated using natural gas, would be economically feasible. As a related topic, we will also discuss different methods for computing process energy and total energy requirements in ironmaking.

Keywords: ironmaking; carbon emissions; energy consumption; flash ironmaking process; alternate ironmaking processes; direct reduction; smelting reduction; iron ore concentrate; natural gas

1. Introduction

The blast furnace (BF), direct reduction (DR), and recently developed smelting reduction (SR) make up major ironmaking processes currently practiced in industry. The BF process currently produces more than 90% of the world production [1]. The balance is produced largely by gas-based direct reduction, with the rotary kiln process accounting for about 1% and smelting reduction contributing approximately 1%. The overarching current issues in the ironmaking industry are energy consumption and carbon dioxide emissions, which drive most of the new developments in ironmaking. National and international efforts in this respect include, among others, the American Iron and Steel Institute (AISI) CO$_2$ Breakthrough Program in the U.S., the ULCOS program in Europe, the COURSE 50 Program in Japan, the development of FINEX in Korea, and the efforts in Chinese steel industry [2]. As part of AISI's CO$_2$ Breakthrough Program, a novel Flash Ironmaking Technology (FIT) has recently been developed at the University of Utah. We will discuss this technology in some detail, vis-a-vis the current processes, from the viewpoint of CO$_2$ emissions and energy consumption.

1.1. The Blast Furnace (BF)

The modern blast furnace is a very large metallurgical reactor with a capacity of 0.50–5.6 million tons of pig iron per year. The largest of them is the No. 1 blast furnace at POSCO's Gwangyang (Korea) Steelworks that has the production capacity of 5.65 million tons per year [3].

In BF ironmaking, the solid feeds are fed from the top of a shaft furnace and preheated air is injected near its bottom. The solid charge consists of iron ore as sinter, pellets or lumps, mainly consisting of hematite (Fe_2O_3) or magnetite (Fe_3O_4), coke, and limestone as a flux. Preheated air is blown through tuyeres near the bottom to burn the coke, generating reducing CO gas and process heat. Coke maintains its strength at elevated temperatures, unlike coal that softens at the same temperatures. Iron ores must be sintered or pelletized to turn them into strong and porous pellets of 1–2 cm sizes to keep the bed permeable to the gas flow and facilitate the reduction.

Often, pulverized coal, natural gas, and/or oil are also injected together with the preheated air to decrease the consumption of expensive coke. The coke and injected fuels burn to produce combustion gas containing CO and CO_2 (>1800 °C). This hot gas flows up while heating the descending solids, causing partial reduction of iron oxides.

Liquid hot metal (pig iron), which typically contains 3.5–4.5 wt% of dissolved carbon, and molten slag collect in the bottom part (hearth) of the furnace.

1.2. Direct Reduction (DR)

In the Direct Reduction (DR) processes, iron ore is reduced to solid sponge iron with coal or a reducing gas mixture made up of H_2 and CO, most often produced from the reforming of natural gas, thus avoiding the use of coke. Worldwide, greater than 90% of the direct reduced iron (DRI) production is based on natural gas, whereas coal is mainly used in India [4].

The best-known technologies for DR are MIDREX [5] and HYL/Energiron [6], with MIDREX accounting for more than 78% of the DRI production in 2016. The shaft furnaces used in these technologies require pellets, which are 10–12 mm in size. Sponge iron produced from DR processes is fed to basic oxygen furnaces (BOF) and electric arc furnaces (EAF), as an alternative to BF because of its low capital cost and often based on the local conditions with respect to raw materials [4].

Other gas-based processes developed for the reduction of iron oxide but have been less adopted are the fluidized-bed processes FINMET, earlier FIOR [7], CIRCORED [8] and SPIREX [9]. These processes use iron ore fines, which are particles of +0.1 mm to −10 mm sizes, and thus provide low production rates because the particles in this size range react slowly. The process cannot be operated at high temperatures because the particles fuse and stick together at high temperatures. Largely because of these reasons, fluidized bed processes have not been very successful. The Flash Ironmaking Technology (FIT) utilizes iron ore concentrates with particle sizes less than 100 µm, which are smaller than fines by up to two orders of magnitude. These particles are reduced in seconds rather than the minutes and hours required in other processes.

The coal-based direct reduction process uses rotary kilns [10]. Although this process is slower and rather cumbersome, it is robust, uses less expensive non-coking coal instead of coke, operates at lower temperatures, and requires less feed preparation than most other ironmaking processes.

1.3. Smelting Reduction (SR)

Successful smelting reduction (SR) processes emerged during the 1990s. The main feature of SR is that pre-reduced iron ore is reduced by char generated from coal by in situ devolatilization to form molten metal and slag. This process bypasses cokemaking and requires less charge preparation, but usually needs a pre-reduction step. COREX, FINEX, and HISMELT are major examples. The more fully commercialized COREX and FINEX may be considered as processes that separate the shaft and the lower smelting sections of a blast furnace [11]. Pre-reduction of iron ore is first carried out, and then smelting is done in a separate melter-gasifier that contains a char bed continuously formed by the devolatilization of coal continuously fed to the vessel. The COREX performs the pre-reduction in a shaft furnace, whereas the FINEX uses fluidized bed reactors in series for the same purpose. The melter-gasifier part is essentially the same in the two processes.

2. Critical Issues in Ironmaking

2.1. Technical Issues

World Steel Association [12] reports that the steel industry world-wide is responsible for 6.7% of the total CO_2 emissions. Further, the steel industry consumes the second largest amount of energy and emits the greatest volume of CO_2 (30%) of any industry. Lowering of energy use and CO_2 emissions in the conventional steelmaking processes is quickly approaching the theoretical bounds. To make significant further reductions in energy usage and CO_2 emissions, steelmaking will require drastically new ideas leading to the development of breakthrough technologies.

The BF operation has been greatly improved over the years with respect to efficiencies in productivity and energy use, coke rate, cokemaking technique, CO_2 emissions, and increased injection of other combustibles such as coal, natural gas, and plastics. Many of such technologies are already in commercial practice. However, the operation of a blast furnace still endures drawbacks related to high energy consumption, greenhouse gas emissions and large infrastructure cost.

The blast furnace is quite efficient from the viewpoint of energy and productivity, but requires pelletizing or sintering of iron ore and in addition must use coke as the main fuel and reducing agent. Cokemaking and sintering/pelletization are energy intensive and pollution prone. In addition, the use of coke generates large amounts of CO_2. Overall, sintering (13%), pelletization (2%), and cokemaking (5%) together produce ~20% of the overall CO_2 emissions in the BF-BOF route and the BF contributes ~70%, with 10% generated by steelmaking [13].

The utilization of iron ore fines or concentrates free of pelletizing and sintering is one of the alternate ways to lower CO_2 emissions and energy consumption [14]. These are important reasons for the development of and increasing attention paid to alternate ironmaking technologies. In addition to the above reasons, alternate processes like DR typically are more versatile and economical than BF at lower production rates. Despite these advantages, the current DR technologies based on shaft furnaces, rotary kilns, and fluidized bed reactors suffer from drawbacks such as a low energy efficiency when applied in a small scale, requirement for pelletization, and the fact that the produced DRI is pyrophoric and tends to re-oxidize. Furthermore, the rates of such processes are not intensive, for they cannot use higher temperatures because of the problems of solid sticking and fusion. The processes that use shaft furnaces, which dominate the DR industry, require the concentrate to be pelletized increasing operating costs and environmental problems. Currently, this process is less economical than the BF process as it requires iron ore of higher quality, limiting its flexibility [5].

Smelting reduction processes address many of the above-mentioned difficulties accompanying DR processes. However, they require more than one stage of operation and are unable to reduce CO_2 generation significantly because of their dependence on the use of coal.

2.2. Energy Requirements

Energy consumption is a critical issue in the steel industry. Table 1 shows a comparison between the four major steelmaking routes in terms of the energy consumed for the iron and steel production and also to generate electricity [15]. These numbers were based on the best practice of modern plants. In terms of just ironmaking, DR consumes the least amount of energy at 12.2 GJ/ton of steel and SR requires the largest amount at 18.7 GJ/ton, with BF positioned in the middle at 15.7 GJ/ton of steel. The use of an electric arc furnace for steelmaking is seen to be energy-intensive whereas the use of BOF for primary steelmaking requires a minimal amount of energy. In DR processes, iron oxide is reduced in solid state and therefore all gangues contained in the iron ore stay in the product iron (DRI) and must be removed to the slag in the EAF. This causes greater electrical energy consumption than to melt scraps [4].

Table 1. Energy requirements by best practices worldwide for iron and steel production (GJ per tonne of mild steel) [15].

Production Step	Process	BF-BOF	DR-EAF	SR-BOF *	EAF-Scrap
Feed preparation	Sintering	2.2	2.2		
	Pelletizing		0.8	0.8	
	Coking	1.1			
Ironmaking		12.4	9.2	17.9	
Steelmaking	Main Step	−0.3	5.9	−0.3	5.5
	Refining	0.4		0.4	
Total (GJ/t)		15.8	18.1	18.8	5.5

* Mainly COREX; does not include FINEX or HISMELT.

Note of Caution Regarding 'Energy Requirement': It is noted here that a caution must be exercised when comparing the presented energy intensity values or energy requirements to clearly understand the energy items included in these values [16]. There are currently different approaches for selecting energy items to include in the overall "energy requirements" when an input fuel also serves as a reactant.

For a process involving heat generation from fuel combustion, the choice of which endothermic reactions to pick to compute the energy requirement can cause confusion. The essential question is whether the heating value of a reactant that is also burned to generate process heat should be added as an item in the 'energy requirement'. Some investigators include this item [17–19] and others do not [20–23]. The choice is arbitrary, but tone should indicate explicitly which approach is used in showing the energy calculations, especially in comparing different processes. The presented 'energy requirement' varies with the selected approach.

The net differences in the energy requirements between technologies are not strongly affected by the selection of calculation approaches if the same method is used for different technologies, but the consumption values for individual processes themselves depend on them. Thus, when presenting 'energy requirements', it is critical to definitively state the applied approach. Whether the heat of combustion of a reactant that can also be burned is included in the energy requirement should be clearly indicated. It will make it much more definitive to indicate the amounts for 'process energy' that contains only the heating value of the fuel and 'reductant energy' ('feedstock energy' in petrochemicals production) that includes the heating value of the substance serving as a reactant.

For detailed description of these different calculation methods and discussion on the subject, the reader is referred to Sohn and Olivas-Martinez [16]. An additional caution that follows from this consideration is that the comparison of different ironmaking processes for their energy requirements should only be done with the values obtained using the same method by the same investigators. In other words, the energy requirement value obtained by one investigator for a certain ironmaking process should not be compared with a value obtained by another investigator for a different ironmaking process without carefully checking the bases of the calculations.

Thus, the energy intensities listed in Table 1 for various technologies for making iron and steel may be used safely to obtain differences among the technologies. However, care should be exercised for example when using the energy intensity value for the BF-BOF route give in Table 1 to compare with the energy intensity of say the DR-EAF combination calculated or reported by a different investigator.

2.3. Carbon Dioxide Emissions

Greenhouse gas is a serious problem facing the world today. Although BF is a highly efficient reactor in terms of energy and chemical reactions, the coke used generates large CO_2 emissions. Steelmaking emits 1.9 tons of CO_2 for every ton of steel produced, accounting for 6.7% of the total man-made CO_2 emissions [12].

Although substantial decreases in emissions have been made, new steelmaking technologies will require drastically new ideas to make much greater reductions in energy consumption and CO_2 emissions.

As seen in Table 2, natural gas-based DRI production generates lower CO_2 emissions, in the range of 0.77–1.1 tons of CO_2 per ton of steel (in contrast to ~1.9 tons of CO_2 per ton of steel for the BF-BOF process), varying with the source of electricity used [1]. Smelting reduction (SR) processes like COREX and FINEX, which are coal-based, produce somewhat more CO_2 than DR processes but less than BF and much less than rotary kilns (SL/RN).

Table 2. CO_2 emissions from various steelmaking technologies in tons per ton of iron.

BF-BOF [a]	Midrex Process [b]	HYL III-Energiron [a]	SL/RN Process [a]	SR [a,c]	Circored [a,d]
~1.9	~1.1	0.77–0.92	~3.2	1.3–1.8	~1.2

[a] Institute for Industrial Productivity [4]; [b] Metius et al. [24]; [c] Hasanbeigi et al. [25]; [d] Husain et al. [8].

3. Development of Novel Flash Ironmaking Technology (FIT)

3.1. Background

Despite the many improvements and technical merits of the currently dominant blast furnace ironmaking process, an essential problem for the steel industry today is the development of a new technology with lower energy consumption, CO_2 emissions and fixed costs than the combined blast furnace and coke oven routes. An optimal process should also be able to produce at least 5000–10,000 tons of metal per day to be able to provide sufficient feed to the steel plants.

Given the limited potential for increased efficiency associated with current technologies and considering the other driving forces presented above, a Flash Ironmaking Technology (FIT) has recently been developed at the University of Utah as an innovative alternate ironmaking process [14,20–22,26–31]. This technology reduces iron ore concentrate in a flash reactor with a suitable reductant gas such as hydrogen or natural gas, and possibly bio/coal gas or a combination thereof. It is the first flash ironmaking process. This technology is suitable for an industrial operation that converts iron ore concentrate (less than 100 microns) to metal without further treatment. This transformative technology produces iron while bypassing pelletization or sintering as well as cokemaking steps that are energy intensive and pollution-prone. Further, the process is intensive due to the fact that the fine particles of the concentrate are reduced at a fast rate at 1200–1600 °C. Thus, the required residence times in this process is of the order of seconds rather than the minutes and hours required for pellets and even iron ore fines.

The fluidized bed processes, which have been less than successful commercially, also use fine iron ore particles. These ore 'fines' are several millimeters in size, compared with concentrate particles of <100 μm sizes. Thus, reduction takes much less time in FIT than in a fluidized bed reactor.

In the FIT concentrate particles are reduced by a hot gas mixture produced by the incomplete combustion of a fuel gas that serves as the source of heat as well as reductant gas (H_2 + CO + H_2O + CO_2 mixture from natural gas and H_2 + H_2O when pure hydrogen is used). A schematic of a possible large-scale flash ironmaking reactor is shown in Figure 1. The partial combustion of the feed gas produces the process heat and at the same time generates a gas mixture that has a sufficient reducing power for the reduction of iron oxide.

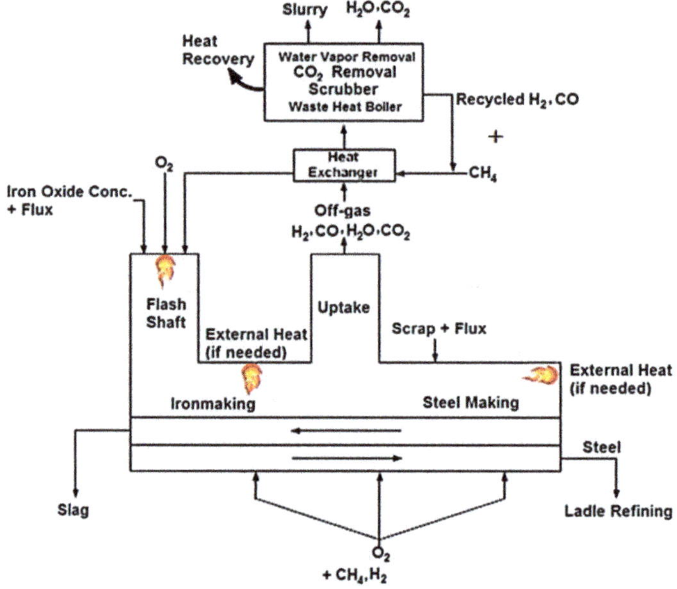

Figure 1. Flash Ironmaking Process. (Concept first proposed by H. Y. Sohn in a research proposal submitted to U.S. Department of Energy through American Iron and Steel Institute (AISI) in November 2003. The figure is adapted from Sohn and coworkers [28,32]).

The operating temperature of this process depends on the target final product. It can be operated at lower temperatures where iron is produced in solid state that can be fed to downstream steelmaking processes such as EAF. In this case, the flash reactor would mainly consist of the shaft portion of the facility shown in Figure 1. In a high-temperature operation, where iron is produced in a molten state, the process can be used as an overall continuous direct steelmaking process, as depicted in Figure 1 [28,32].

The use of concentrate particles and natural gas eliminates the pelletization and cokemaking steps, which lowers the energy consumption by 30% [16] and CO_2 emissions by 39–51% relative to the average blast furnace process [20–22]. If the technology is operated with hydrogen produced from sources with no carbon footprint, the CO_2 emissions will be lowered to 4% (from limestone calcination) of that from the blast furnace process.

In the U.S., iron ore is mostly produced from the Mesabi Range in Minnesota and Michigan, which made up 97% of the iron ore products in the U.S. in 2017 [33]. The ore is largely taconite mineral, a low-grade magnetite ore containing 15–30 wt% iron that requires upgrading for industrial use by crushing and grinding to liberate the iron-bearing minerals and concentrating to remove the gangue materials (mainly silica) by magnetic separation and flotation. The concentrate particles are less than 100 μm in size, whereas the feed to the blast furnace must be coarser than 2 mm. Therefore, the concentrate must be pelletized in the BF process by first forming green pellets [34] and then hardening them by an induration process in which they are dried and fired at ~1300 °C.

3.2. Energy Requirements

Tables 3 and 4 list energy input and output items, respectively, for the flash ironmaking process using in situ partial combustion of natural gas or hydrogen, compared with those for the average blast furnace operation. The energy values in these tables include the heat of combustion of fuel/reductant consumed for the reduction of iron oxide plus that used for process heating. The numbers in these tables are for the production of 1 million tons of iron per year. The values for the flash ironmaking process are based on the flow sheets developed by Pinegar et al. [20,21]. The values for an average blast furnace operation were calculated using published data and applying the same method as for the flash ironmaking process. Sohn and Olivas-Martinez [16] also presented the energy balances for these processes as Sankey diagrams.

Table 3. Energy input for an industrial-scale flash ironmaking process using natural gas or hydrogen vs. an average blast furnace process (for production of 1 million tons of iron). (Adapted from Pinegar et al. [20,21]).

	Process	Reformerless Natural Gas	Hydrogen [d]	Blast Furnace [a,e]
	Fuel combustion [b]	19.22	14.05	13.60
	Heat recovery (sum of next 2)	−4.77	−2.80	−1.32
	(Waste heat boiler)	(−3.39)		
ITEMIZED INPUT	(Steam not used)	(−1.38)		
(GJ/ton Fe)	Sub-total	14.45	11.25	12.28
	Ore/Coke preparation [c]			5.68
	$CaCO_3$ and $MgCO_3$ calcination (external)	0.26	0.26	
	Total	14.7	11.5	18.0

[a] Energy balance was calculated by METSIM based on the published material balance. [b] Higher heating values of the natural gas and hydrogen were used for this calculation. [c] See Sohn and Olivas-Martinez [16] for references. [d] Energy requirement for hydrogen production was not included for this calculation. The energy requirement for hydrogen production depends on the production process such as steam-methane reforming, coal gasification or water splitting. [e] In fairness to excluding the energy requirement of hydrogen production for flash ironmaking, the energy requirement for producing coking coal was not included for the blast furnace.

Table 4. Energy output for an industrial-scale flash ironmaking process using natural gas or hydrogen vs. an average blast furnace process (for production of 1 million tons of iron). (Adapted from Pinegar et al. [20,21]).

	Process	Reformerless Natural Gas	Hydrogen	Blast Furnace [a]
	Reduction [b]	6.68	6.68	7.37
	Sensible heat of iron	1.27 (1773 K)		1.35 (1873 K)
	Sensible heat of slag	0.24 (1773 K)		0.47 (1873 K)
	Slurry (H_2O (l))	2.25 (323 K)	1.93	
	Hot water not used	1.57 (493 K)		
	Flue gas	0.79 (573 K)		0.26 (363 K)
	Removed water vapor		0.01	
	$CaCO_3$ decomposition			0.33
	Slagmaking			−0.17
ITEMIZED OUTPUT	Heat loss in the reactor	0.78	0.78	2.60
(GJ/ton Fe)	Heat loss in the heat exchangers (sum of next 3)	0.73	0.34	0.07
	(Reactor feed gas heater)	(0.40)		
	(Natural gas heater)	(0.21)		
	(WGS reactor feed gas heater)	(0.12)		
	Steam not used (363 K)	0.14		
	Sub-total	14.45	11.25	12.28
	Pelletizing [c]			3.01
	Sintering [c]			0.65
	Cokemaking [c]			2.02
	$CaCO_3$ and $MgCO_3$ calcination (external)	0.26	0.26	
	Total	14.7	11.5	18.0

[a] Energy balance was calculated by METSIM based on the published material balance. [b] For flash ironmaking process, magnetite was used as the iron oxide; hematite was used for the blast furnace, because magnetite in the concentrate is converted to hematite during pelletization required in the blast furnace (BF) process. [c] See Sohn and Olivas-Martinez [16] for references.

In the blast furnace process, the energy needed for sintering, pelletizing and cokemaking operations, which are not necessary in flash ironmaking, represents a large part of the overall energy requirement. The preparation of ore and coke takes up approximately 30% of the total energy input [16].

The 'energy requirement' for ironmaking can be calculated directly from the sums of energy inputs in Table 3 and outputs in Table 4. This is because the recovered heat is listed as a negative input item in Table 3. The energy balance could be written with such an item listed as a positive output term in Table 4, in which case the 'balance' is still attained.

The total value in these tables for BF is somewhat different from that for the ironmaking portion given in Table 1. As noted in that section, energy requirement values depend on the items included in the calculation procedure. However, the differences in the values computed by applying the same calculation procedure remain the same. According to Table 1 a typical current Direct Reduction process for ironmaking consumes approximately 3.5 GJ/t Fe less than BF [15], which would correspond to 14.5 GJ/t Fe if calculated using the same basis for obtaining the results in Tables 3 and 4. This number agrees with the energy requirement for the reformerless mode of the Flash Ironmaking Technology.

3.3. Carbon Dioxide Emissions

When natural gas is the fuel/reductant in the Flash Ironmaking Technology (FIT), the level of carbon dioxide emissions is similar at ~1 ton/ton Fe [21] to those from other natural gas based DR processes listed in Table 2. However, a hydrogen-based flash ironmaking process would emit a lower amount of CO_2, depending on the source of hydrogen.

3.4. Reduction Kinetics of Concentrate Particles

A flash reactor typically provides a residence time measured in seconds, unlike a shaft or fluidized bed reactor that can provide a residence time of minutes and hours. Thus, an essential requirement in the development of a flash ironmaking process is sufficiently fast reduction rate of the iron oxide feed. In a fluidized bed reactor, the iron ore fines require residence times measured in minutes. It is important, however, to differentiate concentrates from iron ore fines. The majority of individual particles in fines are several millimeters in size, while concentrate particles are less than 100 μm. There have been many applications of iron ore fines in ironmaking such as FIOR, FINMET, Circored, Iron Carbide and the recent FINEX processes. On the other hand, there have not been any processes directly using iron ore concentrates in large scale.

Previous researchers have considered the in-flight reduction of iron oxide particles, but the reduction rate was deemed too slow for a flash process that typically provides a residence time of a few seconds. This conclusion was based on data from the reduction of particles ranging from 70 to 42,000 μm at temperatures 600–1000 °C [35]. Sohn [14] examined these data and concluded that there was a potential for concentrate particles to be reduced in a few seconds. He then launched a project to develop the novel Flash Ironmaking Technology (FIT). It was thus important to establish a sufficiently fast rate for the reduction of iron concentrates using H_2 and CO gas mixture for designing the flash reactor.

Therefore, the reduction kinetics were investigated in the temperature range 1150–1600 °C using H_2, CO, or H_2 + CO as the reducing gas by Sohn and coworkers [36–42]. Reduction by H_2 + CO is complicated because of the simultaneous reduction by the two gases as well as the water-gas shift reaction involving CO and H_2 with the product gases H_2O and CO_2.

Due to the melting of the particles above 1350 °C, the rate analysis was done separately in the ranges of 1423–1623 K (1150–1350 °C) and 1623–1873 K (1350–1600 °C). It is seen in Figure 2 that the particles maintain their original solid shape in the lower temperature range, whereas they melt into spheres at higher temperatures, as seen in Figure 3.

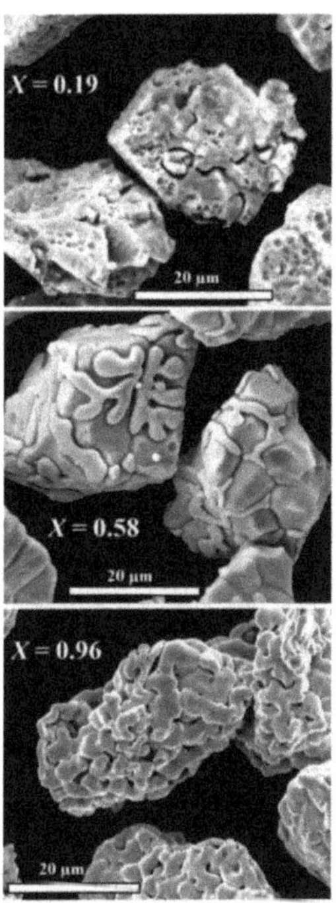

Figure 2. SEM micrographs of particles at 1137–1324 °C. (*X* is fractional conversion.) (The figure is adapted from [39], with permission from John Wiley & Sons, 2017).

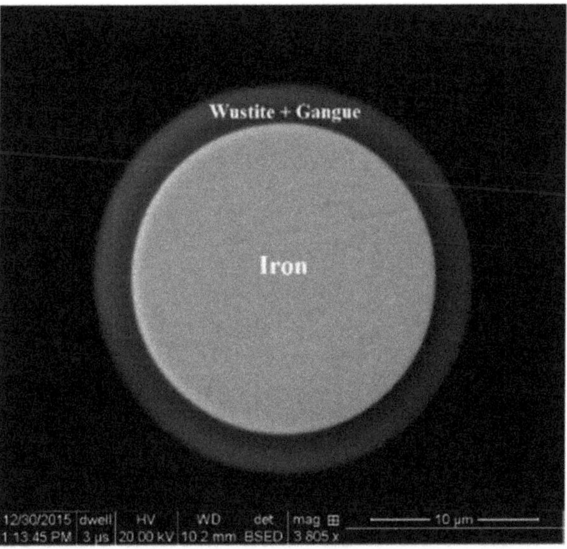

Figure 3. Cross section of a particle reduced to fractional conversion $X = 0.94$ at 1623 K (1350 °C) (The figure is adapted from [43], with permission from authors, 2019).

At these temperatures and particle sizes, it was proved that neither external mass transfer nor pore-diffusion influence the rate of magnetite concentrate reduction, as the reaction would be completed in milliseconds if the rate was diffusion-controlled compared with several seconds observed experimentally. Therefore, the rate equations obtained based on the experimental data are those of the chemical reactions unaffected by mass transfer [28]. The reduction of magnetite to iron proceeds through the formation of $Fe_{0.947}O$ in the temperature range of flash smelting. However, it is not possible to measure the separate kinetics of $Fe_{0.947}O$ and Fe formation for fine particles reacting rapidly. Moreover, different parts of a fine irregular particle react at different rates and thus various oxide phases may coexist at any time, as the SEM and XRD results of quenched samples in Figure 4 show. Thus, a global rate equation was used in which conversion represented the fraction of oxygen removed from the original iron oxide without recognizing individual oxide phases.

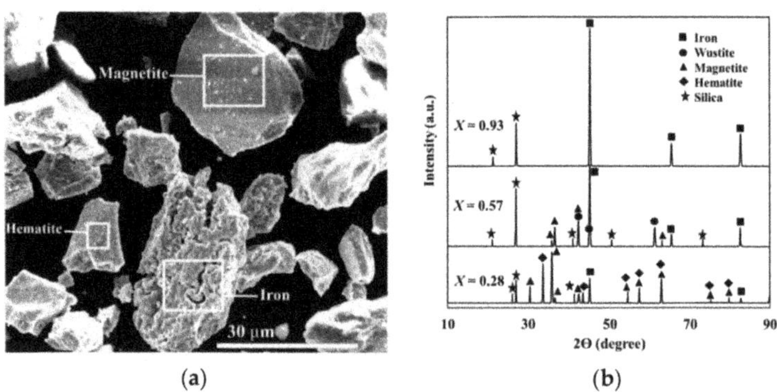

(a) (b)

Figure 4. Existence of different iron oxide phases indicated by: (**a**) SEM micrographs of reduced particles (X = 0.14) and (**b**) XRD of samples at various reduction degrees. (The figure is adapted from [40], with permission from Springer Nature, 2015).

The resulting rate expressions from all the work mentioned above are summarized below:

Reduction of Magnetite Concentrate by H_2

In temperature range 1423–1623 K (1150–1350 °C) [38]:

$$\frac{dX}{dt}\bigg|_{H_2} = 1.23 \times 10^7 \exp\left(-\frac{196,000}{RT}\right) \cdot \left(p_{H_2} - \frac{p_{H_2O}}{K_{H_2}}\right) \cdot (1-X) \tag{1}$$

where t is in seconds, R is 8.314 J/mol K, T is in K, and p is in atm.

In temperature range 1623–1873 K (1350–1600 °C) [43]:

$$\frac{dX}{dt} = 6.07 \times 10^7 \cdot e^{-\frac{180,000}{RT}} \cdot \left(p_{H_2} - \frac{p_{H_2O}}{K}\right) \cdot (d_p)^{-1} \cdot (1-X) \tag{2}$$

where d_p is in µm.

Reduction of Magnetite Concentrate by CO

In temperature range 1473–1623 K (1200–1350 °C) [44]:

$$\frac{dX}{dt} = 5.35 \times 10^{13} \cdot e^{-\frac{451,000}{RT}} \cdot \left(p_{CO} - \frac{p_{CO_2}}{K}\right) \cdot (1-X) \cdot [-\ln(1-X)]^{-1} \tag{3}$$

In temperature range 1623–1873 K (131,600 °C) [44]:

$$\frac{dX}{dt} = 3.225 \times 10^3 \cdot e^{-\frac{88,000}{RT}} \cdot \left(p_{CO} - \frac{p_{CO_2}}{K}\right) \cdot (d_p)^{-1} \cdot (1-X) \cdot [-\ln(1-X)]^{-1} \tag{4}$$

Reduction of Magnetite Concentrate by H$_2$ + CO Mixture [45]

$$\frac{dX}{dt} = \left(1 + 1.3 \cdot \frac{p_{co}}{p_{co} + p_{H_2}}\right) \cdot \left.\frac{dX}{dt}\right|_{H_2} + \left.\frac{dX}{dt}\right|_{CO} \qquad 1423\ \text{K} < T < 1623\ \text{K} \qquad (5)$$

$$\frac{dX}{dt} = \left[1 + (-0.01T + 19.65) \cdot \frac{p_{co}}{p_{co} + p_{H_2}}\right] \cdot \left.\frac{dX}{dt}\right|_{H_2} + \left.\frac{dX}{dt}\right|_{CO} \qquad 1623\ \text{K} < T < 1873\ \text{K} \qquad (6)$$

where $\left.\frac{dX}{dt}\right|_{H_2}$ and $\left.\frac{dX}{dt}\right|_{CO}$ are the instantaneous rates of reaction obtained from the corresponding reaction rates individually by H$_2$ and by CO, respectively, given in Equations (1)–(4).

Reduction of Hematite Concentrate by H$_2$ [36]

$$\frac{dX}{dt} = 8.47 \times 10^7 \times e^{-\frac{218,000}{RT}} \cdot \left[p_{H_2} - \left(\frac{p_{H_2O}}{K_{H_2}}\right)\right] \cdot (1 - X) \qquad 1423\ \text{K} < T < 1623\ \text{K} \qquad (7)$$

Reduction of Hematite Concentrate by CO [36]

$$\frac{dX}{dt} = 5.18 \times 10^7 \times e^{-\frac{241,000}{RT}} \cdot \left[p_{CO} - \left(\frac{p_{CO_2}}{K_{CO}}\right)\right] \cdot (1 - X) \qquad 1423\ \text{K} < T < 1623\ \text{K} \qquad (8)$$

Reduction of Hematite Concentrate by H$_2$+CO Mixture [37]

$$\frac{dX}{dt} = \left[1 + (-0.004T + 7.004) \cdot \frac{p_{co}}{p_{co} + p_{H_2}}\right] \cdot \left.\frac{dX}{dt}\right|_{H_2} + \left.\frac{dX}{dt}\right|_{CO} \qquad 1423\ \text{K} < T < 1623\ \text{K} \qquad (9)$$

with $\left.\frac{dX}{dt}\right|_{H_2}$ and $\left.\frac{dX}{dt}\right|_{CO}$ given by the instantaneous rates of reaction obtained from the reaction rates individually by H$_2$ and by CO, respectively, given in Equations (7) and (8).

A serious problem with the direct reduced iron is its pyrophoric nature due to the morphology and large specific surface area when produced below about 900 °C. Sohn and co-workers [46,47] investigated the re-oxidation rates of the iron produced in the flash ironmaking process under various oxidizing gases. The re-oxidation rate by H$_2$O was determined under various temperatures and water partial pressures at 550–700 °C and H$_2$O concentrations of 40–100%. During the few seconds available in a flash process, the re-oxidation extent of iron particles in water vapor was <0.24%. The investigation of the oxidation of iron particles by an O$_2$-N$_2$ gas atmosphere was also performed and it was concluded that the flash-reduced iron was not pyrophoric, unlike the direct reduced iron. The flash ironmaking takes place at higher temperatures than the direct reduction and thus the surface of the flash reduced iron was passivated, as seen from the micrographs in Figure 5. This figure shows iron particles reduced at a lower temperature, like conventional DRI (Figure 5a), compared with those obtained at a flash ironmaking temperature (Figure 5b). This work established that the oxidation of flash reduced iron does not pose any concern at temperatures below 573 K (300 °C) [46].

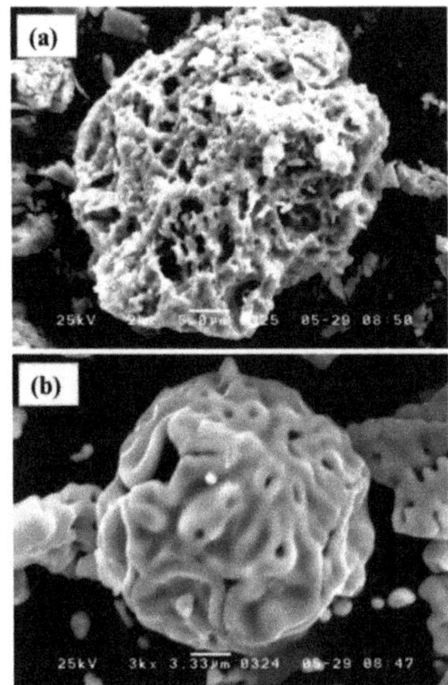

Figure 5. Microstructures of iron particles: (**a**) reduced by hydrogen at 1073 K (800 °C); (**b**) flash reduced at 1623 K (1350 °C). (The figure is adapted from [46], with permission from authors, 2014).

3.5. Laboratory Flash Reactor

An industrial flash reactor would be quite different from the laminar-flow reactor used for the rate measurement, including the fact that an oxy-fuel burner would be the main source of heat and the amount of excess reducing gases would be much lower (20–100%). Therefore, a laboratory flash reactor was installed at the University of Utah [29,31,48] to test the Flash Ironmaking Technology (FIT). The experiments performed in this reactor aimed to establish the optimum conditions for the design of the industrial flash reactor. In this reactor, magnetite concentrate was reduced by a reducing gas mixture generated from the partial oxidation of methane and/or hydrogen with industrial oxygen, which provided heat and produced a reducing gas mixture of H_2 and CO.

The apparatus, shown in Figure 6, consisted of an electrical furnace housing a stainless-steel tube, a gas delivery system, a powder feeding system, a power control system, an off-gas scrubbing system, and an off-gas burner. The electrical furnace housed a 316 stainless-steel tube with 19.5 cm ID and 213 cm length.

The particles were fed into this reactor through openings in the upper flange installed on the top of the reactor tube. Figure 7 shows the two feeding modes tested in this work: (a) feeding through the center of the fuel/oxygen burner; (b) feeding through two ports on opposite sides of the burner.

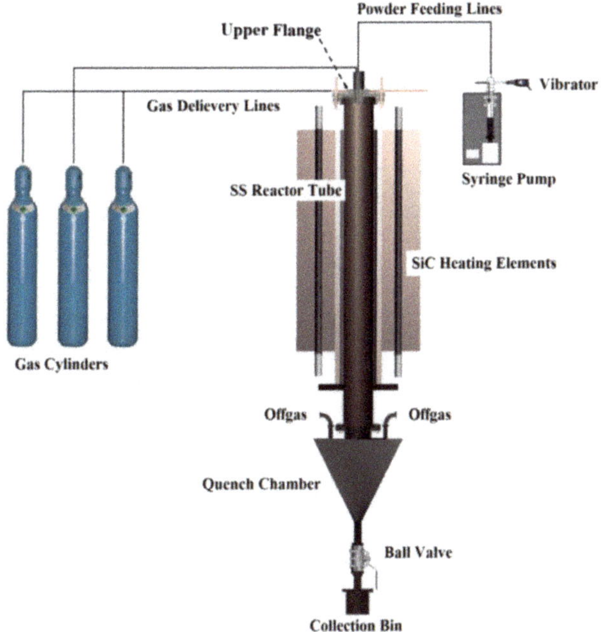

Figure 6. Schematic diagram of the laboratory flash reactor. (The figure is adapted from Sohn et al. [31]).

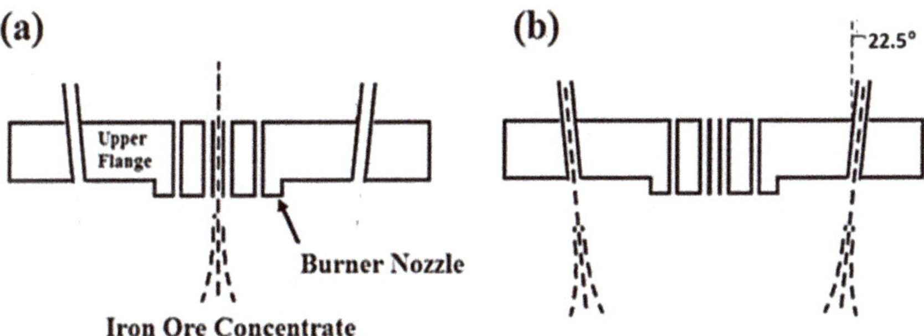

Figure 7. Powder feeding modes: (**a**) Burner feeding port. (**b**) Two Side-Feeding ports. (The figure is adapted from Sohn et al. [31]).

The burner was made of Inconel with crescent-shaped feeding inlets (Slots 1) and cylindrical inlets (Slots 2), as shown in Figure 8. Two different flame configurations were tested by switching the fuel and oxygen injection slots. F-O-F (F = fuel; O = oxygen) was the first configuration where the fuel (hydrogen or methane) was injected through Slots 1 and surrounded the oxygen injected through Slots 2. O-F-O was the second configuration where the oxygen was injected from Slots 1 surrounding the fuel injected from Slots 2.

The variation in the burner configuration changed the temperature distribution in the upper part of the reactor. Computational fluid dynamics (CFD) simulations [48,49] indicated that variation in the flame configuration affected the best feeding modes as the particles experience different temperatures in the different flame configurations, as will be discussed subsequently.

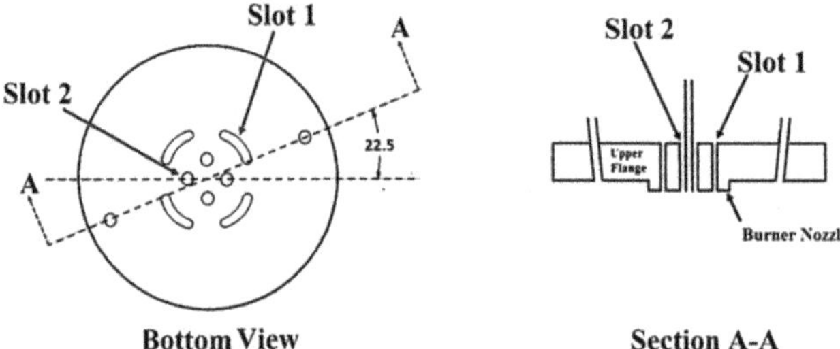

Figure 8. Schematic diagram showing the fuel/oxygen burner: **Left figure**-Bottom view. **Right figure**-Cross section at A-A. (The figure is adapted from Sohn et al. [31]).

3.5.1. Experiments with Hydrogen

A reduction degree of >90% with <100% excess hydrogen at temperature as low as 1175 °C was achieved in a few seconds of residence time. Figure 9 shows the effect of residence time and excess driving force (EDF) on the extent of reduction. EDF, which represents the level of excess hydrogen fed, is defined as follows [28]:

$$\text{EDF} = \frac{\left(\frac{p_{H_2}}{p_{H_2O}}\right)_{\text{actual}} - \left(\frac{p_{H_2}}{p_{H_2O}}\right)_{\text{equ.}}}{\left(\frac{p_{H_2}}{p_{H_2O}}\right)_{\text{equ.}}} = K_H\left(\frac{p_{H_2}}{p_{H_2O}}\right)_{\text{actual}} - 1 \tag{10}$$

where $p_{H2,\text{actual}}$ and $p_{H2O,\text{actual}}$ are the partial pressures of H_2 and H_2O in the gas mixture at complete reduction, respectively, and $p_{H2,\text{equ.}}$ and $p_{H2O,\text{equ.}}$ are the partial pressures of H_2 and H_2O at equilibrium with wüstite and Fe.

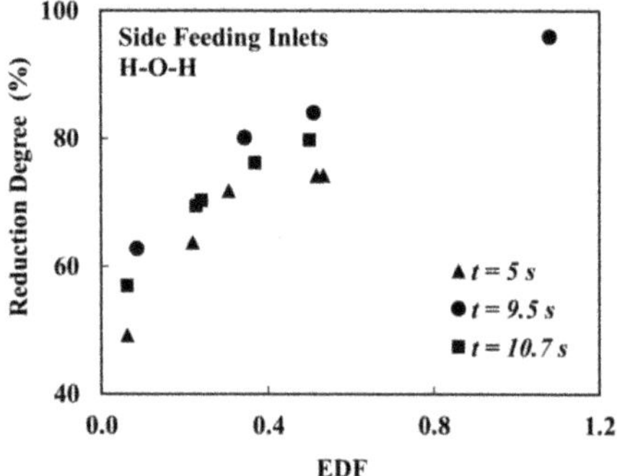

Figure 9. Effect of excess driving force (EDF) on reduction degree (%) when feeding through the two side-ports with the H-O-H configuration [31].

The H-O-H flame generated a higher local temperature in the middle of the flame compared to the O-H-O flame [31]. When the particles were injected through the burner in the H-O-H flame arrangement, they melted and then solidified into spherical particles. Melting reduced the surface area in contrast to side feeding where the particles retained their irregular shape and reactivity that resulted in higher reduction degrees. Therefore, the reduction degree obtained from the burner feeding was lower than that obtained from the side feeding at the same experiment conditions.

The temperature in the middle of the flame with the O-H-O arrangement was lower than with the H-O-H arrangement, 1150 °C and 2577 °C, respectively. Therefore, particle melting did not occur and they kept their irregular shape even when going through the flame, which resulted in a higher reduction degree. Figure 10 illustrates the change in the particles shape with the flame arrangement at otherwise the same conditions.

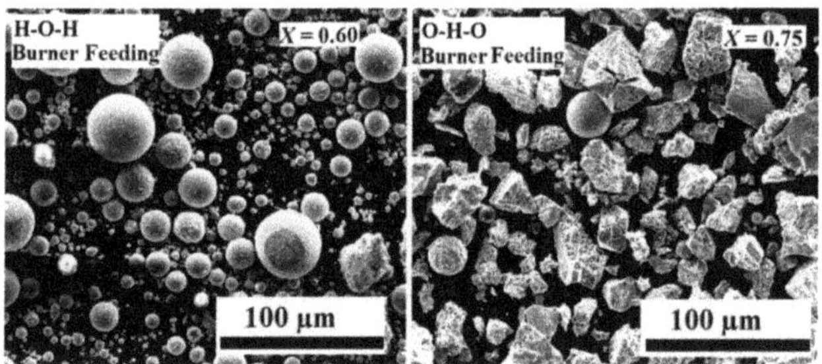

Figure 10. SEM micrographs of the particles collected from experiments with feeding through the burner using H-O-H (Reduction degree = 60%) (**Left**) and O-H-O (Reduction degree = 75%) (**Right**) configurations. (The figure is adapted from Sohn et al. [31]).

3.5.2. Experiments with Methane

Experiments were performed in the laboratory flash reactor in which natural gas was partially burned with industrial oxygen, producing heat and H_2 + CO. It was determined that at above 1150 °C, the reduction in an H_2 + CO gas mixture is done mainly by hydrogen.

To set the experimental conditions, the HSC 5.11 thermodynamics software was used to calculate the equilibrium product composition. Using this calculated equilibrium composition, the nominal residence time of the particles and the excess driving force (EDF) with respect to H_2 were calculated.

The O-M-O configuration was used in the case of solid feeding through the center of the burner to avoid the melting of the concentrate particles. The O-M-O configuration yielded a higher reduction degree compared with the M-O-M configuration, as shown in Figure 11. When feeding through the side ports, the particles experienced higher temperatures in the O-M-O flame compared to the M-O-M flame [31].

The results obtained from this reactor proved that iron oxide could be reduced directly in a flash reactor utilizing natural gas or hydrogen. These results were used to determine the best feeding modes and the flame configuration to be implemented in a larger scale reactor as well as an industrial reactor. Magnetite reduction degree >90% was obtained at a temperature as low as 1175 °C with 100% excess hydrogen driving force or less in a few seconds of residence time.

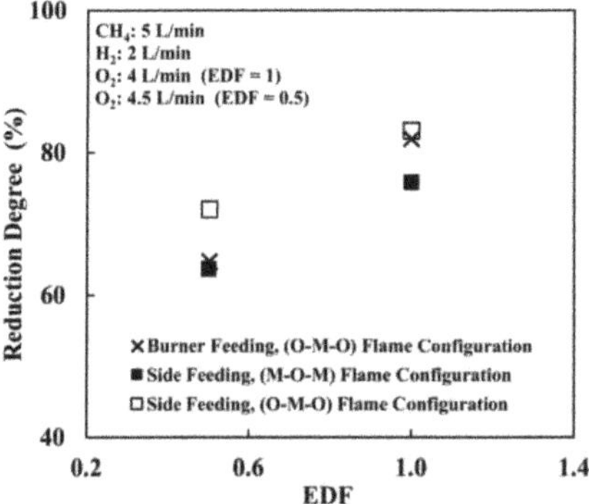

Figure 11. Effect of EDF on reduction with different feeding modes and flame arrangements. (The figure is adapted from Sohn et al. [31]).

3.6. Mini-Pilot Reactor Testing

In this section, we describe a mini-pilot reactor called the Large-Scale Bench Reactor (LSBR) operating at 1200–1600 °C and a concentrate feeding rate of 1–7 kg/h, installed at the University of Utah and shown in Figure 12.

Figure 12. The Large Bench Reactor installed in the University of Utah.

The LSBR consists of a reactor vessel, a vessel roof with various feeding and auxiliary ports, burners, a quench tank, off-gas piping, a flare stack, an off-gas analyzer, a gas valve train, a water cooling system, gas leak detectors, a concentrate feeding system, and human machine interface. Figure 13 shows the main components of the reactor body.

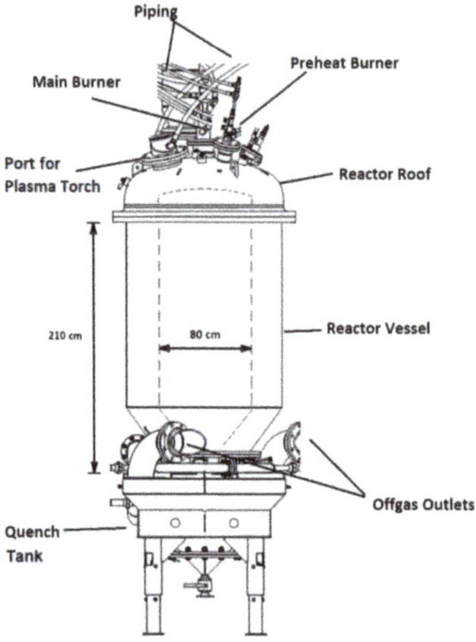

Figure 13. Schematic diagram of the Large-Scale Bench Reactor (LSBR).

The reactor vessel was built with a carbon steel shell and lined with three wall layers: 0.3 cm of alumina-silica fiber blanket, 8 cm layer of high fired and pressed silica, and 18 cm of 99.8% alumina castable refractory layer with high hot strength. The inner diameter of the vessel was 80 cm and the length was 210 cm, as shown in Figures 13 and 14. Figure 14 also shows photos of the main reactor and the burner as well as other details by schematics.

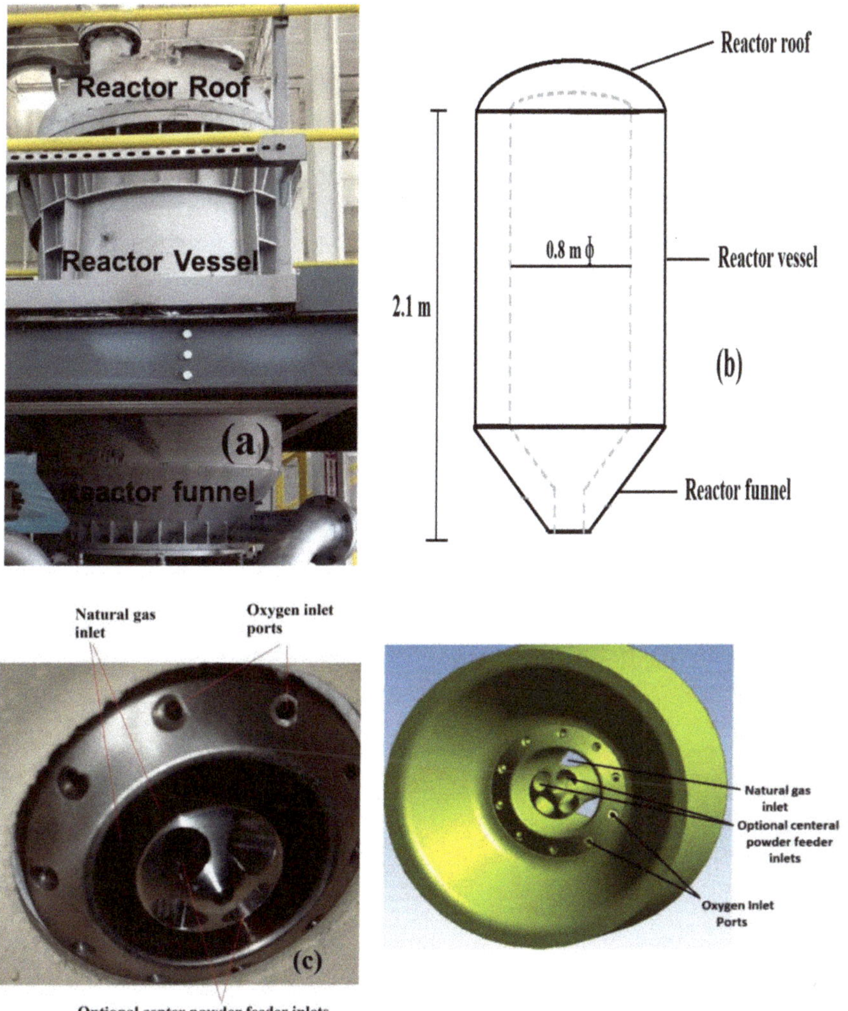

Figure 14. (**a**) The Large-Scale Bench Reactor (LSBR) Vessel, (**b**) Schematic diagram of the LSBR, (**c**) The burner for the LSBR, and Schematic of the burner (**Lower right**).

The LSBR had three burners: a preheat burner, the main burner, and a plasma torch. The preheat burner was used for preheating the reactor to the target temperature using a natural gas and industrial oxygen combustion flame. The preheat burner contained a pilot burner that generated a small flame to ignite the preheat burner. The pilot burner had a fiberglass flame detector that detected the pilot flame and started the flow of natural gas and oxygen through the preheat burner. Through the main burner, natural gas and industrial oxygen were injected to produce a flame and reducing gas mixture by partial oxidation. The plasma torch was installed to provide heat, if needed, without affecting the process gas flow rates.

The Human Machine Interface consisted of the programmable logic controller (PLC) and a computer. The main PLC was connected to all the different parts of the system and to the computer where the operator could monitor the various parts and run the reactor. The programming in the main PLC was responsible for all the safety and emergency steps.

During preheating, combustion of natural gas with industrial oxygen was used to heat the reactor at a ramp rate of 90–95 °C/h to the operating temperature of 1200–1550 °C. The reactor was preheated by burning natural gas with oxygen. The wall temperature was measured by B-type thermocouples imbedded 2.5 cm inside of the inner wall surface. The concentrate was fed to the reactor using a pneumatic feeder with a rate of 1–7 kg/h through 2 feeding ports 0.28 m away from the center of the roof.

The gas analyzer used an NDIR (infrared) detector for measuring the CO, CO_2 and CH_4 contents, a thermal conductivity cell for measuring the H_2 concentration, and an electrochemical sensor for measuring the O_2 content. The partial oxidation of natural gas produced H_2 and CO, but CH_4 and O_2 contents in the off-gas were always less than the detection limits.

The experimental conditions were selected to represent the industrial conditions and to produce a gas mixture that had sufficiently high reducing power and temperature. In addition, the gas flow rates were set to generate sufficient heat and residence time for high degrees of reduction.

The results of the LSBR runs will help in designing an industrial reactor in terms of the identification of the technical hurdles and improvement of the operation. This reactor was simulated by 3-D CFD to optimize the operating conditions for an industrial reactor [50]. Six runs with the LSBR listed in Table 5 were simulated. The results obtained from the CFD model was in satisfactory agreement with the results of the reactor runs, especially considering the complexity of the process and the size of the facility. This work further identified potential safety issues and solutions that are needed in the design and operation of an industrial flash ironmaking reactor.

Table 5. Experimental conditions of the LSBR.

Parameters	Run 1	Run 2	Run 3	Run 4	Run 5	Run 6
Concentrate feeding rate, kg h^{-1}	2.5	4.3	5.0	5.0	4.6	4.0
Particle size range, μm	32–90	less than 90	less than 90	32–90	less than 90	less than 90
Mass average particle size, μm(used for simulation)	45	32	32	45	32	32
Natural Gas flow rate, [a] m^3 h^{-1}	25.16	30.56	20.36	24.80	17.36	15.86
Natural Gas input temperature, K	300					
O_2 flow rate, [a] m^3 h^{-1}	19.85	19.67	14.27	21.53	16.35	14.81
O_2 input temperature, K	300					
Total inlet gas flow rate, [a] m^3 h^{-1}	45.01	50.23	34.63	46.33	33.71	30.67
O_2 to Natural Gas mole ratio	0.79	0.64	0.70	0.87	0.94	0.93
Inner wall temperature, [b] K	1483–1563	1503–1603	1573–1673	1403–1473	1563–1623	1573–1623
Inner wall temperature, K (used for simulation)	1526	1548	1626	1440	1594	1599

[a] Volumetric flow rates are presented at 298 K and 0.85 atm, the atmospheric pressure at Salt Lake City (1 atm = 101.32 kPa). [b] The measured wall temperature varied in the shown range during a run and the average value was selected for Computational fluid dynamics (CFD) simulation.

3.7. Computational Fluid Dynamics Simulation

3-D Computational fluid dynamics (CFD) technique was used to simulate the fluid flow, heat transfer and chemical reaction of the concentrate in the shaft of a flash ironmaking reactor. Flash reactor runs to test the effects of different powder feeding schemes, different flame configurations, and hydrodynamic conditions of an industrial flash ironmaking reactor were simulated using the CFD technique. Temperature and species contours, gas flow patterns, and particle trajectories inside the reactor were computed while incorporating the rate expressions.

The Euler-Lagrange approach was used in this simulation, in which the gas phase was described in the Eulerian frame of reference while the particles were tracked in the Lagrangian framework. Particle spread by turbulence was described using the stochastic trajectory model. Detailed description of the CFD model for LSBR can be found elsewhere [50].

Experimental reduction degrees and the corresponding computed values are presented in Table 6. A reasonable agreement in the reduction degrees is seen in the first three runs. The disagreements for the last three runs were attributed to possible particle agglomeration at higher temperatures, which was not considered in the simulation [50].

Table 6. Experimental vs. computed reduction degrees.

Run	Experimental (pct)	Simulation (pct)
1	94.0	99.8
2	80.0	84.5
3	94.5	99.6
4	74.0	99.8
5	72.5	99.5
6	50.0	85.0

3.8. Economic Analysis

Sohn and coworkers [20–22,51] studied the economic and environmental aspects of the FIT by using the METSIM software to assess different process configurations for a plant that produces 1 million tons per year of solid iron powder. These authors constructed the flow sheet for an industrial-scale plant based on the Flash Ironmaking Technology and carried out detailed material and energy balances. They also calculated the net present value (NPV) after a 15-year operation.

These results suggested that the flash ironmaking process would be economical if it is operated with natural gas. Although the use of hydrogen was not economical at the 2010 price, sensitivity analyses indicated that it could become economical with the development of hydrogen economy with mass production for application as an automobile fuel or with some publicly imposed CO_2 penalty.

Thus, this transformative technology has a significant economic potential in addition to considerable energy saving and reduced CO_2 emissions relative to the current blast furnace process.

4. Concluding Remarks

With the immediate and increasing gravity of global warming caused by anthropogenic CO_2 emissions and similarly serious increasing costs for energy, the steel industry faces a dire need for developing drastically new technologies to respond to these issues. These two issues are coupled in that current energy production is largely dependent on fossil fuels that generates CO_2 emissions and the sequestration of CO_2 requires energy. The amounts of emissions and energy requirements in current ironmaking processes including BF, DR, and SR were compared.

The Flash Ironmaking Technology developed at the University of Utah to address many of the disadvantages of BF and current alternate processes was introduced. This transformative technology removes the energy intensive cokemaking and pelletization/sintering steps by using iron ore concentrate without any further treatments, which allows considerable energy saving and reduced CO_2 emissions. It can be adopted in large enough scales (of the order of millions of tons of iron per year) to compete with the currently available ironmaking technologies or to feed EAF operations for steelmaking. This technology is expected to have an economic advantage over the BF route when it is operated with natural gas as the reducing agent as well as a fuel. Furthermore, the technology may be able to recover iron from fine feed materials other than iron ore concentrates such as dusts, precipitates, and other sources.

Funding: This research was funded by the U.S. Department of Energy under Award Numbers DE-EE0005751 (2012–2018) and DE-FC36-971D13554 (2005–2007) with cost share by the American Iron and Steel Institute (AISI) and the University of Utah, and by American Iron and Steel Institute during 2008–2011.

Acknowledgments: I wish to thank all my undergraduate and graduate assistants plus postdoctoral associates, too many to name all here, who worked with me during the development of the Flash Ironmaking Technology at the University of Utah over the years. Funding from the U.S. Department of Energy, the American Iron, and Steel Institute (AISI), and the University of Utah is gratefully acknowledged.

Conflicts of Interest: The author declares no conflict of interest. The funders had no role in the design of the study; in the collection, analyses, or interpretation of data; in the writing of the manuscript, or in the decision to publish the results.

Disclaimer: This report was prepared as an account of work sponsored by an agency of the United States Government. Neither the United States Government nor any agency thereof, nor any of their employees, makes any warranty, express or implied, or assumes any legal liability or responsibility for the accuracy, completeness, or usefulness of any information, apparatus, product, or process disclosed, or represents that its use would not infringe privately owned rights. Reference herein to any specific commercial product, process, or service by trade name, trademark, manufacturer, or otherwise does not necessarily constitute or imply its endorsement, recommendation, or favoring by the United States Government or any agency thereof. The views and opinions of authors expressed herein do not necessarily state or reflect those of the United States Government or any agency thereof.

References

1. World Steel Association. World Steel in Figures. 2018. Available online: https://www.worldsteel.org/en/dam/jcr:f9359dff-9546-4d6b-bed0-996201185b12/World+Steel+in+Figures+2018 (accessed on 16 December 2019).
2. Sun, W.; Cai, J.; Ye, Z. Advances in energy conservation of China steel industry. *Sci. World J.* **2013**, *2013*, 247035. [CrossRef] [PubMed]
3. Matejak, E.-M.; POSCO S. Ignites the Largest Blast Furnace in the World. Available online: http://e-metallicus.com/en/news/world/posco-ignites-the-largest-blast-furnace-in-the-world.html (accessed on 30 March 2018).
4. Institute for Industrial Productivity. Available online: http://www.ietd.iipnetwork.org/content/direct-reduced-iron (accessed on 30 May 2019).
5. Midrex, World Direct Reduction Statistics. 2016. Available online: https://www.midrex.com/assets/user/news/MidrexStatsBook2016.pdf (accessed on 30 March 2018).
6. Duarte, P.E.; Becerra, J.; Lizcano, C.; Martinis, A. ENERGIRON direct reduction technology-economical, flexible, environmentally friendly. *Acero Latino Americano* **2008**, *6*, 52–58.
7. Hassan, A.; Whipp, R.H. Finmet Process for Direct Reduction of Fine Ore. Available online: https://www.researchgate.net/publication/292554156_Finmet_process_for_direct_reduction_of_fine_ore1996/06/01 (accessed on 30 March 2018).
8. Husain, R.; Sneyd, S.; Weber, P. Circored and Circofer—Two New Fine Ore Reduction Processes. In Proceedings of the International Conference on Alternative Routes of Iron and Steelmaking (ICARISM '99), Perth, Australia, 15–17 September 1999; pp. 123–129.
9. Macauley, D. Options increase for non-BF ironmaking. *Steel Times Int.* **1997**, *21*, 20.
10. IspatGuru. 2017. Available online: https://www.ispatguru.com/coal-based-direct-reduction-rotary-kiln-process/ (accessed on 24 May 2019).
11. Primetals Technologies, Brochure Nr.: T01-0-N001-L4-P-V4-EN, 2015, Primetals Technologies Ltd. and Posco Ltd. Available online: https://www.primetals.com/fileadmin/user_upload/content/01_portfolio/1_ironmaking/finex/The_Finex_process.pdf (accessed on 24 May 2019).
12. World Steel Association. Steel's Contribution to a Low Carbon Future and Climate Resilient Societies. Available online: https://www.worldsteel.org/en/dam/jcr:66fed386-fd0b-485e-aa23-b8a5e7533435/Position_paper_climate_2018.pdf (accessed on 30 May 2019).
13. Orth, A.; Anastasijevic, N.; Eichberger, H. Low CO_2 emission technologies for iron and steelmaking as well as titania slag production. *Miner. Eng.* **2007**, *20*, 854–861. [CrossRef]
14. Sohn, H.Y. Suspension Ironmaking Technology with Greatly Reduced Energy Requirement and CO_2 Emissions. *Steel Times Int.* **2007**, *31*, 68–72.
15. Worrell, E.; Price, L.; Neelis, M.; Galitsky, C.; Nan, Z. *World Best Practice Energy Intensity Values for Selected Industrial Sectors*; LBNL-62806; Lawrence Berkeley National Laboratory: Berkeley, CA, USA, 2008. Available online: http://eaei.lbl.gov/sites/all/files/industrial_best_practice_en.pdf (accessed on 16 December 2019).

16. Sohn, H.Y.; Olivas-Martinez, M. Methods for Calculating Energy Requirements for Processes in Which a Reactant Is Also a Fuel: Need for Standardization. *JOM* **2014**, *66*, 1557–1564. [CrossRef]

17. Remus, R.; Aguado Monsonet, M.A.; Roudier, S.; Delgado Sancho, L. *Best Available Techniques (BAT) in Reference Document for Iron and Steel Production. Industrial Emissions Directive 2010/75/EU (Integrated Pollution Prevention and Control)*; Joint Research Centre (JRC) of the European Union: Seville, Spain, 2013; pp. 304–305.

18. Fruehan, R.J.; Fortini, O.; Paxton, H.W.; Brindle, R. *Theoretical Minimum Energies to Product Steel for Selected Conditions*; Department of Energy: Washington, DC, USA, 2000. Available online: http://www1.eere.energy.gov/industry/steel/pdfs/theoretical_minimum_energies.pdf (accessed on 27 July 2013).

19. Stubbles, J. Energy Use in the U.S. In *Steel Industry: An Historical Perspective and Future Opportunities*; Department of Energy: Washington, DC, USA, 2000. Available online: http://www1.eere.energy.gov/manufacturing/resources/steel/pdfs/steel_energy_use.pdf (accessed on 27 July 2013).

20. Pinegar, H.K.; Moats, M.S.; Sohn, H.Y. Process Simulation and Economic Feasibility Analysis for a Hydrogen-Based Novel Suspension Ironmaking Technology. *Steel Res. Int.* **2011**, *82*, 951–963. [CrossRef]

21. Pinegar, H.K.; Moats, M.S.; Sohn, H.Y. Flowsheet development, process simulation and economic feasibility analysis for novel suspension ironmaking technology based on natural gas: Part 1—Flowsheet and simulation for ironmaking with reformerless natural gas. *Ironmak. Steelmak.* **2012**, *39*, 398–408. [CrossRef]

22. Pinegar, H.K.; Moats, M.S.; Sohn, H.Y. Flowsheet development, process simulation and economic feasibility analysis for novel suspension ironmaking technology based on natural gas: Part 2—Flowsheet and simulation for ironmaking combined with steam methane reforming. *Ironmak. Steelmak.* **2013**, *40*, 32–43. [CrossRef]

23. Burgo, J.A. The Manufacture of Pig Iron in the Blast Furnace. In *The Making, Shaping and Treating of Steel*, 11th ed.; Wakelin, D.H., Ed.; The AISE Steel Foundation: Pittsburgh, PA, USA, 1999; pp. 712–713.

24. Metius, G.E.; McClelland, J.M.; Hornby-Anderson, S. Comparing CO_2 emissions and energy demands for alternative ironmaking routes. *Steel Times Int.* **2006**, *30*, 32–36.

25. Hasanbeigi, A.; Arens, M.; Price, L. Alternative emerging ironmaking technologies for energy-efficiency and carbon dioxide emissions reduction: A technical review. *Renew. Sustain. Energy Rev.* **2014**, *33*, 645–658. [CrossRef]

26. Sohn, H.Y. From Sulfide Flash Smelting to a Novel Flash Ironmaking Technology. In Proceedings of the 143rd TMS Annual Meeting on Celebrating the Megascale, EPD Symposium on Pyrometallurgy, San Diego, CA, USA, 10–20 February 2014; Mackey, P.J., Grimsey, E., Jones, R., Brooks, G., Eds.; TMS/Wiley: Hoboken, NJ, USA, 2014; pp. 69–76.

27. Sohn, H.Y. From Nonferrous Flash Smelting to Flash Ironmaking: Development of an ironmaking technology with greatly reduced CO_2 emissions and energy consumption. In *Treatise on Process Metallurgy, Volume 3 Industrial Processes Part B, Section 4.5.2.2*; Elsevier: Oxford, UK; Waltham, MA, USA, 2014; pp. 1596–1691.

28. Sohn, H.Y.; Mohassab, Y. Development of a Novel Flash Ironmaking Technology with Greatly Reduced Energy Consumption and CO_2 Emissions. *J. Sustain. Metall.* **2016**, *2*, 216–227. [CrossRef]

29. Sohn, H.Y.; Choi, M.E.; Zhang, Y.; Ramos, J.E. Suspension Reduction Technology for Ironmaking with Low CO_2 Emission and Energy Requirement. *Iron Steel Technol.* **2009**, *6*, 158–165.

30. Sohn, H.Y.; Mohassab, Y.; Elzohiery, M.; Fan, D.-Q.; Abdelghany, A. Status of the Development of Flash Ironmaking Technology. In *Applications of Process Engineering Principles in Materials Processing, Energy and Environmental Technologies*; Wang, S., Free, M.L., Alam, S., Zhang, M., Taylor, P.R., Eds.; The Minerals, Metals & Materials Series; Springer: Berlin/Heidelberg, Germany, 2017; pp. 15–23. [CrossRef]

31. Sohn, H.Y.; Elzohiery, M.; Fan, D.-Q. Chapter 2 Development of Flash Ironmaking Technology. In *Advances in Engineering Research*; Petrova, V.M., Ed.; Nova Science Publishers: New York, NY, USA, 2019; Volume 26, ISBN 987-1-715-5.

32. Sohn, H.Y.; Choi, M.E. Steel Industry and Carbon Dioxide Emissions—A Novel Ironmaking Process with Greatly Reduced Carbon Footprint. In *Carbon Dioxide Emissions: New Research*; Carpenter, M., Shelton, E.J., Eds.; Nova Science Publishers: Hauppauge, NY, USA, 2012.

33. U.S. Geological Survey. Iron Ore Statistics and Information. 2018. Available online: https://minerals.usgs.gov/minerals/pubs/commodity/iron_ore/mcs-2018-feore.pdf (accessed on 30 March 2018).

34. U.S. Environmental Protection Agency. Office of Air Quality Planning and Standards, Emission Factor Documentation for AP-42, Section 11.23, Taconite Ore Processing. February 1997. Available online: https://www3.epa.gov/ttnchie1/ap42/ch11/final/c11s23.pdf (accessed on 30 March 2018).

35. Themelis, N.J.; Zhao, B. Flash Reduction of Metal Oxides. In *Flash Reaction Processes*; Davies, T.W., Ed.; Kluwer Academic Publishers: Dordrecht, The Netherlands, 1995; pp. 273–293.

36. Fan, D.-Q.; Sohn, H.Y.; Elzohiery, M. Analysis of the Reduction Rate of Hematite Concentrate Particles by H2 or CO in a Drop-Tube Reactor through CFD Modeling. *Metall. Mater. Trans. B* **2017**, *48*, 2677–2684. [CrossRef]

37. Fan, D.-Q.; Elzohiery, M.; Mohassab, Y.; Sohn, H.Y. A CFD Based Algorithm for Kinetics Analysis of the Reduction of Hematite Concentrate by H_2+CO Mixtures in a Drop Tube Reactor. In *8th International Symposium on High-Temperature Metallurgical Processing*; Hwang, J.-Y., Jiang, T., Kennedy, M.W., Yücel, O., Pistorius, P.C., Seshadri, V., Zhao, B., Gregurek, D., Keskinkilic, E., Eds.; The Minerals, Metals & Materials Series; Springer: Berlin/Heidelberg, Germany, 2017; pp. 61–70. [CrossRef]

38. Fan, D.; Mohassab, Y.; Elzohiery, M.; Sohn, H.Y. Analysis of the Hydrogen Reduction Rate of Magnetite Concentrate Particles in a Drop Tube Reactor through CFD Modeling. *Metall. Mater. Trans. B* **2016**, *47*, 1669–1680. [CrossRef]

39. Elzohiery, M.; Sohn, H.Y.; Mohassab, Y. Kinetics of Hydrogen Reduction of Magnetite Concentrate Particles in Solid State Relevant to Flash Ironmaking. *Steel Res. Int.* **2017**, *88*, 1600133. [CrossRef]

40. Chen, F.; Mohassab, Y.; Jiang, T.; Sohn, H.Y. Hydrogen Reduction Kinetics of Hematite Concentrate Particles Relevant to a Novel Flash Ironmaking Process. *Metall. Mater. Trans. B* **2015**, *46*, 1133–1145. [CrossRef]

41. Chen, F.; Mohassab, Y.; Zhang, S.; Sohn, H.Y. Kinetics of the Reduction of Hematite Concentrate Particles by Carbon Monoxide Relevant to a Novel Flash Ironmaking Process. *Metall. Mater. Trans. B* **2015**, *46*, 1716–1728. [CrossRef]

42. Choi, M.E.; Sohn, H.Y. Development of Green Suspension Ironmaking Technology based on Hydrogen Reduction of Iron Oxide Concentrate: Rate Measurements. *Ironmak. Steelmak.* **2010**, *37*, 81–88. [CrossRef]

43. Elzohiery, M.; Fan, D.-Q.; Mohassab, Y.; Sohn, H.Y. *Kinetics of Hydrogen Reduction of Magnetite Concentrate Particles at 1623–1873 K Relevant to Flash Ironmaking*; University of Utah: Salt Lake City, UT, USA, 2020; Unpublished work.

44. Fan, D.-Q.; Elzohiery, M.; Mohassab, Y.; Sohn, H.Y. *The Kinetics of Carbon Monoxide Reduction of Magnetite Concentrate Particles through CFD Modeling*; University of Utah: Salt Lake City, UT, USA, 2020; Unpublished work.

45. Fan, D.-Q.; Elzohiery, M.; Mohassab, Y.; Sohn, H.Y. *Rate-Amplification Effect of CO in Magnetite Concentrate Particle Reduction by H_2+CO Mixtures*; University of Utah: Salt Lake City, UT, USA, 2020; Unpublished work.

46. Yuan, Z.; Sohn, H.Y. Re-Oxidation Kinetics of Flash Reduced Iron Particles in O2–N2 Gas Mixtures Relevant to a Novel Flash Ironmaking Process. *ISIJ Int.* **2014**, *54*, 1235–1243. [CrossRef]

47. Yuan, Z.; Sohn, H.Y.; Olivas-Martinez, M. Re-oxidation Kinetics of Flash-Reduced Iron Particles in H2-H2O(g) Atmosphere Relevant to a Novel Flash Ironmaking Process. *Metall. Mater. Trans. B* **2013**, *44*, 1520–1530. [CrossRef]

48. Fan, D.Q.; Sohn, H.Y.; Mohassab, Y.; Elzohiery, M. Computational Fluid Dynamics Simulation of the Hydrogen Reduction of Magnetite Concentrate in a Laboratory Flash Reactor. *Metall. Mater. Trans. B* **2016**, *47*, 1–12. [CrossRef]

49. Mohassab, Y.; Elzohiery, M.; Sohn, H.Y. Flash Reduction of Magnetite Concentrate Based on Partial Oxidation of Natural Gas Relevant to a Novel Ironmaking Process. In *AISTech 2016 Proceedings*; AIST: Warrendale, PA, USA, 2016; pp. 713–722.

50. Abdelghany, A.; Fan, D.Q.; Elzohiery, M.; Sohn, H.Y. Flash Ironmaking from Magnetite Concentrate using a Large-Scale Bench Reactor: Experimental and CFD Work. In *AISTech 2018 Proceedings*; AIST: Warrendale, PA, USA, 2018.

51. Pinegar, H.K.; Moats, M.S.; Sohn, H.Y. Flowsheet development, process simulation and economic feasibility analysis for novel suspension ironmaking technology based on natural gas: Part 3—Economic feasibility analysis. *Ironmak. Steelmak.* **2013**, *40*, 44–49. [CrossRef]

 *metals*

Article

Hydrogen Ironmaking: How It Works

Fabrice Patisson *  **and Olivier Mirgaux**

Institut Jean Lamour, CNRS, Université de Lorraine, Labex DAMAS, 54011 Nancy, France;
olivier.mirgaux@univ-lorraine.fr
* Correspondence: fabrice.patisson@univ-lorraine.fr; Tel.: +33-372-742-670

Received: 3 June 2020; Accepted: 30 June 2020; Published: 9 July 2020

Abstract: A new route for making steel from iron ore based on the use of hydrogen to reduce iron oxides is presented, detailed and analyzed. The main advantage of this steelmaking route is the dramatic reduction (90% off) in CO_2 emissions compared to those of the current standard blast-furnace route. The first process of the route is the production of hydrogen by water electrolysis using CO_2-lean electricity. The challenge is to achieve massive production of H_2 in acceptable economic conditions. The second process is the direct reduction of iron ore in a shaft furnace operated with hydrogen only. The third process is the melting of the carbon-free direct reduced iron in an electric arc furnace to produce steel. From mathematical modeling of the direct reduction furnace, we show that complete metallization can be achieved in a reactor smaller than the current shaft furnaces that use syngas made from natural gas. The reduction processes at the scale of the ore pellets are described and modeled using a specific structural kinetic pellet model. Finally, the differences between the reduction by hydrogen and by carbon monoxide are discussed, from the grain scale to the reactor scale. Regarding the kinetics, reduction with hydrogen is definitely faster. Several research and development and innovation projects have very recently been launched that should confirm the viability and performance of this breakthrough and environmentally friendly ironmaking process.

Keywords: ironmaking; CO_2 mitigation; hydrogen; kinetics; direct reduction

1. Introduction

Despite the use of the present tense in the title, using just hydrogen as a reductant for ironmaking is not yet an industrial process. However, it could become one soon according to several recent signs. Important R&D&I (research and development and innovation) programs around the world have recently been launched for this purpose. In Europe, the HYBRIT project, which aims at building a whole demonstration plant in Sweden, including an iron ore direct reduction unit fed with hydrogen by a water electrolysis plant using fossil-free electricity, is one example [1]. The H2FUTURE and GrInHy projects, though directly connected to iron or steelmaking, mostly focus on electrolyzer development [2,3]. Recently, ArcelorMittal announced the start of hydrogen-based ironmaking in its MIDREX direct reduction plant in Hamburg [4]. In parallel, increasing demand from the steel sector is expected in the energy industry: "In the iron and steel industry, where hydrogen can be used to reduce iron ore to iron, we expect the use of clean hydrogen will be demonstrated by 2030 and gain momentum by 2035" [5].

The idea of using hydrogen as a reductant is primarily related to the issue of climate change. The steel industry accounts for between 4% and 7% of global anthropogenic CO_2 emissions [6]. This results from the almost exclusive use of carbon (coal or coke) for both the energy and the chemical reduction needed along the steelmaking route, the major contributor being the blast furnace, in which the solid iron ore, in the form of sinter or pellets, is transformed into liquid pig iron. Most iron ores are oxides (usually hematite Fe_2O_3), and the chemical reduction of an iron oxide to Fe^0 by C (or by CO made from C or from CH_4) produces CO_2. The basic concept of hydrogen ironmaking is to substitute C (or CO) reductant with H_2, replacing

$$Fe_2O_3 + \frac{3}{2}\,C = 2\,Fe + \frac{3}{2}\,CO_2 \tag{1}$$

or

$$Fe_2O_3 + 3\,CO = 2\,Fe + 3\,CO_2 \tag{2}$$

with

$$Fe_2O_3 + 3\,H_2 = 2\,Fe + 3\,H_2O \tag{3}$$

thus releasing harmless H_2O instead of the greenhouse gas CO_2 in the chemical reduction step.

In the steelmaking route, the operations converting the ore into metallic iron are referred to as ironmaking. The majority (90%) of ironmaking is made using the blast furnace, which produces pig iron, i.e., liquid iron saturated in C. The other processes are the so-called direct reduction (DR) processes, whose product is solid iron (DRI-direct reduced iron, also named sponge iron, or HBI–hot briquetted iron). The reduction occurs as a series of gas-solid reactions with the reactant gases CO and H_2. The reactor is generally a vertical shaft furnace, whose reducing gas (a CO-H_2 mixture) is obtained by natural gas reforming. The corresponding industrial processes are MIDREX and HYL-ENERGIRON. Rotary earth furnaces are also employed, using coal as the carbon source.

If this substitution (C by H_2) was carried out in the blast furnace, the substitution rate would remain limited. Indeed, the pulverized coal injected at the tuyeres (one-third of the carbon) could probably be largely replaced by H_2, but the coke (two-thirds of the carbon) needs to remain for proper operation of the blast furnace. The expected benefit in terms of CO_2 emissions is typically reported to be 20% [7]. Conversely, if a direct reduction shaft furnace was used, the possibility of substituting 100% of the carbon (monoxide) with H_2 could be envisaged. This is why most of the current projects mentioned above consider using pure hydrogen in a shaft furnace for ironmaking. Presenting a way that this could be achieved for steelmaking and detailing the underlying physicochemical and thermal issues are the objectives of the present paper. Another recent paper on this topic is that of Vogl et al. [8]. Nevertheless, the scope of this paper was more focused on energy and cost and less focused on physicochemical processes than the present paper.

With different processes, reductants other than hydrogen could be used, such as electrons, thus leading to iron ore electrolysis at high or low temperatures. Interesting projects are currently ongoing, such as MOE at MIT [9] and SIDERWIN in Europe [10], but these lie outside of the scope of our paper.

2. The Hydrogen-Based Route to Steel

To the best of our knowledge, the first comprehensive study on hydrogen-based steelmaking was that undertaken by the 'Hydrogen' subproject of the European program ULCOS (ultra-low CO_2 steelmaking, 2004–2010), in which our research group was involved. A comprehensive overview of ULCOS was recently written [11]. In the Hydrogen subproject, the tasks dealt with massive hydrogen production and the feasibility of using pure hydrogen for ironmaking.

The stoichiometric consumption of hydrogen for reducing hematite is 54 kg per ton of iron. A 1 Mt per year steel plant would require a hydrogen plant capacity of as much as 70,000 m^3_{STP} h^{-1}. Large-scale hydrogen production is currently achieved by steam reforming of methane. This option could be retained and even optimized for hydrogen-based ironmaking, e.g., by targeting a 97–98% purity of H_2 instead of the usual 99.9+. However, since based on a fossil resource, the performance in terms of CO_2 mitigation would overall remain average, unless a CO_2 capture unit was added, which represents a strategy different from the one pursued. The other preferable option is to produce hydrogen by water electrolysis, provided that the electricity is fossil-free. The size of the plant could be achieved by multiplying the electrolytic cells. New, improved technologies have been identified, such as proton exchange membranes and high pressure or high temperature electrolysis. The former two are now being developed in the H2FUTURE and GrInHy projects.

Two ways to reduce iron ore by hydrogen were studied in ULCOS. The first was the reduction of fine ores in a cascade of fluidized beds, as in the FINMET process [12], but with hydrogen instead of reformed natural gas. It must be stressed that the hydrogen reduction of fine ores in a two-stage fluidized bed process, named CIRCORED, was the only direct reduction process using pure hydrogen as a reductant that had ever been commercially operated [13]. Hydrogen was produced by natural gas steam reforming. This process was decommissioned for economic rather than technical reasons. The second process investigated was the direct reduction of iron ore pellets or lumps in a vertical shaft furnace, such as a MIDREX furnace. The latter process, which is detailed in the next section, was eventually selected.

The whole route to steel proposed based on these investigations is depicted in Figure 1. Unsurprisingly, the current hydrogen-based ironmaking projects have retained the same route.

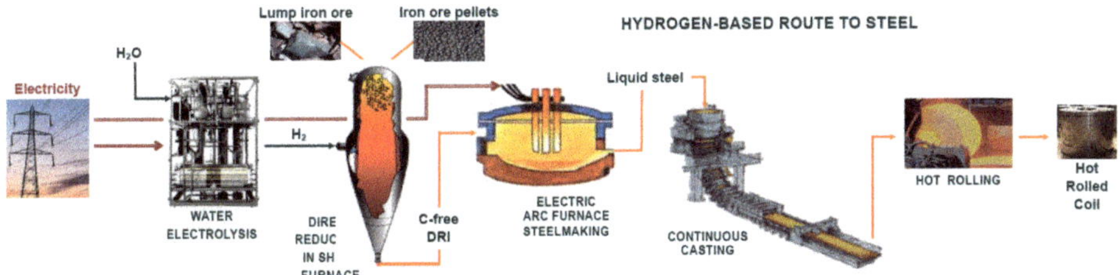

Figure 1. ULCOS hydrogen-based route to steel.

The rest of the route, the steelmaking process that employs an electric arc furnace (EAF), is the same as the usual route for making steel from recycled scrap or from DRI. The only difference lies in the carbon content of the DRI, which is 0% instead of 2–4%. Some EAF steelmakers might be worried about the use of such a DRI since the current practice is to look for DRIs with higher carbon contents. Nuber et al. cogently addressed this question and concluded that carbon-free DRI is not an issue [13]. Plain scrap charges, without carbon, are routinely treated in most EAFs, and the important point is to obtain a good foaming slag by blowing carbon fines and oxygen.

The performance of the whole route, in terms of energy consumption and CO_2 emissions, is indicated in Table 1 (last line) and compared with that of the standard blast furnace-basic oxygen furnace route, as well as with that of the usual direct reduction (MIDREX process) followed by electric arc steelmaking.

Table 1. Energy and CO_2 emissions of various steelmaking routes.

Route	Energy Needed		CO_2 Emissions	
Standard BF-BOF route	18.8 GJ/t_{HRC} (mostly coal)	[14]	1850 kg$_{CO_2eq}$/t_{HRC}	[14]
Direct reduction + EAF	15.6 GJ/t_{HRC} (gas and electricity)	[14]	970 kg$_{CO_2eq}$/t_{HRC}	[14]
	15.4 GJ/t_{HRC}	[15]	196 kg$_{CO_2eq}$/t_{HRC}	[15]
Hydrogen-based route	14.7 GJ/t_{LS} (mostly electricity)	[1]	25 kg$_{CO_2eq}$/t_{LS}	[1]
	13.3 GJ/t_{LS}	[8]	53 kg$_{CO_2eq}$/t_{LS}	[8]

Abbreviations: BF: blast furnace; BOF: basic oxygen furnace; HRC: hot rolled coil; EAF: electric arc furnace; LS: liquid steel.

Regarding the hydrogen-based route, three sources are reported, which give different figures. This discrepancy has to be related to the assumptions and to the boundaries of the system. The ULCOS [14,15] figures reflect a global life cycle approach, from cradle (extraction) to gate (1 ton of hot-rolled coil).

The HYBRIT and Vogl figures are for 1 ton of liquid steel. The energy comparison shows similar energy consumption for the two direct reduction routes, which is slightly lower than that of the standard BF-BOF route, and the hydrogen-based route is at the same level as the natural-gas-based route. The CO_2 comparison is more instructive. While the usual natural gas DR route, with 970 kg_{CO_2eq}/t_{HRC}, halves the CO_2 emissions of the standard BF-BOF route, the hydrogen route reaches far lower levels, 25 to 200 kg_{CO_2eq}/t_{HRC}, i.e., an 89–99% reduction in CO_2 emissions. Here, again, the differences among the sources reflect the systems considered. The upstream pellet-making process and the downstream continuous casting and hot rolling processes are included in [15], whereas HYBRIT's figure excludes the latter and considers neutral pellet making (using biomass), as well as using biomass in the EAF [1]. In any case, the dramatic decrease in CO_2 emissions is definitely the rationale for developing hydrogen ironmaking.

3. Shaft Furnace using just Hydrogen

The shaft furnace is the heart of the hydrogen-ironmaking process, and the crucial question is as follows: can such a shaft furnace be operated under pure H_2 and produce a well-metallized DRI, similar to the current ones—except of course for its carbon content? Some of the projects mentioned above, such as HYBRIT, plan to answer this question by building and operating a pilot/demonstration plant. This is obviously a necessary step. However, another approach can provide an interesting preliminary answer to the question: mathematical modeling. Several mathematical models of the MIDREX and HYL-ENRGIRON processes have been published, and the more detailed models [16–19] provide valuable insights into the detailed physicochemical and thermal behaviors of the reactor. Unfortunately, these models were not used to simulate the case of a shaft furnace operated with hydrogen only. Our research group developed such a model, named REDUCTOR, which, in its first version (v1), was used to study this case [20,21]. The principle of this model is illustrated in Figure 2.

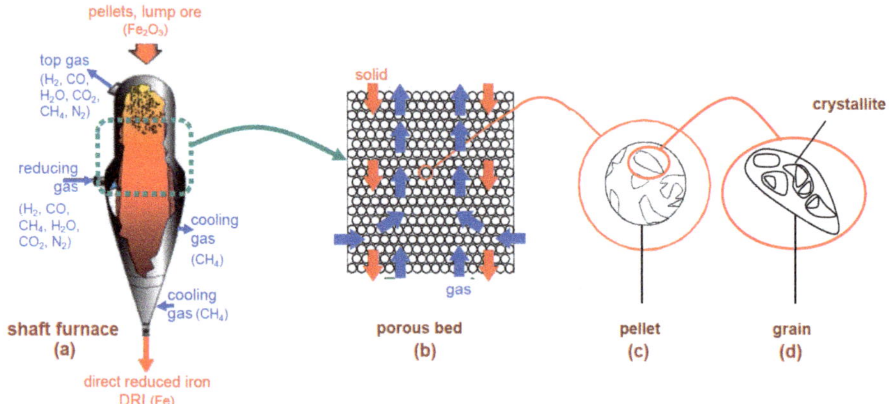

Figure 2. Principle of the REDUCTOR (v1) model; (**a–d**): the four scales considered.

The shaft furnace (Figure 2a) is fed with pellets or lump ore at the top, which descend by gravity and encounter a rising flow of hydrogen, fed laterally at mid-height of the reactor and exiting at the top. The reduction reactions take place in the upper section between the reducing gas outlet and inlet. The conversion to iron is completed at the level of the gas inlet. Below, a conical section can be used to cool the DRI, but preferably using hydrogen instead of methane. The rest of the gas circuit, not shown, is much simpler than that in the MIDREX process: the top gas, consisting of H_2-H_2O, is cooled to condense water, and H_2 is recycled, mixed with fresh H_2 from the electrolysis plant, and reheated to the desired temperature (800–900 °C).

The model simulates the reduction section in two dimensions (radius and height) and in the steady-state regime. It describes the solid and gas flows (Figure 2b), the heat transfer by convection and conduction, and the three reduction reactions

$$3\,Fe_2O_3 + H_2 = 2\,Fe_3O_4 + H_2O \tag{4}$$

$$Fe_3O_4 + \frac{16}{19}\,H_2 = \frac{60}{19}\,Fe_{0.95}O + \frac{16}{19}\,H_2O \tag{5}$$

$$Fe_{0.95}O + H_2 = 0.95\,Fe + H_2O \tag{6}$$

the kinetics of which are calculated from a single-pellet kinetic submodel (see Section 4). Indeed, four scales are considered (Figure 2a–d): the reactor, the pellet bed, the pellets, and the grains and crystallites composing the pellets. The local mass, energy, and momentum balances are rendered discrete and solved iteratively using the finite volume method. The equations were detailed in [20] and are not reproduced here. That paper from our group also presented preliminary results, which were updated in [21] and are shown and discussed below.

We simulated a cylindrical shaft using the geometry and inlet characteristics indicated in Table 2. The main results are presented in Figure 3.

Table 2. Main operating conditions for the reference simulation.

Shaft	Height = 6 m	Radius = 3.3 m	
Pellets	CVRD-DR	Diameter = 14 mm	Porosity = 0.33
Inlet solid	Fe_2O_3	Flowrate = 52 kg s^{-1}	Temperature = 25 °C
Inlet gas	98%H_2, 2%H_2O	Lateral flowrate = 3634 mol s^{-1} Bottom flowrate = 100 mol s^{-1}	Temperature = 800 °C

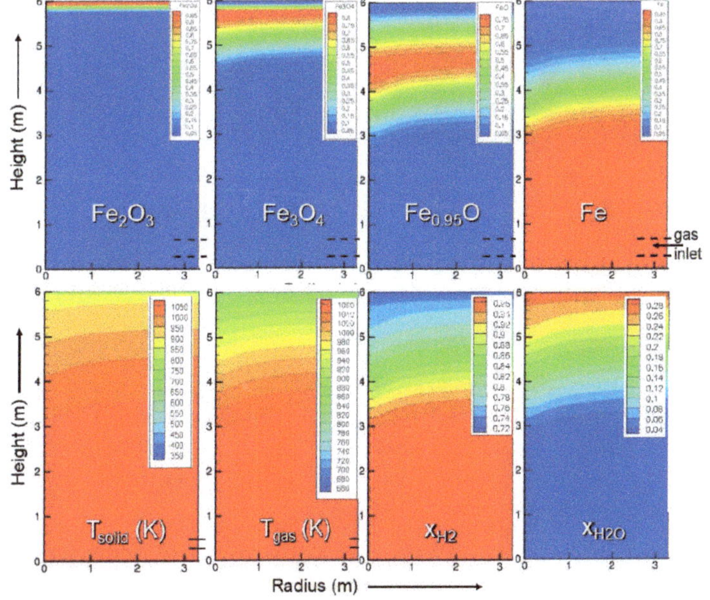

Figure 3. Main results of the shaft model. The calculations were performed under the conditions given in Table 1. Top row: solid mass fractions; bottom row: solid and gas temperatures, gas molar fractions in H_2 and H_2O. The symmetry axis is on the left-hand side of each map, and the wall is on the right-hand side. The main gas inlet is at the wall near the bottom of the reactor.

Regarding the solid mass fractions, these results show that hematite is very rapidly converted to magnetite, followed by the magnetite-to-wustite conversion, which is completed 1.5 m down from the solid inlet. The wustite-to-iron reduction is slower and ends at z = 2.6 m, i.e., at 3.4 m from the top. The solid and gas temperatures are close to each other, being equal to the gas inlet temperature in the lower half of the shaft and between 1073 and 900 K above, due to the endothermic nature of the reactions (5,6) (see Section 5), and fall abruptly approaching the solid inlet, where the solid enters at 25 °C. The H_2 and H_2O molar fraction maps reflect the course of the reductions that consume H_2 and produce H_2O. The thermodynamic conditions produce a reducing environment that can yield 100% iron in the metallic form (Fe^0). The radial profiles are mostly flat with a slight incline towards the symmetry axis. These radial gradients result from the lateral gas inlet at the wall. The principal result of this simulation is that full metallization is reached at 3.4 m, which can be compared to the ≈10 m reduction zone of a MIDREX shaft.

In addition to this reference simulation, the influence of some operating parameters was studied [21]. Figure 4 shows the influence of the inlet gas temperature on the iron mass fraction. At 700 °C, the shaft furnace is not heated enough, and the average iron fraction is 74% with strong radial gradients. At 900 and 950 °C, the reaction is faster than that at 800 °C, with 100% iron obtained in less than 2 m. The profile becomes flat, and the difference is small between 900 and 950 °C. This last trend is explained by the kinetics of the wustite-to-iron reaction in Section 4.

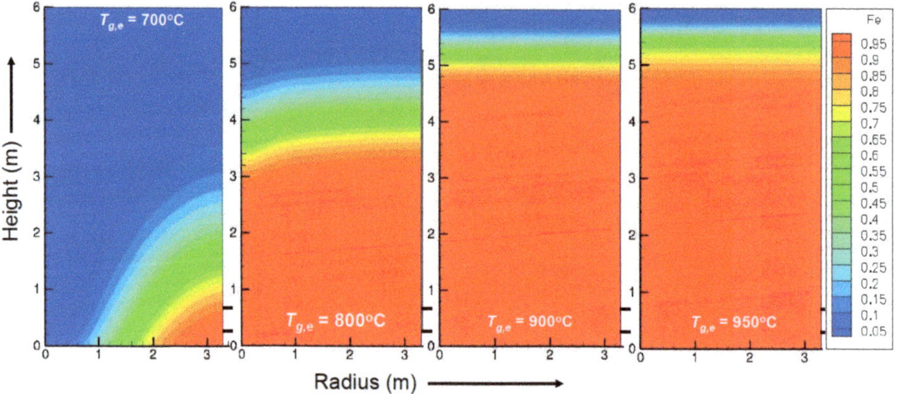

Figure 4. Calculated influence of the inlet gas temperature on the iron mass fraction.

These simulations were not intended for use in designing a reactor. More design features are needed. However, the implications of the main results are clear.

4. Kinetics of Iron Ore Reduction by H_2

Due to its economic importance, the reduction of iron oxides has been thoroughly studied. Thousands of papers have been published over the last century (Table 3). The reduction of iron oxides has been carried out with CO, H_2, and CO-H_2 mixtures. The samples, sized from mg to kg, ranged from pure synthetic oxides, either dense pieces or in powder form, to ores such as hematite and magnetite, in the form of lumps, sinter, and pellets.

Table 3. Publications on the reduction of iron oxides from 1900 to 2020.

Search Terms	Number of Papers
iron (oxide or ore) reduction	20,230
iron (oxide or ore) reduction CO	3692
iron (oxide or ore) reduction H_2	109
iron (oxide or ore) mechanisms	792
iron (oxide or ore) kinetics	741

Source: ISI Web of Science, accessed on 18 May 2020.

All aspects of the reduction reactions were covered: mechanisms, kinetics, and influence of the reducing conditions such as temperature and gas composition. Reviewing this literature is out of the scope of the present paper; a detailed review was presented by Ranzani in 2011 [21], and a very recent review was reported by Zare Ghadi et al. [22]. Only the most important results from the literature and related to the kinetics are recalled here. The whole reduction occurs in two or three steps $(Fe_2O_3 \rightarrow Fe_3O_4 \rightarrow FeO \rightarrow Fe)$, with the occurrence of FeO only above 570 °C. The kinetics of the last reduction are the slowest of the three. Depending on the experimental conditions, the iron phase formed can be dense or porous, which influences the transport process (gas-phase or solid-state diffusion) through the iron layer. The reaction kinetics are said to be controlled by a chemical reaction step, a gas diffusion step, a solid diffusion step, or to be in a mixed regime according to the experimental conditions and the authors. Regarding the reactant, the reaction with H_2 is generally reported to be faster than that with CO.

To obtain a kinetic model that could be integrated into the reactor model (Section 3), our research group performed a number of kinetics experiments on the reduction of iron ore pellets by hydrogen using thermogravimetry. After testing synthetic oxides and then small cubes made from ore pellets, we eventually used whole industrial pellets as samples to obtain the relevant data for modeling. The main parameters studied were the temperature, hydrogen content (H_2 in He), pellet size, and pellet type (Brazilian CVRD pellets of DR and BF grades, Swedish LKAB-KPRS pellets).

Details about the experiments and results can be found elsewhere [20,21]. The main findings, including materials only presented so far in Ranzani's thesis, are given hereafter. The influence of temperature is particularly complex, suggesting different kinetic regimes. The conversion normally accelerates when the temperature increases but not regularly. In particular, a slowing of the kinetics is observed at the end of the reaction at certain temperatures, namely, 700 and 950 °C, as illustrated by the TTT (time-temperature-transformation) diagram of Figure 5. The slowing at ≈950 °C was attributed to the $Fe_\alpha \rightarrow Fe_\gamma$ phase transformation at 912 °C, as the solid-state diffusion through iron is slower in the γ phase. Both types of pellets tested exhibit the same behavior.

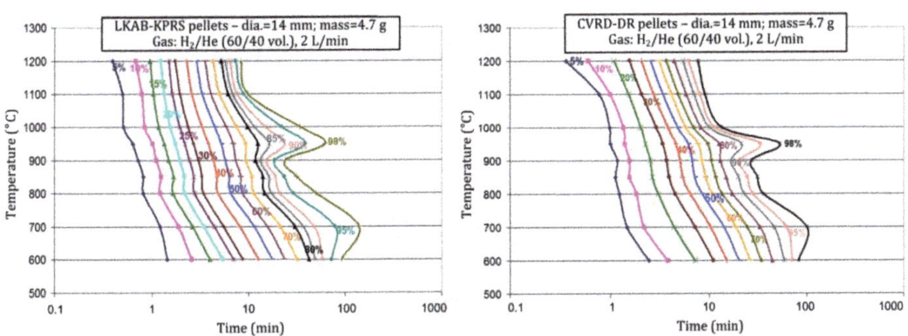

Figure 5. Kinetics of iron ore pellet reduction by H_2 plotted as a TTT diagram.

Scanning electron microscopy observations revealed changes in the pellet internal structure, as shown in Figure 6. The initial structure of a pellet is a porous agglomerate of dense hematite grains (f). It does not change much at the pellet scale after reduction, except for a slight increase in the porosity. Conversely, at the grain scale, the particles transform. The iron grains (a–e) clearly differ from the initial hematite grains, and the metamorphosis depends on the temperature; the iron grains formed at high temperature are larger and smoother. The center column (f–j) depicts the transformation over time. Some pores first appear at the surface of the grains (g–h). Then, at the stage of wustite, the grains tend to break into smaller ones, which we call crystallites (i). Eventually, the iron phase grows internally at the expense of the shrinking wustite cores of the crystallites (k–n) and, at temperatures over 900 °C, spreads over the crystallites and tends to merge them. This last behavior and micrographs similar to (k–n) were reported in [23].

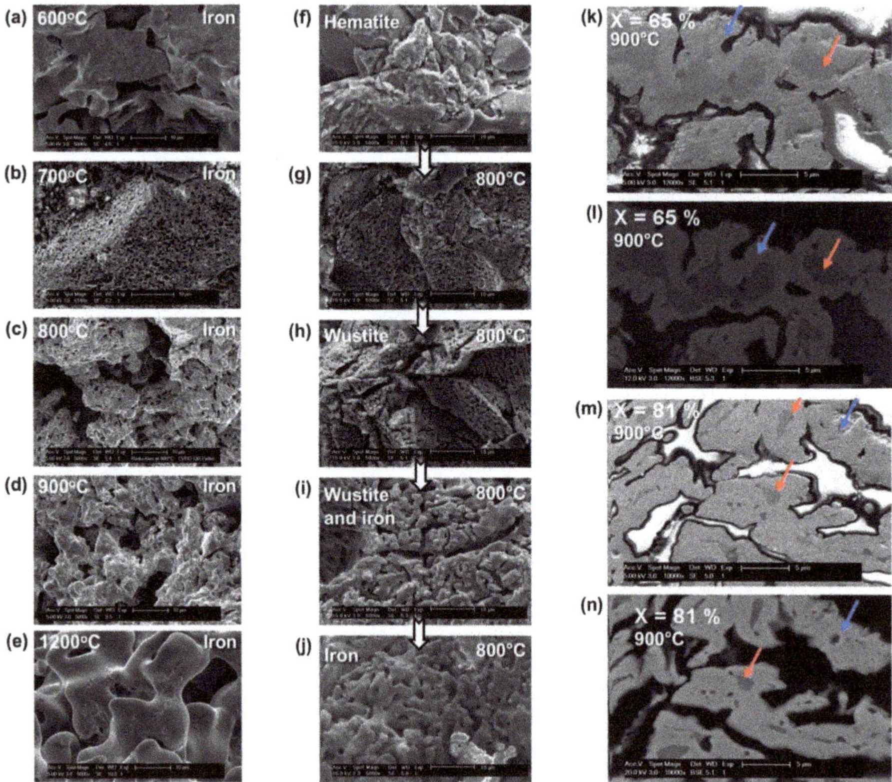

Figure 6. SEM micrographs showing the morphological changes at the granular scale in CVRD-DR pellets; (**a–e**): iron grains after reduction at different temperatures; (**f–j**): grains in different stages of a reduction at 800 °C; (**k–n**): polished cross-sections at conversion degrees of 65% (**k,l**) and 81% (**m,n**) for samples reduced at 900 °C, the red arrows point to shrinking wustite cores (dark gray) surrounded with a spreading layer of iron (light gray), indicated by the blue arrows. All images were taken with SE (secondary electrons), except (**l**) and (**n**), which were taken with BSE (backscattered electrons). The gas used for the reduction was 2 L/min H_2-He (60–40 vol. %) in a thermobalance.

The successive reactions (4–6), the peculiar influence of the temperature, and the morphological evolution led us to build a specific 'kinetic' model to predict the rate of the transformations. Indeed, a simple shrinking core model, even one with three interfaces (interfaces that are not observed at the pellet scale), could not mimic these features. A grain model (pellets made up of grains) is better adapted but has to take into account the three reactions and the grain and pore evolution. It also has to remain sufficiently simple to be later included in a multiparticle reactor model. We thus developed the model illustrated in Figure 7, which physically reflects the structural evolution observed. Mathematically, it is based on an extension of Sohn's law of additive reaction times [24], the advantage of which is its ability to simply take into account multiple possible rate-limiting processes and mixed kinetic regimes. The processes considered here are external transfer of H_2 from the bulk gas to the pellet surface and the reverse for H_2O, intergranular and intragranular transport of H_2 and H_2O by pore diffusion, and solid-state diffusion through the dense iron layer. The porosity and the pore grain diameters vary over the reaction and according to the temperature, the values of which are derived from measurements. The set of corresponding equations was given in [20].

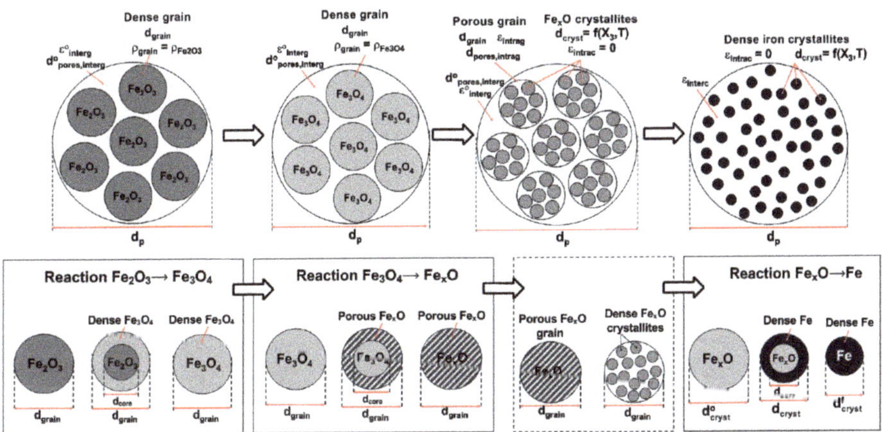

Figure 7. Modeled representation of the pellet (top row) and grain (bottom row) evolution.

Using this single pellet kinetic model, it is possible to calculate the transformation rate as a function of the operating conditions, namely, the gas composition, flow rate and temperature. The sequence of transformation is illustrated in Figure 8a. The reduction of hematite to magnetite is the fastest, and the reduction of wustite to iron is the slowest. The agreement between the measured and calculated values (Figure 8b) is very good at 700, 800, and 900 °C, while the final slowing at 950 °C is less satisfactorily simulated. Finally, the detailed investigation of the calculated reaction times, not reported here, shows that reactions (4,5) start in the chemical regime, i.e., the rate-limiting step is the chemical reaction itself, which then shifts to an intergranular diffusion regime, regardless of the temperature. For reaction (6), the wustite reduction, after a shorter time in the chemical regime, is controlled by intercrystallite diffusion up to 900 °C and in a mixed inter/intracrystallite diffusion regime above 950 °C.

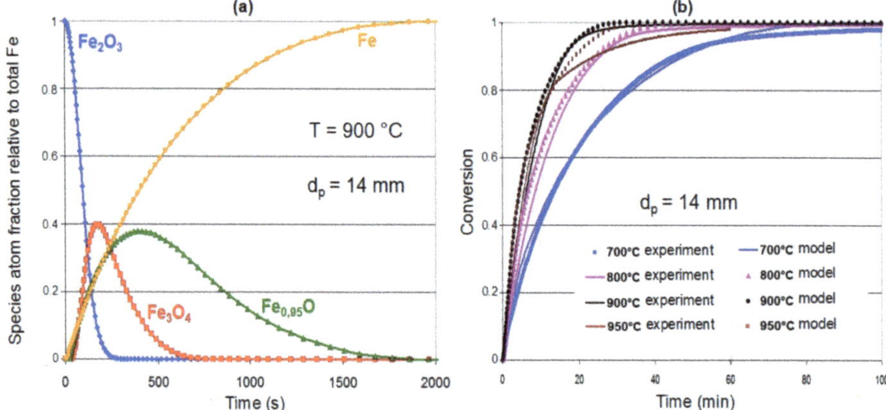

Figure 8. Results of the single pellet model; (**a**): calculated solid fractions as a function of time; (**b**): comparison with experimental data at various temperatures for a CVRD-DR pellet reduced by 2 L/min of H_2-He (60–40 vol. %) in a thermobalance.

5. CO or H_2 for the Reduction, the Main Differences

For ironmakers who are considering the transition to hydrogen, it is important to anticipate the change in the behavior of the reactors when transitioning from H_2-CO mixtures to H_2 only. Several factors can interact in different ways, such as kinetics, thermodynamics, heat transfer, and gas flow.

Regarding kinetics, laboratory studies on the reduction of iron oxides with CO, H_2, and CO–H_2 mixtures have clearly shown that, all else being equal, the kinetics with H_2 are faster (up to 10 times) than those with CO [22,25,26], as illustrated below (Figure 9).

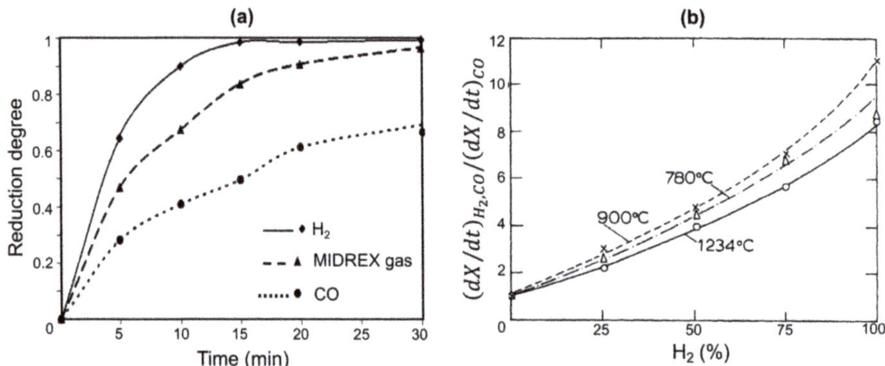

Figure 9. Comparison of the reduction kinetics of hematite pellets with H_2, CO, and H_2-CO mixtures. (**a**): reduction curves at 850 °C (the MIDREX gas contains approx. 56%H_2 and 34% CO), data from [26]; (**b**): relative reduction rates at 55% conversion at 780, 900 and 1234 °C as a function of the H_2 content in the H_2-CO mixture, data from [25].

However, such laboratory experiments do not integrate the influence of thermal and thermodynamic characteristics, since the experiments are generally performed in isothermal conditions and without the presence of CO_2 or H_2O in the gas and are thus far from equilibrium. In all the current iron ore reduction processes, H_2 and CO are present together, and both contribute to the reduction. Their utilization in the reduction reaction depends not only on their relative concentrations but also on the temperature and the reactor configuration. We carried out simulations of a MIDREX shaft furnace using different H_2/CO ratios at the reducing gas inlet [27]. Figure 10 shows that the higher the H_2 content is, the lower the metallization degree of the DRI, which at first glance is contrary to the kinetic behavior.

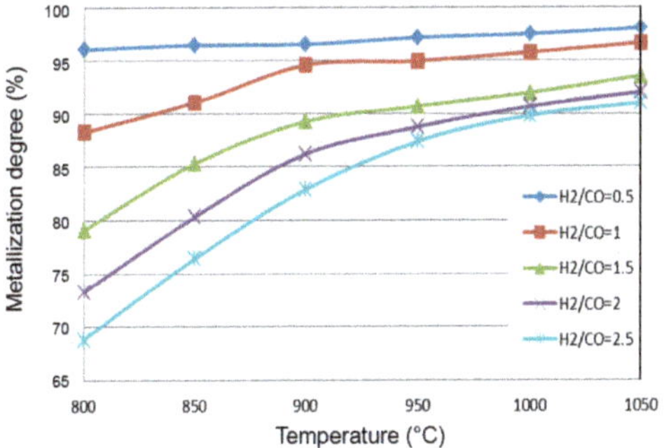

Figure 10. Metallization degree of the DRI at the exit of a MIDREX shaft furnace as a function of temperature and H_2/CO ratio.

The explanation for this contradiction relies on several factors. The first is the thermodynamics, which favor CO at low temperatures, as evidenced by the Chaudron diagram (Figure 11). The vertical blue arrows represent the driving force for the wustite-to-iron reduction, which increases with temperature with H_2 and decreases with temperature with CO.

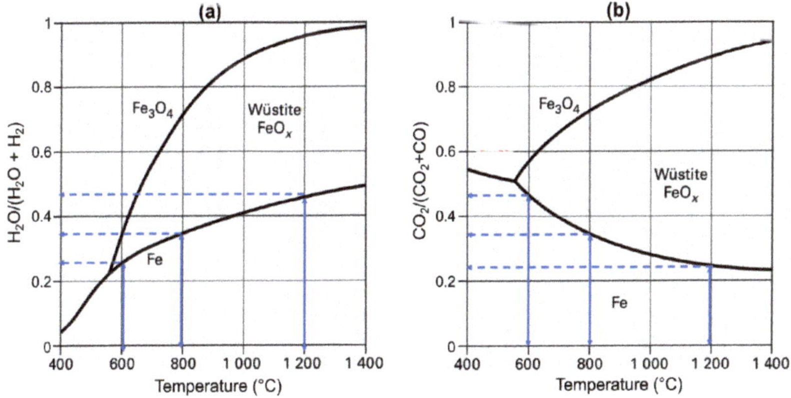

Figure 11. Chaudron (or Baur-Glaessner) phase diagram of the iron phase domains as a function of the temperature and oxidizing power of the gas, (**a**) in the case of a H_2-H_2O atmosphere and (**b**) CO-CO_2 atmosphere. The vertical blue arrows represent the driving force for the wustite-to-iron reduction.

The second factor is the heat of the reduction reactions (Table 4). The hematite-to-magnetite reaction is less exothermic with H_2 than with CO, the magnetite-to-wustite reaction is more endothermic, and chiefly, the wustite-to-iron reaction is endothermic with H_2 and exothermic with CO. Globally, the balance is an endothermic reduction with H_2 and an exothermic reduction with CO.

Table 4. Heat values of the reduction reactions. A minus sign indicates an exothermic reaction.

Reaction	$\Delta_r H_{800\,°C}$ ($J\ mol^{-1}$)
$3\,Fe_2O_3 + H_2 = 2\,Fe_3O_4 + H_2O$	-6020
$3\,Fe_2O_3 + CO = 2\,Fe_3O_4 + CO_2$	$-40{,}040$
$Fe_3O_4 + H_2 = 3\,FeO + H_2O$	$46{,}640$
$Fe_3O_4 + CO = 3\,FeO + CO_2$	$18{,}000$
$FeO + H_2 = Fe + H_2O$	$16{,}410$
$FeO + CO = Fe + CO_2$	$-17{,}610$

As a result, the temperature and compositions in the shaft greatly change with the inlet gas composition, as depicted by Figure 12. When leaving the gas injection zone, the temperature decreases due to methane cracking, but with a higher CO content, the bed is maintained at a higher temperature (a) as a result of the exothermic heat of the reduction reactions, whereas the temperature is lower with more H_2 (c). Plots (d) and (e) show that the wustite reduction in the peripheral zone is completed in less than 5 m versus the 8 m in (f). In the central zone, the low temperature (due to the cooling effect of the methane rising from the cooling zone) hampers the reduction by H_2 (thermodynamics), and wustite reduction in this zone is only possible by CO and thus goes further with more CO (d). Even if in all cases, more H_2 than CO is globally utilized for the 3 reductions (a consequence of the kinetics), the latter effect, i.e., the reduction by CO in the central zone, is decisive regarding the final metallization degree, as noted in Figure 10.

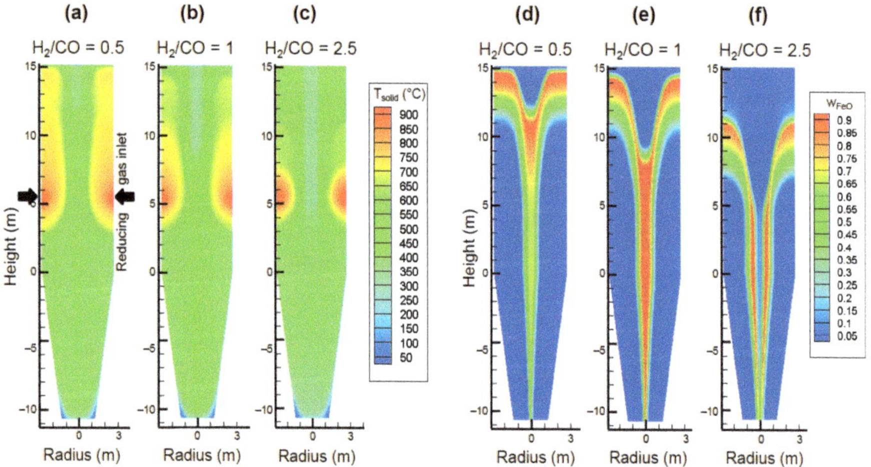

Figure 12. Calculated solid temperature (**a**–**c**) and iron mass fraction (**d**–**f**) throughout a shaft furnace fed with different inlet reducing gas compositions: $H_2/CO = 0.5$ (**a**,**d**), $H_2/CO = 1$ (**b**,**e**), $H_2/CO = 2.5$ (**c**,**f**), the other species being CH_4 (9%), H_2O (4%) and CO_2 (2%).

Finally, recall that when using only H_2 (both at the reducing gas inlet and at the bottom inlet) the colder central zone does not exist, the temperatures are more uniform radially, and the reduction, due to efficient kinetics, goes to completion (100% metallization).

6. Conclusions

In the context of reducing the CO_2 emissions of steelmaking, the hydrogen-based route is currently receiving much attention. This paper presented the principles and characteristics of this breakthrough steelmaking route from the plant scale to the granular scale. As replacements for coal, coke, and gas, hydrogen can be used for ironmaking, and electricity can be used for steelmaking. The expected

CO_2 emissions of this new route would be reduced by 89–99% compared to those of the current blast furnace-basic oxygen furnace route.

Hydrogen production needs to be fossil-free, and thus, the appropriate production method is water electrolysis with CO_2-lean electricity, i.e., renewable or nuclear electricity. Although water electrolysis is a well-known technology, some developments are needed to reach the target of massive amounts of CO_2-lean and, above all, affordable hydrogen for ironmaking. Some projects have been launched by the steelmaking industry, but other projects could emerge from other sectors, such as the transportation and power industries.

Regarding the core process of this new route, the reduction of iron ore by hydrogen, we selected direct reduction in a shaft furnace similar to the MIDREX and HYL-ERNERGIRON reactors. First, compared to other breakthrough technologies with similar levels of CO_2 mitigation, such as the direct electrolysis of iron ore, this process is much closer to industrialization, since the current DR shaft furnaces already work with H_2-CO mixtures that "only" need to be replaced with 100% H_2. Moreover, a DR plant using only hydrogen would be much simpler than MIDREX and HYL-ERNERGIRON, because the gas loop is shorter and methane reforming is not required. Third, compared to fluidized bed processes, it has the advantage of being able to treat lump ores and pellets, not fines. Finally, compared to intermediate options, such as the partial use of hydrogen in a blast furnace, it results in much higher CO_2 mitigation.

Downstream, the steelmaking occurs in an electric arc furnace, as is currently practiced for making steel from scrap and from DRI. The only difference is that the hydrogen-produced DRI would be 0% carbon instead of the usual 2–4% content. Although real technical difficulties are not anticipated, as some carbon will be added in the EAF, this point merits experimental confirmation.

From the mathematical modeling of the reduction zone of a shaft furnace operated with 100% H_2, we found that, due to the fast reduction kinetics with H_2, complete metallization could theoretically be achieved faster than that with H_2-CO, opening avenues to reactors smaller than the current DR shafts. The results have to be verified experimentally, which will be possible in some of the planned demonstrators [1]. Further work should also be performed to precisely determine the reactor geometry and dimensions, the configuration of the cooling section, and the details of the recycling gas loop, as well as to optimize the operating conditions for the selected configuration. Mathematical models such as REDUCTOR could be helpful.

Such mathematical reactor models rely on a proper evaluation of the reduction reaction kinetics at the pellet scale as a function of the temperature and local gas composition. Much knowledge has been acquired so far. Three reactions ($Fe_2O_3 \rightarrow Fe_3O_4 \rightarrow FeO \rightarrow Fe$) successively take place in grains but can coexist at the pellet scale. The last reaction is the slowest, and at approximately 700 and 950 °C, a final slowing can be observed. We proposed a pellet kinetic model based on the structural pellet and grain evolution in the course of the reaction, which very satisfactorily simulates the reduction up to 900 °C but could be further improved for higher temperatures.

Finally, the main differences between iron oxide reduction with H_2 and that with CO were discussed. At the scale of a pellet, the kinetics are faster with H_2. At the scale of a MIDREX shaft furnace using a H_2-CO-CH_4 mixture, due to the roles of thermodynamics, thermal gradients, and gas flow effects, CO plays a decisive role by better reducing wustite in the colder central zone. Using only H_2, the temperature and composition profiles are flatter and full metallization is predicted.

Author Contributions: Conceptualization, F.P. and O.M.; methodology: F.P. and O.M.; investigation: F.P. and O.M.; writing: F.P. and O.M.; supervision: F.P. and O.M.; Project Administration and Funding Acquisition: F.P. All authors have read and agreed to the published version of the manuscript.

Funding: Most of this research was supported by the European Commission within the 6th Framework Programme, project No: 515960 ULCOS (Ultra-low CO_2 steelmaking).

Acknowledgments: The authors warmly thank their Ph.D. students (at that time) Damien Wagner [15] and Andrea Ranzani Da Costa [20,21] for the H_2 studies and Hamzeh Hamadeh for the CO-H_2 study [19,27].

References

1. Hybrit (Hydrogen Breakthrough Ironmaking Technology) Brochure. Available online: https://ssabwebsitecdn. azureedge.net/-/media/hybrit/files/hybrit_brochure.pdf (accessed on 28 April 2020).
2. H2FUTURE. Verbund Solutions GmbH. Available online: https://www.h2future-project.eu (accessed on 28 April 2020).
3. GrInHy. Green Industrial Hydrogen. Salzgitter Mannesmann Forschung GmbH. Available online: https://www.green-industrial-hydrogen.com (accessed on 28 April 2020).
4. ArcelorMittal. Hydrogen-Based Steelmaking to Begin in Hamburg. Available online: https://corporate. arcelormittal.com/media/case-studies/hydrogen-based-steelmaking-to-begin-in-hamburg (accessed on 28 April 2020).
5. Hydrogen Scaling Up. The Hydrogen Council. Available online: https://hydrogencouncil.com/wp-content/ uploads/2017/11/Hydrogen-scaling-up-Hydrogen-Council.pdf (accessed on 28 April 2020).
6. European Commission. Energy Efficiency and CO_2 Reduction in the Iron and Steel Industry. Available online: https://setis.ec.europa.eu/system/files/Technology_Information_Sheet_Energy_Efficiency_and_CO2_ Reduction_in_the_Iron_and_Steel_Industry.pdf (accessed on 28 April 2020).
7. Yilmaz, C.; Jens Wendelstorf, J.; Turek, T. Modeling and simulation of hydrogen injection into a blast furnace to reduce carbon dioxide emissions. *J. Clean. Prod.* **2017**, *154*, 488–501. [CrossRef]
8. Vogl, V.; Ahman, M.; Nilsson, L.J. Assessment of hydrogen direct reduction for fossil-free steelmaking. *J. Clean. Prod.* **2018**, *203*, 736–745. [CrossRef]
9. Molten Oxide Electrolysis. Boston Metals. Available online: https://www.bostonmetal.com/moe-technology (accessed on 28 April 2020).
10. Siderwin. Development of New Methodologies for Industrial CO_2-Free Steel Production by Electrowinning. Available online: https://www.siderwin-spire.eu/content/objectives (accessed on 28 April 2020).
11. Abdul Quader, M.; Ahmed, S.; Dawal, S.Z.; Nukman, Y. Present needs, recent progress and future trends of energy-efficient Ultra-Low Carbon Dioxide (CO_2) Steelmaking (ULCOS) program. *Renew. Sustain. Energy Rev.* **2016**, *55*, 537–549. [CrossRef]
12. Plaul, F.J.; Böhm, C.; Schenk, J.L. Fluidized-bed technology for the production of iron products for steelmaking. *J. S. Afr. Inst. Min. Metall.* **2009**, *108*, 121–128.
13. Nuber, D.; Eichberger, H.; Rollinger, B. Circored fine ore direct reduction-the future of modern electric steelmaking. *Stahl und Eisen* **2006**, *126*, 47–51.
14. Birat, J.P. Society, materials, and the environment: The case of steel. *Metals* **2020**, *10*, 331. [CrossRef]
15. Wagner, D. Etude Expérimentale et Modélisation de la Réduction du Minerai de fer par L'Hydrogène. Ph.D. Thesis, Institut National Polytechnique de Lorraine, Nancy, France, 2008. Available online: https://tel.archives-ouvertes.fr/tel-00280689/ (accessed on 28 April 2020).
16. Parisi, D.R.; Laborde, M.A. Modeling of counter current moving bed gas-solid reactor used in direct reduction of iron ore. *Chem. Eng. J.* **2004**, *104*, 35–43. [CrossRef]
17. Valipour, M.S.; Saboohi, Y. Numerical investigation of nonisothermal reduction of haematite using syngas: The shaft scale study. *Model. Simul. Mater. Sci. Eng.* **2007**, *15*, 487–507. [CrossRef]
18. Shams, A.; Moazeni, F. Modeling and Simulation of the MIDREX Shaft Furnace: Reduction, Transition and Cooling Zones. *JOM* **2015**, *67*, 2681–2689. [CrossRef]
19. Hamadeh, H.; Mirgaux, O.; Patisson, F. Detailed Modeling of the Direct Reduction of Iron Ore in a Shaft Furnace. *Materials* **2018**, *11*, 1865. [CrossRef] [PubMed]
20. Ranzani da Costa, A.; Wagner, D.; Patisson, F. Modelling a new, low CO_2 emissions, hydrogen steelmaking process. *J. Clean. Prod.* **2013**, *46*, 27–35. [CrossRef]
21. Ranzani da Costa, A. La Réduction du Minerai de fer par L'Hydrogène: Etude Cinétique, Phénomène de Collage et Modélisation. Ph.D. Thesis, Institut National Polytechnique de Lorraine, Nancy, France, 2011. Available online: https://tel.archives-ouvertes.fr/tel-01204934/ (accessed on 28 April 2020).
22. Zare Ghadi, A.; Valipour, M.S.; Vahedi, S.M.; Sohn, H.S. A Review on the Modeling of Gaseous Reduction of Iron Oxide Pellets. *Steel Res. Int.* **2020**, *91*, 1900270. [CrossRef]
23. Gransden, J.F.; Sheasby, J.S.; Bergougnou, M.A. Defluidization of iron ore during reduction by hydrogen in a fluidized bed. *Chem. Eng. Progress Symp. Ser.* **1970**, *66*, 208–214.

24. Sohn, H.Y. The law of additive reaction times in fluid-solid reactions. *Metall. Trans.* **1978**, *9B*, 89–96. [CrossRef]
25. Towhidi, N.; Szekely, J. Reduction kinetics of commercial low-silica hematite pellets with CO-H$_2$ mixtures over temperatures range 600–1234 °C. *Ironmak. Steelmak.* **1981**, *6*, 237–249.
26. Bonalde, A.; Henriquez, A.; Manrique, M. Kinetic analysis of the iron oxide reduction using hydrogen-carbon monoxide mixtures as reducing agent. *ISIJ Int.* **2005**, *45*, 155–1260. [CrossRef]
27. Hamadeh, H. Modélisation Mathématique Détaillée du Procédé de Réduction Directe du Minerai de fer. Ph.D. Thesis, Université de Lorraine, Nancy, France, 2017. Available online: https://tel.archives-ouvertes.fr/tel-01740462 (accessed on 28 April 2020).

Communication

Toward a Fossil Free Future with HYBRIT: Development of Iron and Steelmaking Technology in Sweden and Finland

Martin Pei [1],*, Markus Petäjäniemi [2], Andreas Regnell [3] and Olle Wijk [4]

[1] SSAB AB, 10121 Stockholm, Sweden
[2] LKAB (Luossavaara-Kiirunavaara Aktiebolag), Box 952, 97128 Luleå, Sweden; markus.petajaniemi@lkab.com
[3] Vattenfall AB, 16992 Stockholm, Sweden; andreas.regnell@vattenfall.com
[4] Hybrit Development AB, 10724 Stockholm, Sweden; olle.wijk@hybrit.se
* Correspondence: martin.pei@ssab.com; Tel.: +46-8454-5780

Received: 10 June 2020; Accepted: 13 July 2020; Published: 18 July 2020

Abstract: The Swedish and Finnish steel industry has a world-leading position in terms of efficient blast furnace operations with low CO_2 emissions. This is a result of a successful development work carried out in the 1980s at LKAB (Luossavaara-Kiirunavaara Aktiebolag, mining company) and SSAB (steel company) followed by the closing of sinter plants and transition to 100% pellet operation at all of SSAB's five blast furnaces. However, to further reduce CO_2 emission in iron production, a new breakthrough technology is necessary. In 2016, SSAB teamed up with LKAB and Vattenfall AB (energy company) and launched a project aimed at investigating the feasibility of a hydrogen-based sponge iron production process with fossil-free electricity as the primary energy source: HYBRIT (Hydrogen Breakthrough Ironmaking Technology). A prefeasibility study was carried out in 2017, which concluded that the proposed process route is technically feasible and economically attractive for conditions in northern Sweden/Finland. A decision was made in February 2018 to build a pilot plant, and construction started in June 2018, with completion of the plant planned in summer 2020 followed by experimental campaigns the following years. Parallel with the pilot plant activities, a four-year research program was launched from the autumn of 2016 involving several research institutes and universities in Sweden to build knowledge and competence in several subject areas.

Keywords: fossil-free steel; hydrogen direct-reduced iron (H_2DRI); melting of H_2DRI in EAF (Electric Arc Furnace); hydrogen production by water electrolysis; hydrogen storage; grid balancing; renewable electricity

1. Introduction

The three Swedish companies SSAB (steel company), LKAB (mining company), and Vattenfall AB (energy company) aim at creating the first fossil-free value chain from mining to finished steel. The initiative is based on the importance of steel as a material for the development of the modern society and the big impact of today's production lines on the emission of fossil-based carbon dioxide.

Steel is 100% recyclable and has eternal applications including transportation, energy, drinking water, food production, buildings, and infrastructure. Volume wise, steel is the dominating material, and the global steel demand is estimated to increase from 1800 in 2018 to 2500 million tons in 2050 [1].

In the Paris agreement from 2015 [2], countries of the world agreed on a new global climate strategy that provides the framework for future long-term climate efforts. The global annual CO_2 emission from the steel industry is currently around 2.8 billion ton corresponding to 7% of the global total CO_2 emission. As illustrated in Figure 1, to reach the goals of global temperature increases of maximum 2 °C defined in the Paris agreement, the emission from the steel industry should be decreased to a level of 400–600 million ton per year in 2050 at the same time as the forecasted production volumes will increase.

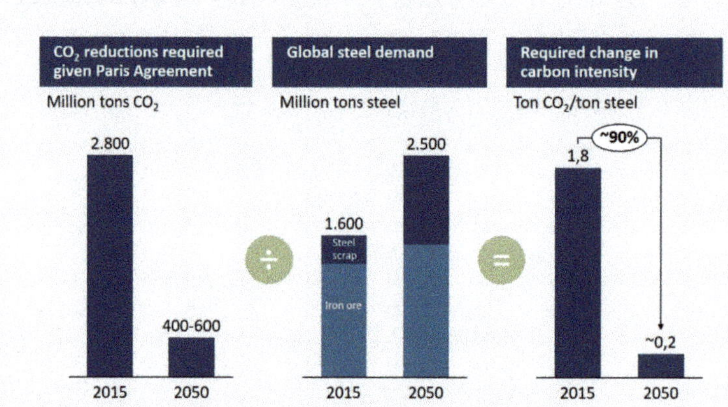

Figure 1. Global steel demand and required decrease of CO$_2$ emissions in 2050, data from [1].

In Sweden, following the Paris agreement, the parliament has adopted a climate law with the target that Sweden should have no net emissions of greenhouse gases by the year 2045. Furthermore, a long-term agreement was reached between the majorities of the left- and right-wing political parties, where it was confirmed that the Swedish electricity production should aim at 100% renewable energy sources by the year 2040.

In Finland, a climate plan was presented in 2019 with the target that Finland shall be carbon-neutral by 2035 with the ambition to be "the first fossil-free welfare society".

SSAB has at its Nordic production units a total annual CO$_2$ emission of around 9 million tons, as shown in Figure 2, of which 5 million tons are emitted in Sweden corresponding to 10% of the total emissions in the country. The corresponding figures for the Finnish production are 4 million tons and 7% respectively of total Finnish emission.

CO$_2$ emissions from SSAB's blast furnaces

Figure 2. CO$_2$ emissions in SSAB's (steel company's) steel works with BF-BOF (Blast Furnace-Basic Oxygen Furnace) in Sweden and Finland.

In the present paper, the HYBRIT (Hydrogen Breakthrough Ironmaking Technology) project, a Swedish initiative between the three companies SSAB, LKAB, and Vattenfall AB, is described.

2. The Blast Furnace Route

Scandinavia has a very long history in iron and steelmaking based on abundant natural resources in forests, hydro-power, and high-grade iron ores. Globally as well as in Scandinavia, the steel industry in periods has met big challenges that have resulted in changes of the industrial landscape, as shown in Figure 3 [3].

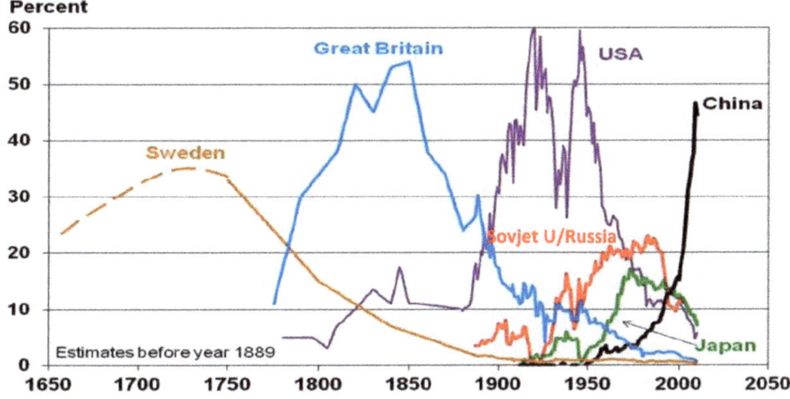

Figure 3. World steel production, historical data from [3].

In the beginning of the 18th century, Sweden dominated global steel production. During a few decades in the late 1700s and the beginning of the 1800s, Sweden's share of the global production was around 35%. This leading position changed rapidly to below 10% when expensive charcoal could be replaced by low-cost natural coal to produce commodity products. Many of the small iron and steelworks went into bankruptcy. A similar development could be seen in the second half of the 19th century. Around 1860, Sweden had 225 blast furnaces in operation. The introduction of processes such as Bessemer, Thomas, and Martin replaced different kinds of small-scale steelmaking processes, resulting in the closing of many iron and steelworks.

During the 20th century, several efforts were made in Sweden to replace the blast furnace. The common aim of these developments was to decrease or eliminate the need for the agglomeration of iron ore fines and to avoid dependence on coking coal. In some cases, the aim was also to find effective solutions to process the Swedish high-phosphorus iron ores. These initiatives peaked in the 1970s when several processes [4] were under development such as INRED, ELRED, Plasmasmelt, and the KTH-converter smelting reduction process. All of these were run on a pilot scale, and the KTH-converter process was also used for coal gasification in a converter combined with iron production.

Extensive research was performed around 1980 in a Swedish national program—Steel Production from High-phosphorus Iron Ore [5], where the focus was on the LKAB iron ores and the challenges facing the steel industry. The production process from mining to the production of the finished product was thoroughly studied in cooperation between academia and the industry. In this program, which was performed both on a national and international level, knowledge and competence was developed, which contributed to the competitive iron and steelmaking technology still used in Sweden.

Today, the Swedish high-phosphorus iron ores are upgraded to pellets with very low phosphorus content, which together with the development of both pellet properties and blast furnace technology have given world-leading results and low emissions. These achievements are based on a long-term close cooperation between LKAB and SSAB.

With an annual Nordic iron ore-based steel production of around 6 million tons, the focus today is on niche products for special purposes and high customer values. This strategy has made the industry competitive despite, from a global perspective, limited production volume. In addition to a very strong focus on product development both regarding iron ore products and finished products, there has been extensive activities regarding process development to improve productivity and decrease fuel consumption in the different production steps. The most significant development step is the joint development effort at LKAB and SSAB in the beginning of the 1980s with the self-fluxing olivine pellet, which resulted in the significant efficiency improvement of the combined LKAB–SSAB production value chain from iron ore to crude steel.

In Figure 4, the development of fuel consumption at LKAB for iron–ore agglomeration from 1960 until today is given in oil equivalents. The decrease is from 50 L per ton in 1960 to 8 L per ton pellet in 2013. The CO_2 emissions have decreased from 192 kg/ton product in 1960 to 31 kg in 2013. The step from shaft furnace to grate kiln resulted in a significant improvement, and new generations of pelletizing processes during the following decades resulted in further improvements, as shown in the figure below. To further decrease/eliminate these emissions, special actions must be taken.

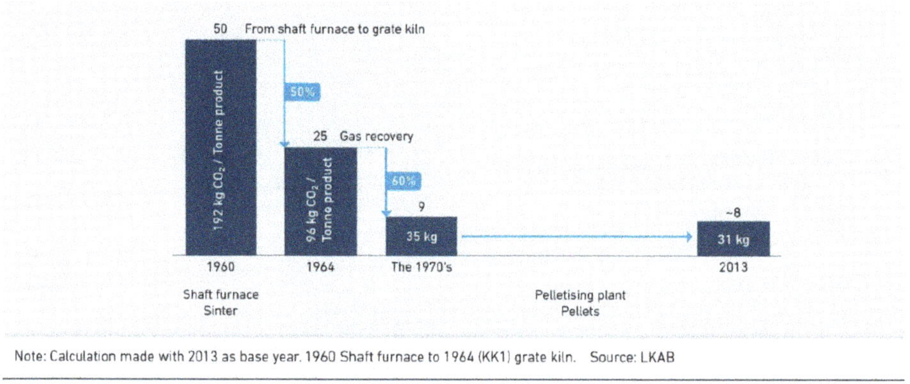

Note: Calculation made with 2013 as base year. 1960 Shaft furnace to 1964 (KK1) grate kiln.   Source: LKAB

Figure 4. Agglomeration of iron ore development in fuel consumption (oil equivalents per ton of product) and CO_2 emissions since 1960.

The modern blast furnace is a very effective and advanced unit with sophisticated process control systems. SSAB, which today has a world-leading position in blast furnace fuel rate, has in total five blast furnaces in Sweden and Finland. The blast furnaces are all run with 100% self-fluxing iron ore pellets, and pulverized coal injection (PCI) is performed. Figure 5 gives the development during the last decades at the SSAB blast furnace in Luleå, Sweden. The fuel rate today is on average 450–460 kg per ton of hot metal, of which around 150 kg comes from PCI. The slag volume is very low: around 120 kg per ton of hot metal.

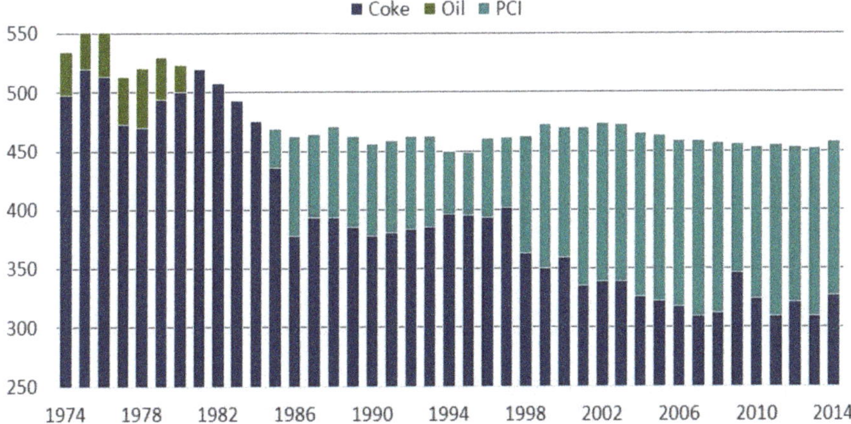

Figure 5. Development of the fuel rate in the blast furnace at SSAB in Luleå, Sweden.

Several initiatives [6] have been taken globally during the last decades to decrease CO_2 emissions and to further develop the blast furnace and to utilize the process as a flexible tool for different fuels. Examples are replacing PCI with bio-coal or recycled plastic materials, top gas recycling, the plasma heating of blast, the injection of hydrogen from coke oven gas or hydrogen produced with electricity, the further increased injection of oxygen, etc. There are also ongoing investigations, including Carbon Capture and Storage (CCS) for the storage of produced CO_2 and utilization of the produced CO_2 as a raw material to produce different chemicals (CCU). It is obvious that several initiatives are needed to solve the very demanding situation for the industry.

Pilot plant experiments within the European ULCOS (Ultra–Low Carbon dioxide (CO2) Steelmaking) project [7] carried out at LKAB's experimental blast furnace at the research institute SWERIM in Luleå Sweden showed that when modifying the blast furnace with top gas recirculation combined with CCS, a reduction of approximately 50% CO_2 emission is achievable.

Within SSAB´s Raahe unit in Finland, full-scale experiments were performed during 2019 aiming at replacing fossil coal with bio-coal [8]. It was shown that replacing 5–20% of the PCI coal with bio-coal is technically feasible and would, with the highest rates of bio-coal injection in the blast furnace in question, give a CO_2 emission reduction of 200,000 ton/year. The limiting factor is the availability of bio-coal.

3. The HYBRIT Initiative

The use of hydrogen produced by water electrolysis using fossil-free electricity to reduce iron ore pellets in a shaft furnace was the main alternative chosen for the HYBRIT initiative. A conversion to a fossil-free value chain from the mine to the finished steel includes many issues to be developed where also local market and geographical conditions must be taken into consideration. Sweden has a unique situation with overcapacity in electrical power in the northern part of the country, vicinity to iron ore mines, good access of biomass and steelworks, and a strong network between industry, research institutes, and universities. Three key areas important for the HYBRIT initiative should be mentioned:

Hydrogen reduction of iron ore has been studied for many decades. There were pioneering investigations in Sweden by Wiberg [9] and Edström [10] already in the 1950s, which have been followed by a number of investigations [11–13] describing the kinetics of the process as well as the reactivity of the produced product. The first commercial scale hydrogen direct reduction of iron ore (H_2DRI) plant based on fluidized bed technology, Circored [14], was built in 1998 by Cliffs and Associates Ltd. at Point Lisas Industrial Complex in Trinidad in 1998. The plant had a designed capacity of 500,000 tons per annum of Hot Briquetted Iron (HBI). The plant did not succeed commercially and was closed down in 2016.

The overall reaction for the reduction of hematite with hydrogen is:

$$Fe_2O_3 + 3H_2 = 2Fe + 3H_2O. \tag{1}$$

Reaction (1) is endothermic with a heat of reaction, $\Delta H_{298} = 95.8$ kJ/mol [15], which is negative for the energy balance of the process and demands an addition of energy with the injected reduction gas/gas mixture. The focus in developing the production line is optimization based on the reduction temperature, kinetics of the reaction, pellet composition, and technology for the preheating of the reduction gas.

Hydrogen storage will give benefits from an electricity sourcing/pricing perspective, but it is not as such critical for the process concept. Today, the most cost-efficient alternative for hydrogen storage is underground pressurized storing, where there the most cost-efficient alternatives are storing in underground salt formations, which today is the only technology for hydrogen storage tested on an industrial scale [16]. Other solutions attracting a lot of interest nowadays include utilizing natural gas pipelines and conversion to ammonia or hydrocarbons as intermediate hydrogen storage. Initial evaluation of the Lined Rock Cavern (LRC) technology used currently in Sweden for natural gas is considered to be promising also [17–19]. Other alternative methods for hydrogen storage under development are storage in metal hydrides and in porous materials.

Since the electric arc furnaces (EAFs) for the melting of steel scrap were introduced on an industrial scale in the 1920s, the technology has developed tremendously. For many years, the EAFs were used for both melting and refining, resulting in a low productivity. Today's furnaces, based on ultra-high power, water cooled panels, and foaming slag practices, are used as melting units, and they are combined with refining in the ladle before casting. Tap to tap times, the time period between sequential tappings of the furnace into the casting ladle, have developed from being several hours to typically below 60 min. The introduction of a foaming slag practice in many steel shops in the 1970s has been crucial for this development. The injection of coal powder and oxygen creates a formation of CO/CO_2 which combined with a proper slag composition results in a foaming slag that protects the furnace walls from the electric arcs. By doing so, it is possible to have a high-power input also at the end of the melting process when the melt temperature is increased before tapping.

The worldwide production of direct reduced iron based on natural gas was 104 million tons in 2018 [20], and the melting in EAF shops is a well-established technology. A melting practice for increased productivity was developed already in the 1970s [5,21–23] and includes typically continuous feeding of DRI and the application of a foaming slag practice. Commercial direct reduced iron contains normally 1–2% carbon, which is beneficial for the melting process and for obtaining a foaming slag.

A hydrogen-based reduction process results in a low carbon iron product, indicating that a fossil-free carbon source is needed to obtain a final commercial attractive product for the steelmaking step.

On 4 April 2016, the three Swedish companies—SSAB, LKAB, and Vattenfall AB—launched a project aimed at investigating the feasibility of a hydrogen-based sponge iron production process, with CO_2 emission-free electricity as the primary energy source—HYBRIT (Hydrogen Breakthrough Ironmaking Technology). A joint venture company was formed, HYBRIT Development AB, with the three companies being owners. This gives full access to top competence in the entire value chain from energy production, mining, ore beneficiation and pellet production, direct reduction, melting, and the production of crude steel.

A prefeasibility study on hydrogen-based direct reduction was carried out in 2017. The study concluded that the proposed process route is technically feasible and, in view of future trends on costs for CO_2 emissions and electricity, it is also economically attractive for conditions in northern Sweden/Finland.

The principle layout of the HYBRIT production process and as a comparison the blast furnace production process is given schematically in Figure 6. The main characteristics are the following:

- Non fossil fuels are used in pellet production
- Hydrogen is produced with electrolysis using fossil-free electricity
- Storage of hydrogen in a specially designed unit is used as a buffer to the grid
- A shaft furnace is used for iron ore reduction
- Tailor-made pellets are used as iron ore feed
- The reduction gas/gas mixture is preheated before injection into the shaft
- The product can either be DRI or HBI free of carbon or carburized
- The DRI/HBI is melted together with recycled scrap in an electric arc furnace

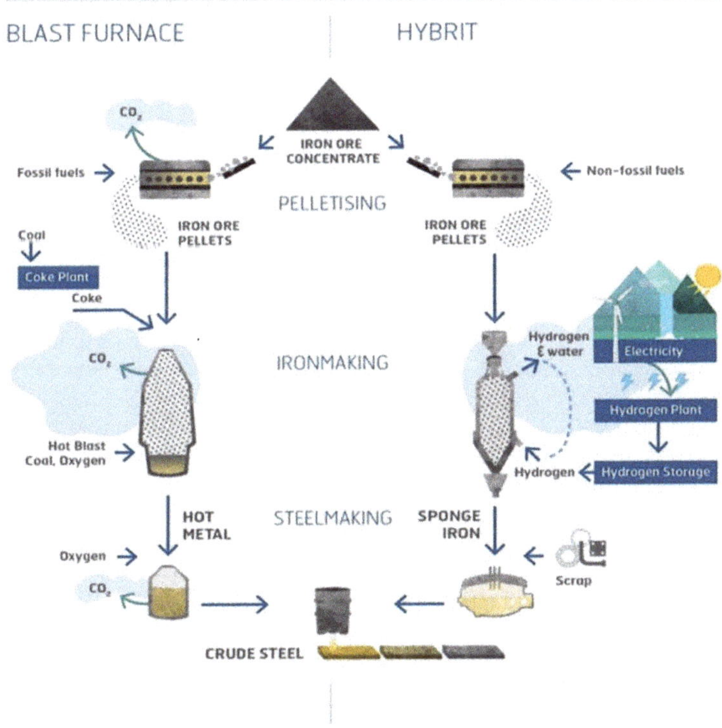

Figure 6. The blast furnace route and HYBRIT—Hydrogen Breakthrough Ironmaking Technology.

A comparison of some consumption figures between the blast furnace route and the HYBRIT route are given in Figure 7 [24]. In the HYBRIT initiative, fossil-free energy sources are used in the whole production line from the pellet production to the produced crude steel. The only direct fossil carbon load comes from the consumption of graphite electrodes and from carbon added in the EAF melt shop. In the example, the fossil CO_2 emission decreases from 1600 kg/ton of crude steel to 25 kg/ton. In this example, in northern Sweden, the availability of renewable electricity (such as hydropower and wind power) together with good access to biomass used in the HYBRIT route replaces the fossil fuel used in the Blast Furnace route, thus eliminating most of the fossil CO_2 emissions.

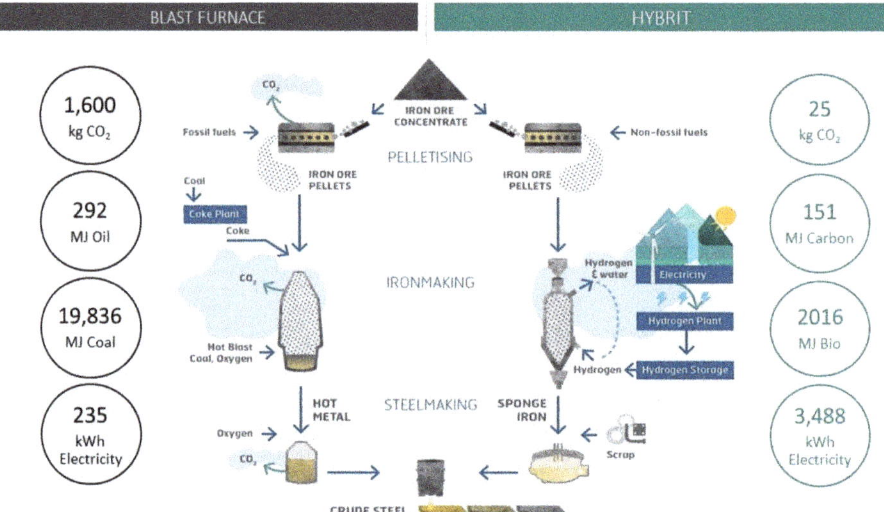

Figure 7. Comparison of some emission and consumption figures for the SSAB blast furnace route and the HYBRIT concept, data from [24].

A critical parameter for the production economy is the cost of electricity. By utilizing fossil-free energy sources such as wind power and optimizing the production time in the electrolysis and the size of the hydrogen storage, simulations have shown that the price for electricity can be lowered by 6–7%.

A production cost comparison, January 2018, based on greenfield units for Swedish/Finnish conditions resulted in increased production costs of 20–30% compared with the blast furnace route. In the future, the HYBRIT concept is estimated to be more attractive due to higher costs for the emission of CO_2 and lower renewable electricity costs. The industry will also prevent being dependent on the limited resources of coking coal. It was also concluded that a scale up to pilot and demo plants is critical for the development.

A decision was made in February 2018 to build a pilot plant with a capacity of 1 ton/h at the SSAB site in Luleå, Sweden, and the construction started in June 2018. Completion of the plant is planned for the summer 2020 followed by experimental campaigns during the coming years, as shown in Figure 8.

The plant is designed for using hydrogen as a reduction gas but will have flexibility where other gases such as biogas can be used also.

Since 2016, the HYBRIT project has expanded, and four different pilot projects are today included in the initiative, as shown in Figure 9. Since 2016, a four-year research program involving several research institutes and universities in Sweden has been run also to build competence and knowledge in different areas, such as the fundamental reduction mechanisms of hydrogen direct reduction of iron ore (H_2DRI), the melting behavior of H_2DRI, new heating and sintering techniques without fossil fuel, hydrogen production and storage techniques, and electricity grid balancing, as well as research in policy aspects as an enabler for the steel industries' transition to carbon neutrality. Around 100 engineers and researchers from industry, research institutes, and academia are today active in the HYBRIT initiative.

Figure 8. The pilot plant in Luleå, Sweden, under construction.

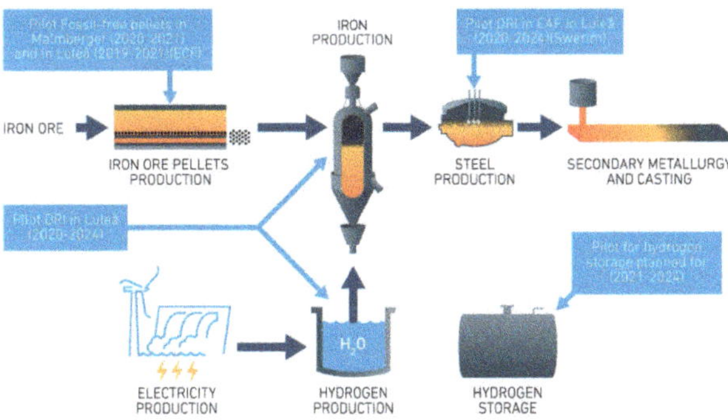

Figure 9. Pilot projects included in the HYBRIT initiative: Pilot for fossil-free pellets, pilot for hydrogen storage, direct reduction of iron ore (DRI) pilot, and electric arc furnace (EAF) pilot.

A roadmap for the transformation of the production line from pellet production to crude steel is given in Figure 10. During 2019, a decision was made to start the planning for the first demonstration plant for HYBRIT to be put into production in 2025, which is the same year that SSAB's Oxelösund plant is planned to be converted from the blast furnace route to the electric arc furnace route, to enable SSAB to be the first steel company in the world providing fossil-free steel to the market.

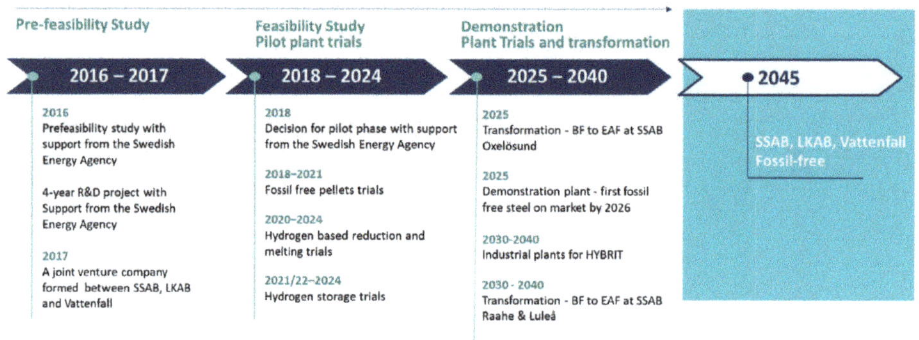

Figure 10. Roadmap to replace coal/coke-based blast furnace operations with fossil-free electricity in Sweden and Finland.

4. Final Comments

The climate change has put a lot of focus on global CO_2 emissions. All sectors of the society including energy, agriculture, transportation, construction, and industry are affected by the situation. Regarding the steel industry, several initiatives are ongoing, which is very important depending on the technical risks and geographical differences regarding the availability of energy and raw materials.

In the present paper, which gives a general overview of the HYBRIT project, some findings and a future strategy are described. So far, the results look promising, and the very broad competence areas including energy supply and hydrogen production, mining, and the development of pellets for direct reduction and iron and steelmaking strengthens this belief. The planned pilot plant experiments will be critical for the scaling up of this technology.

Author Contributions: Conceptualization, M.P. (Martin Pei), M.P. (Markus Petäjäniemi), A.R. and O.W.; Data curation, M.P. (Martin Pei); Formal analysis, M.P. (Martin Pei) and O.W.; Funding acquisition, M.P. (Martin Pei), M.P. (Markus Petäjäniemi) and A.R.; Investigation, M.P. (Martin Pei) and O.W.; Methodology, M.P. (Martin Pei) and O.W.; Project administration, M.P. (Martin Pei) and O.W.; Resources, M.P. (Martin Pei), M.P. (Markus Petäjäniemi) and A.R.; Supervision, M.P. (Martin Pei) and O.W.; Validation, M.P. (Martin Pei) and O.W.; Visualization, M.P. (Martin Pei) and O.W.; Writing—original draft, M.P. (Martin Pei) and O.W.; Writing—review and editing, M.P. (Martin Pei), M.P. (Markus Petäjäniemi), A.R. and O.W. All authors have read and agreed to the published version of the manuscript.

Funding: The Swedish Energy Agency, through the Swedish Industrial Leap program, together with the three companies involved, have provided financial support to HYBRIT Development AB.

Conflicts of Interest: The authors declare no conflict of interest.

References

1. Material Economics. The Circular Economy—A Powerful Force for Climate Mitigation. Available online: https://www.climate-kic.org/insights/the-circular-economy-a-powerful-force-for-climate-mitigation-2/ (accessed on 18 April 2019).
2. Paris Agreement. *Agreement between 197 nations after negotiations in Paris within the United Nations Framework Convention on Climate Change*; Paris Agreement: Paris, France, 2015.
3. *Swedish Iron and Steel Makers Association*; Jernkontoret: Stockholm, Sweden, 2019.
4. Edström, J.O. *Råjärnsprocesser—Sveriges FoU-Behov*; STU-Utredning nr 74: Stockholm, Sweden, 1977.
5. Edström, J.O. *Stålframställning ur Fosforrik Järnmalm*; STU Information 256: Stockholm, Sweden, 1981.
6. Roland Berger. *The Future of Steelmaking—How the European Steel Industry Can Achieve Carbon Neutrality*; Roland Berger GMBH: Munich, Germany, 2020.
7. Burström, E. *ULCORED SP12, Concept for Minimized CO_2 Emission*, 4th ed.; ULCOS Seminar: Essen, Germany, 2008.
8. Lilja, J. SSAB Europe, Private Communication, 2019.

9. Wiberg, M. Some new ways of producing Sponge Iron. *Jernkontorets Ann.* **1956**, *142*, 289–355.

10. Edström, J.O. The Mechanism of Reduction of Iron Oxides. *J. Iron Steel Inst.* **1953**, *175*, 289–304.

11. Tsay, Q.T.; Ray, W.H.; Szekely, J. The modelling of hematite reduction with hydrogen plus carbon monoxide mixtures. *AlChE J.* **1976**, *22*, 1064–1072. [CrossRef]

12. Wagner, D.; Devisme, O.; Patisson, F.; Ablitzer, D. A Laboratory Study of the Reduction of Iron Oxides by Hydrogen. In Proceedings of the Sohn International Symposium, San Diego, CA, USA, 27–31 August 2006.

13. Kazemi, M.; Pour, P.S.; Sichen, D. Experimental and Modeling Study on Reduction of Hematite Pellets with Hydrogen Gas. *Metall. Mater. Trans. B* **2017**, *48*, 1114–1122. [CrossRef]

14. Nuber, D.; Eichberger, H.; Rollinger, B. Circored Fine Ore Direct Reduction. *Stahl Eisen* **2006**, *126*, 4.

15. Jörgensen, C.; Thorngren, I. *Thermodynamic Tables for Process Metallurgist*; Almqvist och Wiksell: Stockholm, Sweden, 1969.

16. Kruck, O.; Crotogino, F.; Prelicz, R.; Rudolph, T. *Overview on All Known Underground Storage Technologies for Hydrogen*; Grant Agreement No.: 303417, Deliverable No. 3.1; Hydrogen Europe: Bruxelles, Belgium, 2013.

17. Mansson, L.; Marion, P. The LRC Concept and the Demonstration Plant—A New Approach to Commercial Gas Storage. In Proceedings of the IGU World Gas Conference, Tokyo, Japan, 1–5 June 2003.

18. Tengborg, P. Stora bergrum I Sverige LRC Gaslager I Halmstad. 2006. Available online: http://www.dftu.dk/Faelles/Modereferater/St%20bergrum%20060228%20Per%20Tengborg.pdf (accessed on 1 March 2017).

19. Tengborg, P.; Johansson, J.; Durup, J.G. Storage of highly compressed gases in underground Lined Rock Caverns–More than 10 years of experience. In Proceedings of the World Tunnel Congress 2014—Tunnels for a better Life, Foz do Iguacu, Brazil, 9–15 May 2014.

20. WorldSteel Association. *Steel Statistical Yearbook*; WorldSteel Association: Brussel, Belgium, 2019.

21. Hutchinson, T.P. Profitable flat Products from Electric Arc Furnace Steel. *Steel Times Int.* **1980**, *208*, 40.

22. Ottmar, H.; Ameling, D. Über einige besondere Problemebei der Verarbeitung von Eisenschwamm im Hochleistungs- Lichbogenofen. In *Internationaler Eisenhuttentechnischer Kongress*; Düsseldorf: Sohnstraße 65, Düsseldorf, Germany, 1974.

23. Plöckinger, E.; Etterich, O. *Electrostahlerzeugung*; Verlag Stahleisen, M.B.H., Ed.; Düsseldorf: Sohnstraße 65, Düsseldorf, Germany, 1979.

24. HYBRIT—Fossil Free Steel. *Summary of Findings from HYBRIT Pre-Feasibility Study 2016-HYBRIT Development AB*; HYBRIT—Fossil Free Steel: Stockholm, Sweden, 2018.

 *metals*

Article

Metals Production, CO$_2$ Mineralization and LCA

Ron Zevenhoven 

Process and Systems Engineering, Åbo Akademi University, FI-20500 Turku, Finland; ron.zevenhoven@abo.fi;
Tel.: +358-2-215-3223

Received: 2 February 2020; Accepted: 27 February 2020; Published: 4 March 2020

Abstract: Modern methods of metal and metal-containing materials production involve a serious consideration of the impact on the environment. Emissions of greenhouse gases and the efficiency of energy use have been used as starting points for more sustainable production for several decades, but a more complete analysis can be made using life cycle assessment (LCA). In this paper, three examples are described: the production of precipitated calcium carbonate (PCC) from steelmaking slags, the fixation of carbon dioxide (CO$_2$) from blast furnace top gas into magnesium carbonate, and the production of metallic nanoparticles using a dry, high-voltage arc discharge process. A combination of experimental work, process simulation, and LCA gives quantitative results and guidelines for how these processes can give benefits from an environmental footprint, considering emissions and use and reuse of material resources. CO$_2$ mineralization offers great potential for lowering emissions of this greenhouse gas. At the same time, valuable solid materials are produced from by-products and waste streams from mining and other industrial activities.

Keywords: metals; metallic products; environmental impact; carbon capture and storage; CO$_2$ mineralization; steelmaking slags; nanoparticles; life cycle assessment (LCA)

1. Introduction

Metal production and subsequent production, use, and disposal of metallic (or metal-containing) products have a significant and worldwide environmental impact. Besides the immediate impact of extraction of rock and ores, a large contribution to emissions of carbon dioxide (CO$_2$) resulting from a vast use of energy, a very significant use of fresh water and the effect of disposing enormous amounts of tailings, other mining residues and by-products/wastes of metal processing can be mentioned. On top of this, an increasing world population requests ever increasing amounts of materials and products while facing the challenges of climate change and environmental pollution in the world that has finite resources.

The greatest potential for reversing the trends is offered by methods and assessment tools that address several problems at the same time. Three of these methods from the field of metal and metal-containing products manufacturing are addressed in this paper, describing and summarizing the work done in Finland, mostly as PhD thesis works supervised by the author, during the last fifteen to twenty years. These three methods are as following:

1.1. Steelmaking Slags Valorization

Steelmaking slags, more specific steel converter (basic oxygen furnace, BOF) slags can be converted with CO$_2$ into precipitated calcium carbonate (PCC) with market value. This route also known as the slag2PCC concept [1–9].

Figure 1 gives an overview of how the sectors of iron- and steelmaking, paper/plastics and the mining of limestone are interconnected with typical global annual mass flows indicated [6]. With significant fixed amounts of CO$_2$, steel converter (BOF) slags can be diverted from landfill and processed into valuable PCC, which reduces the need for the mining and calcination of limestone. (It may be argued that producing PCC binds the same amount of CO_2 as what was released during limestone calcination, but it should be noted that CO_2 emissions from fuels used for limestone calcination, which stands for approximately 1/3 of the CO_2 emissions, are avoided when reusing calcium-rich slags.)

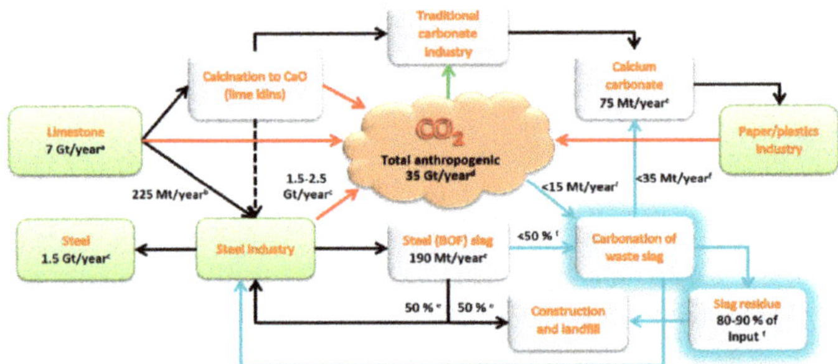

Figure 1. Typical annual mass flows of anthropogenic CO_2 and steelmaking slags, paper/plastics and limestone rock from a viewpoint of precipitated calcium carbonate (PCC) production.

Figure 2 summarizes the slag2PCC concept that may operate on CO_2-containing gas directly (if the target PCC quality allows for it) without a need for a separate CO_2 capture step. This opportunity is one of the main strengths of CO_2 mineralization as a CO_2 capture and storage (CCS) technology, since the capture step gives a very significant energy penalty (see [10]). The overall chemistry can be summarized by Reactions (R1)–(R3), where ion "X" is nitrate—NO_3^-, chloride—Cl^-, or acetate—CH_3COO^-:

$$CaO(s) + 2NH_4X(aq) + H_2O(l) \leftrightarrow CaX_2(aq) + 2NH_4OH(aq), \tag{R1}$$

$$CO_2(g) + 2NH_4OH(aq) \leftrightarrow (NH_4)_2CO_3(aq) + H_2O(l), \tag{R2}$$

$$(NH_4)_2CO_3(aq) + CaX_2(aq) \leftrightarrow CaCO_3(s) + NH_4X(aq). \tag{R3}$$

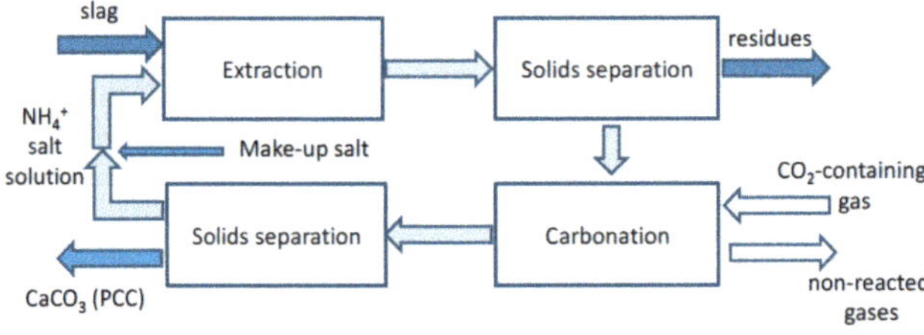

Figure 2. Schematic overview of the slag2PCC process concept.

As shown, the solvent solution is recycled for reuse, although some makeup salt is needed depending on PCC separation from the solvent and subsequent washing.

1.2. Blast Furnace Top Gas Processing

The second concept implies fixing CO_2 from a blast furnace (BF) top gas into solid carbonate using magnesium extracted from serpentinite rock. This is based on the first one of several so-called Åbo Akademi (ÅA) routes for step-wise carbonation of serpentinite rock. Mineral sequestration offers a large-scale CCS option for Finland and many other countries, where the underground storage of CO_2 is not possible [10]. While a significant volume of the literature using ÅA routes for CO_2-containing exhaust gases from heat and power production exists, metal production or mineral processing [11–18] a special opportunity arises for the processing of a BF top gas with $Mg(OH)_2$ that can be obtaineded from thevast(more than needed) natural resources of magnesium silicate rock available worldwide [19]. BF top gas, containing CO_2 as well as CO in more or less equal amounts of approximately 20 vol.%, can be processed by sequential mineralization of CO_2 followed by conversion of CO into CO_2 and H_2 via the CO/water shift reaction [20]:

$$Mg(OH)_2(s) + CO_2(g) \leftrightarrow MgCO_3(s) + H_2O(g), \tag{R4}$$

$$CO(g) + H_2O(g) \leftrightarrow CO_2(g) + H_2(g), \tag{R5}$$

$$Overall : \ CO(g) + Mg(OH)_2(s) \leftrightarrow MgCO_3(s) + H_2(g), \tag{R6}$$

which (depending on H_2O pressure and temperature) competes with the following reaction:

$$Mg(OH)_2(s) \leftrightarrow MgO(s) + H_2O(g). \tag{R7}$$

Two different conversion efficiencies can be defined from the CO_2 conversion and the carbonation of $Mg(OH)_2$, respectively, shown as:

$$CO_2 \text{ fixation efficiency} = 1 - \frac{(CO + CO_2)_{out}}{(CO + CO_2)_{in}}, \tag{1}$$

$$\text{carbonation efficiency} = \frac{MgCO_{3out}}{Mg(OH)_{2in}}. \tag{2}$$

As shown, the overall chemistry gives not only the fixation of CO_2 in thermodynamically stable magnesite, but $MgCO_3$ produces hydrogen at the same time. The latter can be easily separated from a gas stream using a membrane. Magnesite may be hydrated to nesquehonite, $MgCO_3 \cdot 3H_2O$, which may be used as a thermal energy storage (TES) material in a cyclic magnesite/nesquehonite conversion process [21]. As the first analysis of this potential, an optimization study was made for BF top gas processing as a function of temperature, pressure, and $Mg(OH)_2$ vs. $CO + CO_2$ conversion [20]. A similar approach was followed in experimental work in China using calcium oxides for carbonation in parallel with CO/water shift reaction in a fluidized bed reactor [22]. While this method has the advantage of atmospheric pressure operation and somewhat higher temperatures than used with $Mg(OH)_2$, the fixed CO_2 is eventually released in order to regenerate CaO, which gives it rather a calcium looping character.

1.3. Metallic Nanoparticle Production

The third concept involves production of metallic nanoparticles (NPs) using a production route with a potentially lower environmental impact than conventional methods. Development of a list of metals and alloys (Ag, Al, Au, Cu, Ni, Zn and FeCr, and NiCu) in the unit of kilograms per day was the task of the recent EU FP7 project BUONAPART-E [23]. One objective was to reduce the environmental impact of metallic NP production by avoiding the use of complex mixes of toxic or hazardous chemical solutions (for reducing oxidized metallic salts), resulting in a significant postprocessing of by-products and wastes.

The BUONAPART-E NP production route involves the use of high-voltage arc discharges for melting and evaporation of metal, followed by condensation and solidification in an inert gas atmosphere (N_2, Ar, Ne, and N_2/H_2 mixture (95%/5% vol./vol.)). A simplified process diagram is shown in Figure 3a, Figure 3b displays a 2×8 operational units set-up at the University of Duisburg-Essen (UDE), Germany. The units were shown to be capable of producing (using 15 of the 16 electrode pairs) Cu NPs with a primary particle size of 79 nm at a production rate of 69 g/h at a specific electricity consumption (SEC) of 170 kWh/kg in a N_2 carrier gas. NP sizes down to 14 nm were obtained by adjusting gas composition, gas flow, or applied electric power [24].

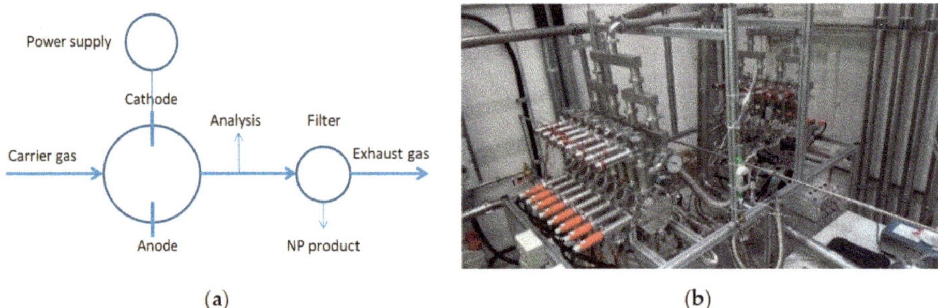

Figure 3. (a) Schematic overview of the BUONAPART-E concept for metallic nanoparticle production. (b) Multiple operating units (2 × 8) system at the University of Duisburg-Essen (UDE), Duisburg, Germany.

At ÅA, the efficiency of energy (i.e., electricity) use and the environmental footprint of the dry, arc discharge NP production route were analysed and compared with those of more conventional production routes based on aqueous solution chemistry. One feature that makes the energy efficiency assessment very interesting is the fact that below a size of approximately 50 nm the thermodynamic properties of (metallic) NPs start to change: melting and evaporation temperatures as well as heat (i.e., enthalpy) for that were lowered. This is illustrated by Figure 4 for the melting/solidification of NPs with both sizes, D_1 and D_2, of <<100 nm ($D_2 < D_1$), compared with the properties of bulk materials, which was confirmed by the data for 5 nm Ag NPs in [25].

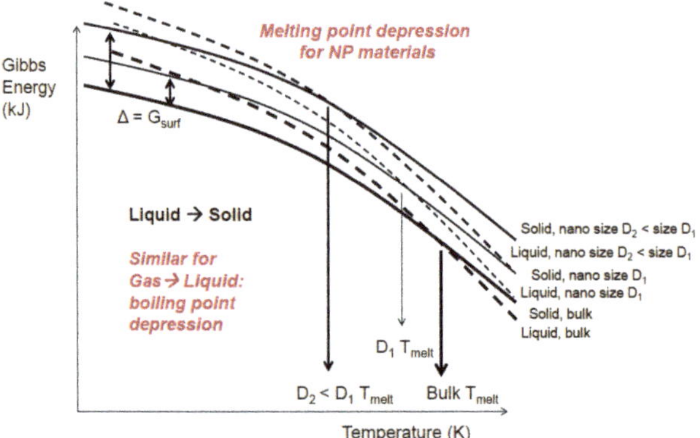

Figure 4. Gibbs energies of nanosized droplets and solid particles vs. those of bulk size materials, showing melting point depression for the diameter D_1 of <<100 nm and $D_2 < D_1$.

This can be found in [26], where an analysis of energy needs for (metallic) NP production is based on the surface free energy (SFE) of atoms at the surface compared to those in the bulk volume of a material. (For liquid–gas and liquid–liquid interfaces, SFE is commonly referred to as surface tension.) As in fact work is needed, here in the form of electricity, to increase or form an interface with the surroundings, energy efficiency assessments based on exergy (see below for more details) are straightforward, once data on SFE are found. The work by Xiong et al. [27,28] reported on diameter-dependent thermodynamic properties of metallic NPs, such as temperatures and enthalpies for melting and evaporation and their temperature dependences.

Using the scarce literature, for six metallic elements, the relations between NP size and melting temperature and melting heat could be calculated as given in Figure 5 (see also [26]). It should be noted that entropy change as the ratio melting heat/melting temperature is unchanged. The same was found for the evaporation/condensation process. As reported below, this had significant consequences for the energy use of the dry, arc discharge process for (metallic) NP production and the life cycle footprint of this production route.

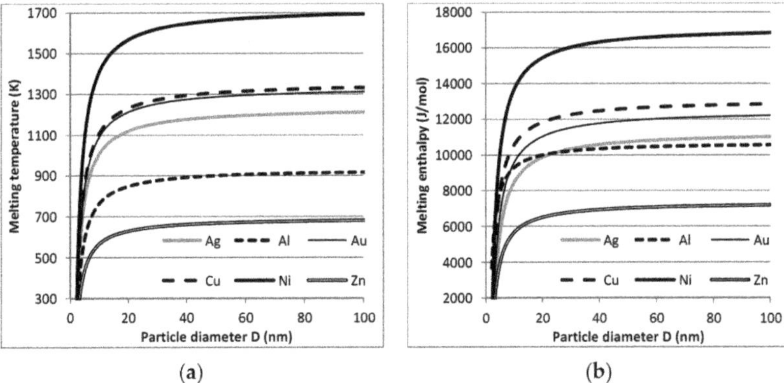

(a)   (b)

Figure 5. Size-dependent melting points (**a**) and melting enthalpy (**b**) for six metallic nanoparticles (NPs).

These three concepts will be evaluated for their potential when aiming at reduced CO_2 emissions, less production of wastes or an overall smaller environmental footprint when compared to "business as usual" or other profitable production. Prohibitive legislation and/or regulation may be on the horizon. Nonetheless, however promising a "better" method may seem and regardless of evidence that these methods are from many points of view superior to existing methods, large-scale implementation will not occur until industry recognizes an effective business model. Minimizing or lowering costs is not sufficient, since profit must be made at all times while guaranteeing some level of growth, with new products promptly available in the market place.

The red line of this paper is new technologies and tools for production of metals and metal-containing products, resulting in a smaller environmental footprint. This is illustrated by three project-driven and independent studies with each having a narrower focus. For this reason, readers are encouraged to consult the original papers as referenced for more details on background, methods and results.

2. Materials and Methods

Since 2005, the development of the slag2PCC concept has involved experimental work later followed by process simulation (Aspen Plus) on steel converter (BOF) slags from Finnish iron- and steelmaking industry. A range of possible solvent salts was tested in aqueous solutions at ambient pressure and low temperature (20–30 °C). In a few cases, heating was used until up to 70 °C [2–8], in order to selectively leach calcium from slags (see Figure 6 [4]).

Successful experiments were done with ammonium salts, mostly ammonium chloride. Initial tests were done with 1 L solutions, and later tests were conducted with 2 × 28 L vessels (made of Perspex) connected with separators and circulation pumps [5,6]. At the same time, Aalto University in Espoo Finland put up a demonstration plant with 200 L reactor vessels [7,8]. Later, the work involved life circle assessment (LCA) simulations using SimaPro with Ecoinvent and other databases [9] and process scale-up and integration with a focus on separation methods for PCC product solids and spent slags from aqueous dispersions [1,6].

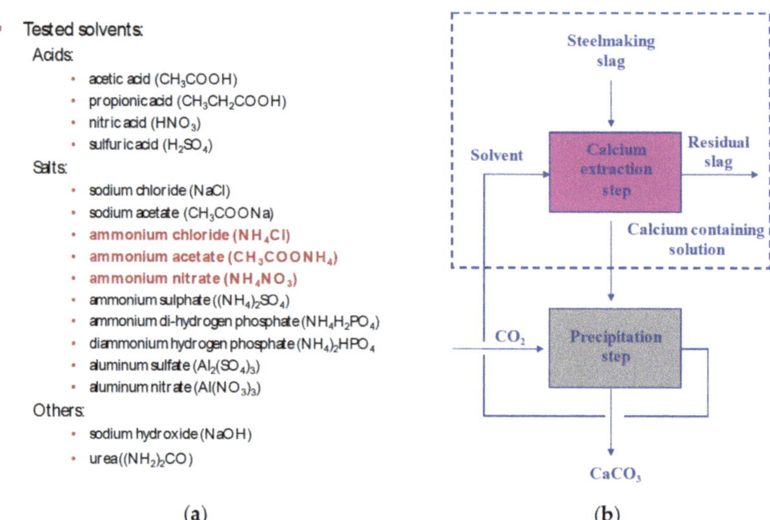

Figure 6. (**a**) Solvents tested as aqueous solutions for selective leaching of calcium from steel converter (basic oxygen furnace (BOF)) slags with three successful solvents highlighted. (**b**) Carbonation to produce PCC with the recycling of the leaching solvent.

For the analysis of carbonation of $Mg(OH)_2$ with CO_2 in the BF top gas, where in parallel the CO/water shift reaction supplies CO_2 for subsequent mineralization, a simulation study was made using Aspen Plus (AspenTech, Bedford, MA, USA) (v.8.2 and later v.9.0). Pressures, temperatures, and $Mg(OH)_2$ feed vs. gas feed were optimized for maximum CO_2 as well as combined $CO_2 + CO$ conversion efficiencies while minimizing production of reactive MgO and energy input requirements. LCA studies that include the production of $Mg(OH)_2$ as part of serpentinite carbonation for treatment of flue gases from a power plant or a lime kiln are given elsewhere [14,17].

For the analysis of the efficiency of energy use with a dry, arc discharge (BUONAPART-E) process route for metallic NP production, the production of waste heat was measured and quantified at the 2×8 operational single-unit production facility at the UDE, Germany, using an infrared camera (Fluke Ti9, Fluke Europe, Eindhoven, The Netherlands) and an infrared thermometer (Testo Quicktemp 860-2, Sensorcell, Helsinki, Finland) (see [14,29]). Exergy analysis was used for assessing how much electricity was used for creating NP surface energy and how much was dissipated as waste heat. While for electric power P the exergy Ex(P) is P, the exergy of heat Q depends on its absolute temperature T and that of the surroundings $T°$ and can be written as $Ex(Q) = Q \cdot (1 - T°/T)$. Szargut et al. [30] gave more details on exergy analysis, a concept that follows the second law of thermodynamics. The (Carnot) factor $(1 - T°/T)$ quantifies the quality of heat, and thus the exergy of energy sources normalizes these to the capacity of doing work.

Besides electricity, also materials and other resources are used or consumed by NP production processes. Here, LCA simulations were made using SimaPro with Ecoinvent and other databases for comparing NP production routes and the use of metallic NPs in consumer products. These benefits were from special properties obtained already when using very small amounts of NPs. Examples are silver NPs in hospital cotton (bed sheets and lab coats), gold NPs in solar energy collectors, nickel NPs for catalysts, zinc NPs as flame redardants in plastics like PP, or copper NPs in water giving a nanofluid with improved cooling performance. Besides the summary in [14], more details on LCA work were given in [29] and [31] with focuses on energy use and comparisons of conventional methods (wet chemistry) for metallic NP production, respectively.

3. Results and Discussion

3.1. Steelmaking Slags Carbonation: the Slag2PCC Concept

3.1.1. Experimental Results and Process Scale-up

Steel converter slag leaching tests with a range of solvents as shown in Figure 6 gave for most cases no good results for calcium leaching and its selectivity. A good leaching (~90% after 1 h, 1 g per 50 mL at ambient conditions for particle sizes of 74–125 μm) without significant formation of precipitates was obtained with acetic acid, but unfortunately with significant leaching of other species was obtained as well, primarily silicon, iron, and manganese, as shown in Figure 7a. Leaching with ammonium salts gave much more selective leaching of calcium, with only very little silicon, as can be seen from Figure 7b. Leaching with ammonium nitrate gave a 40–45% leaching efficiency for calcium under the same conditions as for the acetic acid tests. Although the obtained leached calcium was less than with acidic solutions, the selectivity achieved with ammonium salts solvents further benefited from the alkalinity of the final solution. This made the precipitation of calcium carbonate possible with a pH buffering effect when adding CO_2. After the precipitation, the solvent solution was returned to the leaching reactor. More details are given elsewhere [4].

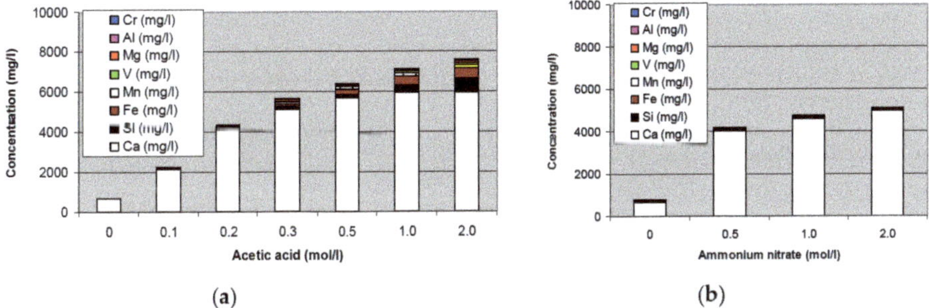

Figure 7. Selective leaching of calcium and other elements from steel converter (BOF) slags at ambient conditions using acetic acid (**a**) and ammonium nitrate (**b**) solutions with different strengths.

Having identified a suitable and selective solvent that allows for PCC production at (near) ambient conditions, the work proceeded with scale-up and operation in a continuous process mode. Compared to conventional PCC production, which typically implies a batch process (starting with lime, CaO), a continuous process offers more flexibility for varying composition, calcium content, and calcium compounds of industrial steel converter slags. A schematic of the process set-up at ÅA, constructed of Perspex parts, is given in Figure 8, with two 28 L reactors for extraction and carbonation, respectively. The schematic also shows the two tubular-inclined (45°) settlers that removed ~99% of the dispersed solids from the solution (inner diameter: 0.1 m; length: 1.1 m for PCC removal; length: 0.5 m for spent slag removal) upstream of candle filters for further cleaning. The inset photo in Figure 8 shows the reactors with white PCC and black slag dispersions, respectively. A set of pumps circulated the dispersions between the reactor vessels at a rate of 1.5 L/min, while temperatures and pH values of the dispersions were measured on-line and logged.

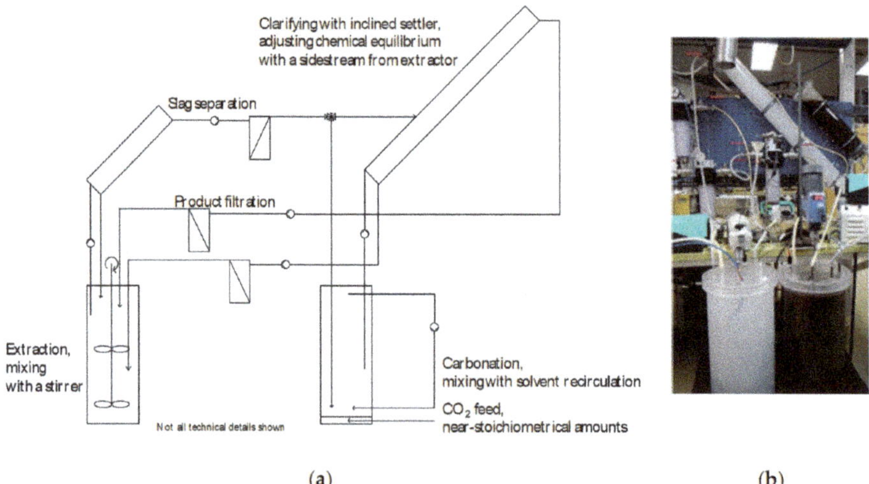

(a) (b)

Figure 8. (**a**) Schematic of the continuous lab-scale slag2PCC process set-up with two 28 L reactors and inclined settlers followed by candle filter separators for particle product and residue removal. In (**b**) the reactor vessel with black liquids is the extractor with dispersed steel converter slags, and the reactor with a white solution is the carbonator with dispersed PCC particles.

Using this set-up, PCC products with varying particle size and crystal shape were produced, depending on pH levels of the dispersions. Examples are given as SEM photos in Figure 9.

(a) (b)

(c)

Figure 9. PCC products produced from steel converter (BOF) slag using the slag2PCC concept with a 1 M ammonium chloride solvent: (**a**) rhombohedral calcite particles; (**b**) cubic calcite particles; and (**c**) spherical vaterite particles. Scale bar in (**b**): 1 μm.

The results of the lab-scale experimental work combined with the parallel work at Aalto University resulted in the design and construction of a pilot-scale test set-up at Aalto University. The reporting by Said et al. [7,8] gives a description of this set-up, composed of 200 L reactor vessels, pumps, a feeding silo, and candle filters for the separation of particles from aqueous dispersions. Figure 10 gives an impression of the test facility. The objectives of ongoing work are scale-up to yet a larger scale than ~100 kg/h this pilot unit can handle, debottlenecking with respect to solid particles separation and increasing the amount of calcium extracted from the slags while guaranteeing the continuous production of PCC with preselected properties. This shall eventually take the technology into large-scale use and commercialization (following the patent in 2008).

Figure 10. Slag2PCC pilot-scale test facility at Aalto University, Espoo Finland. The photo was taken by the author.

3.1.2. LCA of PCC Production

A wide-scope analysis of whether a certain product or process is more preferable than another from an environmental footprint point of view can be made using LCA. A comparison of the slag2PCC process route with conventional PCC production was made, as illustrated by Figure 11. SimaPro software (v. 7.3.2) with several life cycle inventory (LCI) datasets (Ecoinvent v3, European Life Cycle Data (ELCD), EU and Danish Input Output Library, and Swiss Input Output) were used for the calculations (see [6,9] for more details). Four impact categories being climate change, human health, ecosystem quality, and resources were considered. A cradle-to-gate assessment was made, excluding the product use and the end-of-life phase, as these would be the same as for conventional PCC production.

The results are given in Figure 12, initially for a comparison between conventional PCC production and slag2PCC operating with a 0.65 M ammonium solvent solution and a 0.1 kg/L slag/liquid loading, which from a process performance point of view was found to be optimal. However, this gave a larger environmental impact than the conventional PCC production, as the solvent leaving the process attached to the PCC product required a significant energy input (steam) for its recovery. The outcome was better using a "system expansion" consideration that took into account the fact that the PCC market has a limited volume. Producing PCC using an alternative route lowered the impact of the conventional process route, most likely with a location elsewhere, which was partly taken out of business (see [9]). This is illustrated by the dotted rectangles in Figure 12.

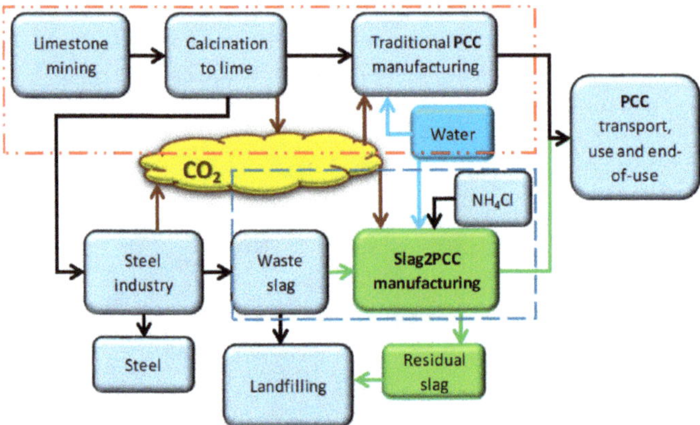

Figure 11. System boundaries for PCC production using traditional processing (red striped line box) and the use of the slag2PCC concept (blue striped line box).

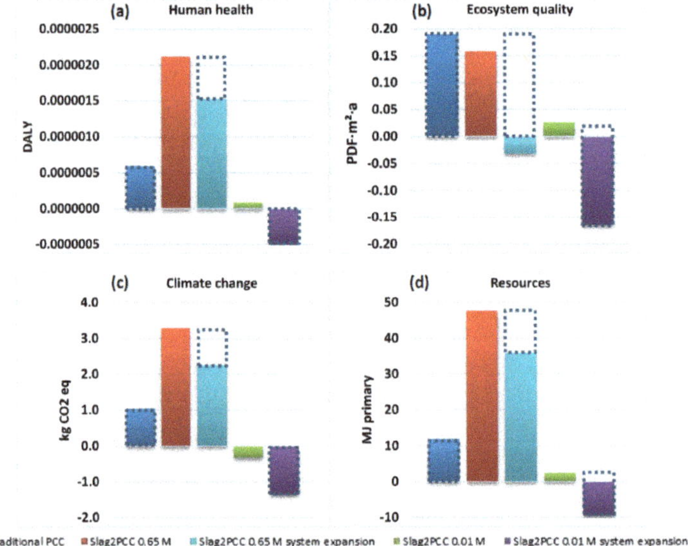

Figure 12. Life cycle assessment (LCA) results for PCC production using a traditional method versus the use of the slag2PCC concept at different solvent concentrations and/or as a system expansion concept (see the text) for four impact categories: Human health (**a**), Ecosystem quality (**b**), Climate change (**c**), Resources (**d**) DALY: disability adjusted life years. PDF·m^2·a: potentially disappeared fraction of species times area and time.

Nonetheless, a strong environmental benefit was obtained by diluting solutions in the slag2PCC process to 0.01 M NH$_4$Cl while keeping the slag loading similar to 0.65 M operation, after which much less product washing was needed. A significant negative environmental footprint was obtained for the system expansion approach, replacing the market volume and the environmental footprint of conventionally produced PCC. In practice, operating the slag2PCC route at lower solvent concentrations implies that dispersions may make several cycles to obtain full conversion. More details on the consequences of operating the slag2PCC process under such conditions are given in [6,9].

3.2. Mineralization of CO_2 from BF Top Gas Using Magnesium Hydroxide

For the assessment of the feasibility of integrated mineralization of CO_2 from BF top gas using $Mg(OH)_2$ produced from magnesium silicate rock and CO/water shift reaction supplying CO_2 from CO, a BF top gas composed of 20.7 vol% CO_2, 20.7vol% CO, 3.8 vol% H_2, 6 vol% H_2O, and 48.8 vol% N_2 was assumed [32]. The flow sheet for the Aspen Plus model used for the simulations is shown in Figure 13, with SOLIDIN as the $Mg(OH)_2$ supply that eventually gives solid products COLDCARB and BF top gas stream feed GASIN that is compressed to process conditions eventually giving product gas FINALGAS after expansion. Three gas/solid heat exchangers, a gas/solid (chemical equilibrium) reactor, and a gas/solid separator made up the final process equipment [20]. This model may be readily added to or integrated with Aspen Plus models for BFs or other sections of iron- and steelmaking processes, such as for example given in [33,34].

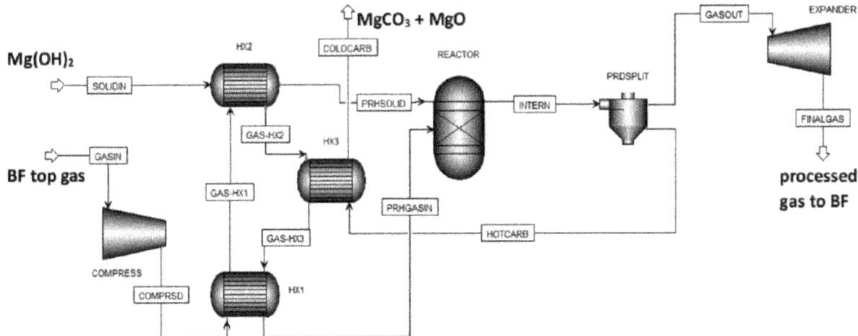

Figure 13. The Aspen Plus flow sheet for BF top gas processing.

A gas feed of 1 kmol/s with the abovementioned composition was used with an excess of 17% $Mg(OH)_2$ for the conversion, corresponding to 0.3 mol Mg per mol input gas. Temperature and pressure for the reactor were varied in the ranges of 400–500 °C and 40–100 bar, respectively. This was based on preliminary calculations using Gibbs energy minimization and the earlier work at ÅA on $Mg(OH)_2$ carbonation using CO_2-containing exhaust gases from other processes. The simulation results showed that in all cases $Mg(OH)_2$ was converted not only to $MgCO_3$ but also to MgO. At 40 bar, the carbonation efficiency dropped from 93% at 400 °C to 66% at 460 °C. For 100 bar, the carbonation efficiency changed to 97% at 400 °C and 87% at 460 °C. The values for the CO + CO_2 conversion efficiency were similar to the values for the carbonation efficiency. Interestingly, there was little effect of process temperature and pressure on the amount of H_2 produced, which eventually left the system at approximately six times the amount entering with the feed gas. One feature of the process is that with increasing temperature the CO/water shift reaction equilibrium caused CO to be more stable while the corresponding higher H_2O partial pressure could not prevent MgO formation. Experimental work under the preferable process conditions can be a next step, and the required suitable equipment (e.g., a pressurized fluidized bed as in [11]) is available at ÅA.

The study was finalized by making an analysis of energy input/output and exchanger duties, with results as shown below in Figure 14. It can be seen that power requirements may be compensated for by heat that is produced by the overall process with the surroundings temperature T° and the process unit temperatures T_i and the exergies of the combined heat outputs, Q_i, calculated by $\Sigma Q_i \cdot (1 - T°/T_i)$, was larger than the required netto power input. See [20] for more details on these results and [30] for the use of exergy analysis, which based on the second law of thermodynamics allows recalculating energy flows of different forms (here are power and heat) into the equal denominator of useful work.

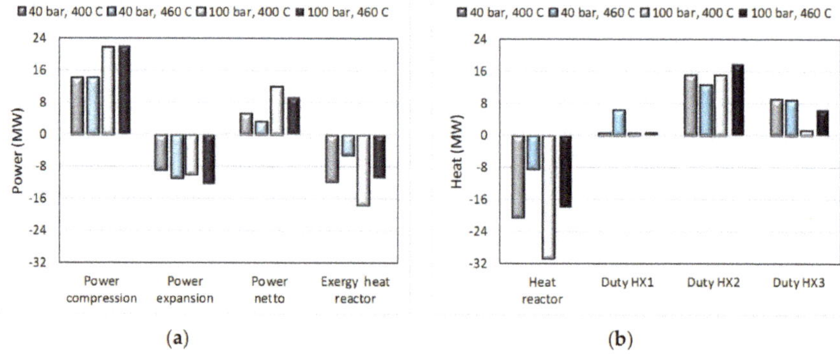

Figure 14. Energy input and output and heat exchanger duties for process conditions as indicated for the process shown in Figure 13. Power in/out and exergy of reaction heat (**a**) and heat of reaction and heat duties of heat exchangers (**b**).

3.3. Metallic NP Production Using an Arc Discharge Route

3.3.1. Specific Electricity Consumption (SEC)

Analyzing the SEC for the production of metallic NPs leads to a significant penalty resulting from the NP size dependence and the energies (enthalpies) of melting/solidification, evaporation/condensation and the temperatures for these. Melting and evaporation (volatilization) requires heats Q_m and Q_v at temperatures T_m and T_v, respectively, while later the heat released by condensation Q_c is smaller than Q_v for condensation temperature $T_c < T_v$ and the heat released by solidification Q_s is smaller than Q_m for solidification temperature $T_s < T_m$. Input exergies were significantly larger than output exergies calculated as $Q \cdot (1 - T°/T)$ for the heating and cooling processes involving the material that produces NPs.

For copper NPs, the results are illustrated by Figure 15, showing the SEC for several NP sizes as a result of exergy input for evaporation not being returned as exergy of condensation and exergy input for melting not being returned as exergy of solidification. This comes on top of the exergy needed for producing increased surface energies of atoms in NPs compared to those of atoms inside the material. It was clearly shown that the evaporation/condensation exergy consumption is by far the most important one of the three processes, which come on top of the SEC that arises from pumping around the carrier gas [14,29]. The values given in Figure 15 are several orders of magnitude lower than the 170 kWh/kg reported for experimental production of 79 nm Cu [29].

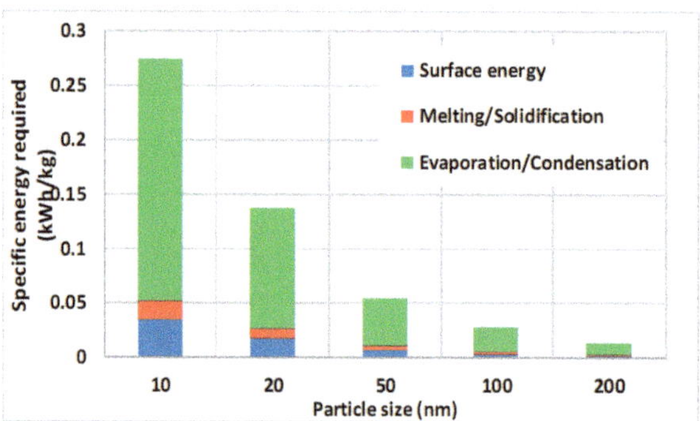

Figure 15. Specified electricity consumption (SEC) for copper NP production considering surface energy and losses due to melting/solidification and evaporation/condensation.

3.3.2. LCA of NP Production

Besides the energy efficiency analysis, a wider analysis of the environmental footprint was made of metallic NP production via the dry, arc discharge process versus more conventional wet, aqueous solution metallic oxide salt reduction. For copper NPs, the results are given in Figure 16, illustrating that the BUONAPART-E concept is preferable only if more of the input metal material is obtained as NP products rather than remaining as deposits inside the production facility. Although NP metal production and LCA results were reported, metal NP yield needs to be improved [24].

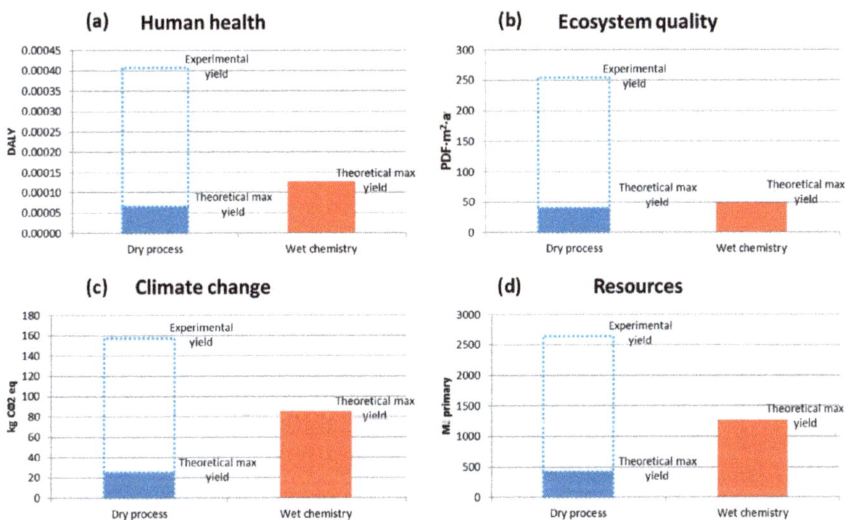

Figure 16. Life cycle impact (LCI) comparisons of dry, arc discharge NP production and chemical reduction methods for copper particles for four impact categories: Human health (**a**), Ecosystem quality (**b**), Climate change (**c**), Resources (**d**). See Figure 12 for the explanations of abbreviations.

The LCA studies on metal NP production were expanded to products that contain metallic NPs, such as silver NPs in cotton used in hospitals and copper NPs in water, so as to give a nano-cooling fluid. These assessments can be found elsewhere [14,35] and showed great potential and special properties of consumer product materials containing metallic NPs. The increased demand for NP materials obviously calls for scale-up of production routes, while at the same time negative impacts need to be addressed, as was the objective of the work reported here. Note, however, that the LCA studies were cradle-to-gate assessments, i.e., from producing metal from ore to products containing NP particles leaving production. Lack of data on end-of-life product handling makes a full cradle-to-gate LCA impossible, although a first step in this direction was recently presented [36].

4. Conclusions

The work reported addresses modern trends seen in development of more sustainable process routes for iron- and steelmaking as well as nonferrous materials and products in which these are used. CO_2 emissions reduction is obviously on top of many industrial production agendas together with energy efficiency, followed by water use and waste and by-product disposal. This faces the facts of limited resources and an increasing need for more circular economies for materials, resources, and processes that allow producing them. Experimental methods combined with modern theoretical tools such as exergy assessment for energy efficiency analysis and LCA give quantitative information on whether improved processes or products actually are an improvement vis-à-vis profitability. Three examples were given in this paper.

Slag2PCC is a proven concept developed in Finland for valorization of steelmaking (BOF) slags, binding CO_2 while producing high-value PCC. Scale-up and commercialization is ongoing, after the lab-scale proof-of-concept was demonstrated followed by successful operation of a pilot plant. The LCA results showed that water use may be a critical factor.

CO_2 (and CO) from/in BF top gas can be converted with $Mg(OH)_2$ that can be produced from abundant serpentinite rock to $MgCO_3$, hydrogen and steam, integrated with CO/water shift. The process simulation studies showed under the process conditions (pressure: 40 bar; temperature: 400 °C) good conversion levels were obtained, yet to be experimentally verified.

The dry production of metallic NPs using high-voltage (arc discharge) evaporation has many benefits, including life cycle impact, compared to with wet methods for NP production, and appears to allow for easier scale-up to production levels of kilograms per hour. The LCA calculations showed the overall benefit from an environmental footprint of metallic NP production as well as the production of products that, with very small amounts of NP, have a variety of beneficial properties.

5. Patents

The slag2PCC concept described in this paper has been patented under Finnish patent 122348 and US patent 8603428.

Author Contributions: The author presented this work at the International Process Metallurgy Symposium IPMS2019 event in Espoo Finland 5–6 November 2019 and produced this manuscript based on that presentation.

Funding: The development work of the slag2PCC concepts was possible, during 2005–2016 with primarily Tekes funded projects Slag2PCC, Slag2PCC+ and Cleen Oy/Clic Oy CCSP and Finland's Graduate School for Chemical Engineering (2010–2014). The work on metal nanoparticulates was conducted during 2012–2016 funded by EU FP7 project BUONAPART-E (grant agreement: 280765) and Finland's earlier Graduate School of Energy Engineering and Systems (2012–2015).

Acknowledgments: Daniel Lindberg of Aalto University, Espoo Finland is acknowledged for inviting the author to present the work at International Process Metallurgy Symposium IPMS2019, and Lauri Holappa of Aalto University is acknowledged for the invitation to produce this manuscript. The author acknowledges his earlier co-workers, currently postdoctoral researchers in most cases, for intensive and productive cooperation, a great deal of which is summarized in this paper.

Conflicts of Interest: The author declares no conflicts of interest.

References

1. Teir, S.; Kotiranta, T.; Pakarinen, J.; Mattila, H.-P. Case study for production of calcium carbonate from carbon dioxide in flue gases and steelmaking slag. *J. CO$_2$ Util.* **2016**, *14*, 37–46. [CrossRef]
2. Teir, S. Fixation of Carbon Dioxide by Producing Carbonates from Minerals and Steelmaking Slags. Ph.D. Thesis, Helsinki University of Technology, Espoo, Finland, 2 June 2008. Available online: http://lib.tkk.fi/Diss/2008/isbn9789512293537/ (accessed on 19 February 2020).
3. Eloneva, S.; Said, A.; Fogelholm, C.-J.; Zevenhoven, R. Preliminary assessment of a method utilizing Carbon dioxide and steelmaking slags to produce precipitated calcium carbonate. *Appl. Energy* **2012**, *90*, 329–334. [CrossRef]
4. Eloneva, S. Reducing CO_2 Emissions by Mineral Carbonation: Steelmaking Slags As Raw Material with a Pure Calcium Carbonate End-Product. Ph.D. Thesis, Aalto University, Espoo, Finland, 26 November 2010. Available online: http://lib.tkk.fi/Diss/2010/isbn9789526034577/ (accessed on 19 February 2020).
5. Mattila, H.-P.; Zevenhoven, R. Designing a continuous process setup for precipitated calcium carbonate production from steel converter slag. *ChemSusChem* **2014**, *7*, 903–913. [CrossRef] [PubMed]
6. Mattila, H.-P. Utilization of Steelmaking Waste Materials for Production of Calcium Carbonate ($CaCO_3$). Ph.D. Thesis, Åbo Akademi University, Turku, Finland, 10 October 2014. Available online: https://www.doria.fi/handle/10024/99011 (accessed on 19 February 2020).
7. Said, A.; Laukkanen, T.; Järvinen, M. Pilot-scale experimental work on carbon dioxide sequestration using steelmaking slag. *Appl. Energy* **2016**, *177*, 602–611. [CrossRef]

8. Said, A. CO$_2$ Sequestration by Steelmaking Slags for the Production of Precipitated Calcium Carbonate—From Laboratory to Demonstration Stage. Ph.D. Thesis, Aalto University, Espoo, Finland, 12 May 2017. Available online: https://aaltodoc.aalto.fi/handle/123456789/25401 (accessed on 19 February 2020).

9. Mattila, H.-P.; Hudd, H.; Zevenhoven, R. Cradle-to-gate life cycle assessment of precipitated calcium carbonate production from steel converter slag. *J. Clean. Prod.* **2014**, *84*, 611–617. [CrossRef]

10. Zevenhoven, R.; Romão, I.S. CO$_2$ mineralisation as a route to energy-efficient CO$_2$ sequestration or materials with market value. In *CO$_2$ Sequestration by Ex-Situ Mineral Carbonation*; Sanna, A., Maroto-Valer, M.M., Eds.; World Scientific Publ. Co.: London, UK, 2017; pp. 41–90.

11. Fagerlund, J. Carbonation of Mg(OH)2 in a Pressurized Fluidized Bed for CO$_2$ Sequestration. Ph.D. Thesis, Åbo Akademi University, Turku, Finland, 2 March 2012. Available online: https://www.doria.fi/handle/10024/74477 (accessed on 19 February 2020).

12. Nduagu, E.I. Production of Mg(OH)$_2$ from Mg-Silicate Rock for CO$_2$ Mineral Sequestration. Ph.D. Thesis, Åbo Akademi University, Turku, Finland, 13 December 2012. Available online: https://www.doria.fi/handle/10024/86170 (accessed on 19 February 2020).

13. Soares Romão, I.S. Production of Magnesium Carbonates from Serpentinites for CO$_2$ Mineral Sequestration: Optimisation towards Industrial Application. Ph.D. Thesis, Åbo Akademi University, Turku, Finland, University of Coimbra, Coimbra, Portugal, 17 December 2015. Available online: https://www.doria.fi/handle/10024/117766 (accessed on 19 February 2020).

14. Slotte, M. Two Process Case Studies on Energy Efficiency, Life Cycle Assessment and Process Scale-up. Ph.D. Thesis, Åbo Akademi University, Turku, Finland, 11 January 2017. Available online: https://www.doria.fi/handle/10024/130097 (accessed on 19 February 2020).

15. Lavikka (n. Sjöblom), S. Geological and Mineralogical Aspects on Mineral Carbonation. Ph.D. Thesis, Åbo Akademi University, Turku, Finland, 3 January 2017. Available online: https://www.doria.fi/handle/10024/130096 (accessed on 19 February 2020).

16. Koivisto, E. Membrane Separations, Extractions and Precipitations in Aqueous Solutions for CO$_2$ Mineralisation. Ph.D. Thesis, Åbo Akademi University, Turku, Finland, 11 October 2019. Available online: https://www.doria.fi/handle/10024/171340 (accessed on 19 February 2020).

17. Zevenhoven, R.; Slotte, M.; Koivisto, E.; Erlund, R. Serpentinite carbonation process routes using ammonium sulphate and integration in industry. *Energy Technol.* **2017**, *5*, 945–954. [CrossRef]

18. Zevenhoven, R.; Slotte, M.; Åbacka, J.; Highfield, J. A comparison of CO$_2$ mineral carbonation processes involving a dry or wet carbonation step. *ENERGY* **2016**, *117*, 604–611. [CrossRef]

19. Lackner, K.S. A guide to CO$_2$ sequestration. *Science* **2003**, *300*, 1677–1678. [CrossRef] [PubMed]

20. Zevenhoven, R.; Virtanen, M. CO$_2$ mineral sequestration integrated with water-shift reaction. *ENERGY* **2017**, *141*, 2484–2489. [CrossRef]

21. Erlund, R.; Zevenhoven, R. Thermal energy storage (TES) capacity of a lab-scale magnesium hydrocarbonates/silica gel system. *J. Energy Storage* **2019**, *25*, 100907. [CrossRef]

22. Liu, Y.; Li, Z.; Xu, L.; Cai, N. Effect of sorbent type on the sorption enhanced water gas shift process in a fluidized bed reactor. *Ind. Eng. Chem. Res.* **2012**, *51*, 11989–11997. [CrossRef]

23. EU FP7 Project Better Upscaling and Optimization of Nanoparticle and Nanostructure Production by Means of Electrical Discharges BUONAPART-E (2012–2016). Available online: http://www.buonapart-e.eu/ (accessed on 3 January 2020).

24. Stein, M.; Kruis, F.E. Scaling-up metal nanoparticle production by transferred arc discharge. *Adv. Powder Technol.* **2018**, *29*, 3138–3144. [CrossRef]

25. Hu, W.; Xiao, S.; Deng, H.; Luo, W.; Deng, L. Thermodynamic properties of nano-silver and alloy particles. In *Silver Nanoparticles*; In-Tech: Vukovar, Croatia, 2010; pp. 1–34.

26. Zevenhoven, R.; Beyene, A. The exergy of nano-particulate materials. *Int. J. Thermodyn.* **2014**, *17*, 145–151. [CrossRef]

27. Xiong, S.Y.; Qi, W.H.; Cheng, Y.J.; Huang, B.Y.; Wang, M.P.; Li, Y.J. Modeling size effects on the surface free energy of metallic nanoparticles and nanocavities. *Phys. Chem. Chem. Phys.* **2011**, *13*, 10648–10651. [CrossRef] [PubMed]

28. Xiong, S.Y.; Qi, W.H.; Cheng, Y.J.; Huang, B.Y.; Wang, M.P.; Li, Y.J. Universal relation for size dependent thermodynamic properties of metallic nanoparticles. *Phys. Chem. Chem. Phys.* **2011**, *13*, 10652–10660. [CrossRef] [PubMed]

29. Slotte, M.; Zevenhoven, R. Energy efficiency and scalability of metallic nanoparticle production using arc/spark discharge. *ENERGIES* **2017**, *10*, 1065. [CrossRef]

30. Szargut, J.; Morris, D.; Steward, F.R. *Exergy Analysis of Thermal, Chemical and Metallurgical Processes*; Hemisphere Publishing Co.: New York, NY, USA, 1988.

31. Slotte, M.; Mehta, G.; Zevenhoven, R. Life cycle indicator comparison of copper, silver, zinc and aluminum nanoparticle production through electric arc/spark evaporation or chemical reduction. *Int. J. Energy Environ. Eng. (IJEEE)* **2015**, *6*, 233–243. [CrossRef]

32. Saxén, H.; Åbo Akademi University, Thermal and Flow Eng., Turku, Finland. Personal Communication, 13 October 2015.

33. Schultmann, F.; Engels, B.; Rentz, O. Flowsheeting-based simulation of recycling concepts in the metal industry. *J. Clean. Prod.* **2004**, *12*, 737–751. [CrossRef]

34. Porzio, G.F.; Colla, V.; Fornai, B.; Vannucci, M.; Larsson, M.; Stripple, H. Process integration analysis and some economic-environmental implications for an innovative environmentally friendly recovery and pre-treatment of steel scrap. *Appl. Energy* **2016**, *161*, 656–672. [CrossRef]

35. Slotte, M.; Zevenhoven, R. Energy requirements and LCA of production and product integration of silver, copper and zinc nanoparticles. *J. Clean. Prod.* **2017**, *148*, 948–957. [CrossRef]

36. Zevenhoven, R. Energy requirements for recovery of (metallic) nanoparticulate material from waste. In Proceedings of the 2019 World Resources Forum (WRF) Conference, Geneva, Switzerland, 22–24 October 2019; Available online: http://users.abo.fi/rzevenho/SS7-5-Zevenhoven.pdf (accessed on 31 January 2020).

 *metals*

Review

Reuse and Recycling of By-Products in the Steel Sector: Recent Achievements Paving the Way to Circular Economy and Industrial Symbiosis in Europe

Teresa Annunziata Branca [1], Valentina Colla [1,*] , David Algermissen [2], Hanna Granbom [3], Umberto Martini [4], Agnieszka Morillon [2], Roland Pietruck [5] and Sara Rosendahl [3]

[1]   Scuola Superiore Sant'Anna, TeCIP Institute, 56124 Pisa, Italy; teresa.branca@santannapisa.it
[2]   FEhS-Institut für Baustoff-Forschung e.V., 47229 Duisburg, Germany; d.algermissen@fehs.de (D.A.); a.morillon@fehs.de (A.M.)
[3]   SWERIM, 97125 Lulea, Sweden; hanna.granbom@swerim.se (H.G.); Sara.Rosendahl@swerim.se (S.R.)
[4]   RINA Consulting-Centro Sviluppo Materiali S.p.A. (CSM), 00128 Castel Romano (Rome), Italy; umberto.martini@rina.org
[5]   VDEh-Betriebsforschungsinstitut GmbH, 40237 Dusseldorf, Germany; Roland.pietruck@bfi.de
*   Correspondence: valentina.colla@santannapisa.it; Tel.: +39-348-071-8937

Received: 8 February 2020; Accepted: 3 March 2020; Published: 5 March 2020

Abstract: Over the last few decades, the European steel industry has focused its efforts on the improvement of by-product recovery and quality, based not only on existing technologies, but also on the development of innovative sustainable solutions. These activities have led the steel industry to save natural resources and to reduce its environmental impact, resulting in being closer to its "zero-waste" goal. In addition, the concept of Circular Economy has been recently strongly emphasised at a European level. The opportunity is perceived of improving the environmental sustainability of the steel production by saving primary raw materials and costs related to by-products and waste landfilling. The aim of this review paper was to analyse the most recent results on the reuse and recycling of by-products of the steelmaking cycles as well as on the exploitation of by-products from other activities outside the steel production cycle, such as alternative carbon sources (e.g., biomasses and plastics). The most relevant results are identified and a global vision of the state-of-the-art is extracted, in order to provide a comprehensive overview of the main outcomes achieved by the European steel industry and of the ongoing or potential synergies with other industrial sectors.

Keywords: by-products; circular economy; industrial symbiosis; reuse; recycling; steel industry

---

1. Introduction

The increasingly stringent European regulation and the ever-higher disposal costs currently affect manufacturing industries, leading them to strengthen their efforts in order to improve the recycling rate of their by-products and waste [1]. Iron- and steelmaking by-products result from the processes producing steel by two main routes: the iron ore-based steelmaking and the scrap-based steelmaking. In total, 70% of the world steel is produced utilising the first one, based on Blast Furnace (BF), where iron ore is reduced to pig iron, which is afterwards converted into steel in the Basic Oxygen Furnace (BOF). Input of this route are mainly iron ore, coal, limestone and steel scrap. Through the second route, mainly based on the Electric Arc Furnace (EAF), 30% of the world steel is produced, using scrap steel as input as well as electricity as energy source. Other raw materials can often be used, such as Direct Reduced Iron (DRI) and pig iron.

During the iron- and steelmaking processes, several by-products are produced, such as slags, dusts, mill-scales and sludges. By providing some figures on the by-product productions, on average, for one tonne of steel 200 kg (in the scrap-based steelmaking) and 400 kg (in the iron ore-based steelmaking) of by-products are produced [2].

Slags represent the iron- and steel by-products, which are produced in the greatest quantity with a worldwide average recovery rate from over 80% (steelmaking slag) to nearly 100% (ironmaking slag). They mainly contain silica, calcium oxide, magnesium oxide, aluminium and iron oxides. Slags derive from the smelting process, where some fluxes (e.g., limestone, dolomite, silica sand) are introduced to the different furnaces, such as BF, BOF, EAF. Slags tasks include removing impurities, which are present in iron ore, steel scrap and other feeds, as well as protecting the liquid metal from oxygen and maintaining temperature inside the furnace. The slag density is indeed lower than that of liquid metal and this implies the slag itself floats over the metal surface, hindering the contact of the steel with an external atmosphere. Finally, the difference of density between slag and molten metal facilitates the slag removal at the end of the process. During the iron- and steelmaking processes, dusts and sludges are also produced. In particular, sludges derive from dust or fines in different processes, such as steelmaking and rolling, and they contain a high moisture percentage. Dusts and sludges are collected in the abatement plants, equipped with filters. After they are removed from the gases, they contain high amounts of iron oxides and also carbon, which can be used for internal purposes [3]. On the other hand, iron containing residues which are not internally recycled can be externally sold and used by other sectors, in different applications, such as Portland cement or electric motor cores. Mill-scales is mainly produced during the continuous casting and rolling mill processes in oxidising atmospheres. Iron oxides layer is formed at the surface of steel. It can be reused as a raw material in the sintering plants as well as for briquettes and pellets. Finally, iron- and steelmaking gases are mainly coke oven gas, BF gas and BOF gas. Usually they are cleaned and they are internally used (or externally sold), in order to produce steam and electricity, providing between 60% to 100% of the plant power [4].

The emerging concept of Circular Economy (CE) has driven several industrial sectors to cooperate in the reuse and recycling of by-products in order to globally reach the ambitious "zero waste" goal. On this subject, significant efforts and commitments are undertaken within both European and worldwide activities. In particular, the European steel industry, in order to increase its competitiveness, is committed to introducing innovative actions on high performance products and to increasing process efficiency, by also reducing its environmental impacts [5–8] (see Figure 1).

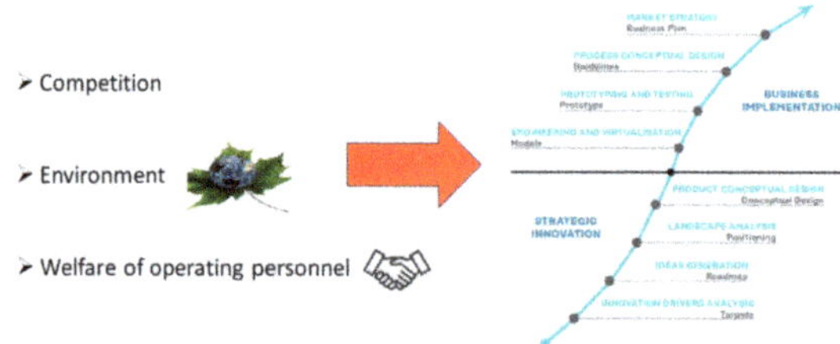

Figure 1. Circular Economy for a cleaner industry.

This strategy needs further investigations in different conditions for new implementations [9], such as the recycling method of red mud, comprised of the carbon-bearing red mud pellets roasting in the rotary hearth furnace and smelting separation in the EAF [10]. Furthermore, the development of new technologies aims not only at reusing by-products in manufacture of conventional products but also at converting them into new products [11], through new destination of secondary materials in a CE perspective, subtracting large quantities of them from the destination to landfills and saving extraction of new raw materials [12]. The previous examples show the role played by the steel sector in the CE context, based on the 4 "R" [13,14]: Reduce, Reuse, Recycle and Restore. Some aspects of this approach are shown in Figure 2. Reduce represents the concept based on avoiding or minimising the environmental impact. Reuse concerns the internal recycling of by-products, such as the sludges reuse (through the thermal technologies, eliminating or reducing its Zn content) as well as the slag reuse (focused on the lime content reduction, resulting in its direct recycling). The Recycling concept concerns also the creation of Industrial Symbiosis (IS), which aims at developing the synergies among different sectors, identifying new business opportunities for underutilised resources outside the boundary of the production chain [15]. Through the IS concept implementation, by-products from one sector (e.g., the steel industry) can be valuable inputs to other sectors. Restore mainly involves the impact reduction of steel products. For instance, currently, for every tonne of CO_2 produced during the steelmaking process, six tonnes of CO_2 are saved during the application of the steel product. However, the aim of the steel sector is to achieve further improvements in the impact reduction of its products [16]. Indeed, these results do not represent total aspects of novelty in this sector. For instance, the main by-product, slag, is currently internal partially recycled (due to its high content of valuable elements, such as iron) or applied in different fields (e.g., cement production, road building and restoration of marine environments), according to the national legislations. However, there are further potential applications of the steelworks by-products [2], both inside and outside the steel production cycle [17]. Over the last few decades, iron- and steelmaking by-products recovery and use have led to a material efficiency rate of 97.6% worldwide. For instance, in 2013, 81% of production residues in the steel company ArcelorMittal were reused or recycled as by-products, and only 9% went to landfills. On the other hand, in 2013 24% of residues produced in its mining activities were disposed in landfill [18].

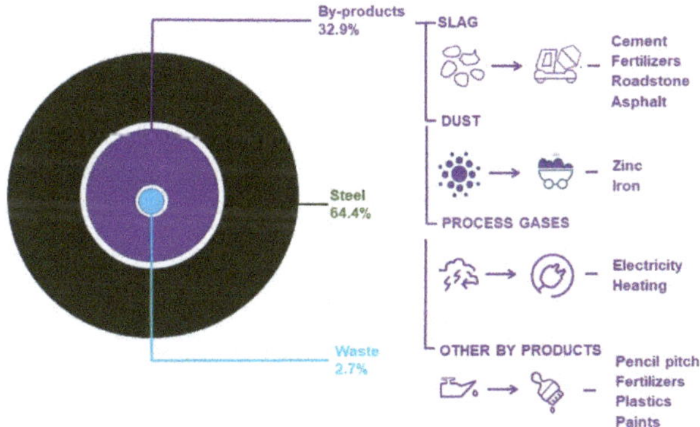

Figure 2. Reuse of steelmaking by-products.

In order to achieve better results, over the last few years, both industries and academic researchers have undertaken several initiatives and activities in order to apply new approaches and techniques aiming at by-product management finalised at increasing their recycling. For instance, the internal recycling of some by-products in the pelletization process has been recently investigated, taking into account achieving a high quality of pellets and reducing environmental impacts and operating costs [19]. In addition, dust recovered from EAF gas treatment has been used for substituting clays in traditional brick manufacturing, by producing energy savings, environmental impact reduction and possible economic benefits [20]. On the other hand, the valorisation of wastes or by-products from different industrial sectors as Thermal Energy Storage (TES) materials has been investigated in depth [21]. Furthermore, simulation models development has allowed identifying the BOF, EAF and Ladle Furnace (LF) slag quality to be internally reused and to provide significant economic and environmental improvements, compared to the current slag use in the steelworks [22,23]. However, there is still significant room for improvement for increasing the recovery rate of by-products, achieving environmental and economic benefits, also according to the principles of IS, as shown in Figure 3.

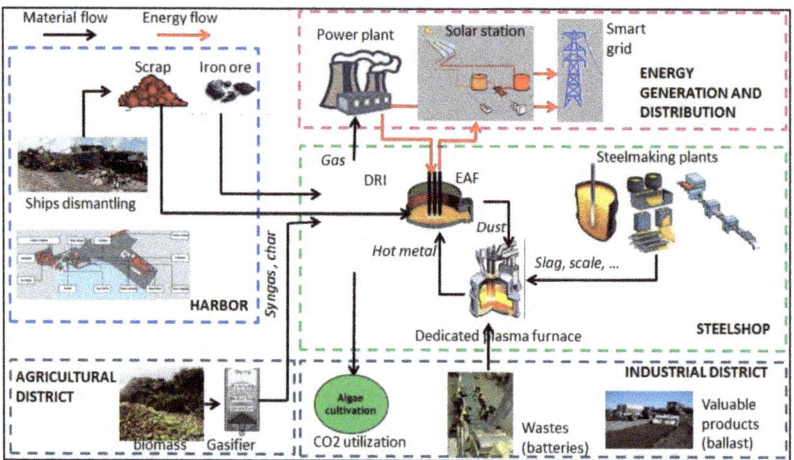

Figure 3. Conceptual design of steel industry-based industrial symbiosis.

This review paper was derived from an analysis performed in the dissemination project entitled "Dissemination of results of the European projects dealing with reuse and recycling of by-products in the steel sector (REUSteel)", co-founded by the Research Fund for Coal and Steel (RFCS). The aim of the project is to identify, organise, combine and integrate the most relevant and promising outcomes from a large number of previous and running European projects, focused on the reuse and recycling of by-products. Furthermore, some areas with promising results in the same topics will be identified and suggestions for future research will be made.

The paper is organised as follows: Section 2 introduces the context of IS, providing some important examples. In Section 3 the recent achieved developments in slags reuse are depicted. Section 4 describes the main achievements in by-products reuse in cement production, while Section 5 concerns their reuse in road construction. Section 6 discusses the liming and amending properties of by-products, and Section 7 is devoted to the reuse of other by-products not fully analysed in the previous Sections. Finally, Section 8 provides some concluding remarks.

2. Industrial Symbiosis in the Steel Sector

Relevant examples about the recent achieved results in IS implementation, including the steel and other sectors, can be found in the literature. In general, by-products generated in the iron- and steelmaking processes can be used in different sectors. On the other hand, by-products from other industrial sectors can be applied in the steel industry as secondary and recycled materials. For instance, some iron- and steelmaking by-products, such as slag, EAF dust, mill scale and zinc sludge, can be used in a cement plant and a zinc smelter plant as raw materials. In particular, sludge can be used as raw material for zinc ingots, which in turn are used in steelmaking as raw material for producing wire rod in the galvanising process [24] (this process consists of an electroplating process, which coats the wire with zinc, in order to prevent corrosion). Currently, gas from iron- and steel processes are cleaned and internally used, for instance, for producing electricity. Hydrogen, contained in the coke oven gas (about 55%), can provide up to 40% of the power for the steelmaking plant, and ammonium sulphate can be used as fertiliser. In addition, BTX (benzene, toluene and xylene) can be applied in plastic products, and tar and naphthalene for producing electrodes for the aluminium industry, plastics and paints. On the other hand, iron oxides and slags can be used for external applications, such as Portland cement; zinc oxides, produced in the EAF route, can be used as a raw material mainly through the Waelz process with over 85% of the market (source: World Steel Association: https://www.worldsteel.org/). On this subject, alternative processes are available, particularly focused on technological solutions for

Zn recovery from EAF dust [25]. These processes are promising and in continuous evolution, based on different approaches. In particular, to recover zinc from EAF dust, present in the form of franklinite (60%), a study based on ultrasound-assisted leaching process has been carried out [26]. In addition, to investigate the selective zinc removal from EAF dust, the microwave heating oven has been used as a heat source [27]. Recovery of zinc from the pre-treatment of coated steel scrap before it is fed to the EAF has also been deeply investigated [28]. Moreover, some examples can be provided about the recovery of valuable metals from by-products [29] as well as the selective leaching processes tested or applied to reclaim pure zinc compounds or metallic zinc from EAF secondary steelmaking [30]. Furthermore, although research towards recycling of the high-zinc fraction of BF sludge is limited, recent studies have been performed in order to incorporate this fraction in self-reducing cold-bonded briquettes and pellets [31] (see some examples in Figure 4).

Figure 4. Exemplar of cold-bonded briquettes, agglomerates and pellets produced at VDEh-Betriebsforschungsinstitut GmbH.

Other significant examples of IS have been recently performed, based on by-products from sources outside the steel sector that can be used in the steel industry as secondary and recycled materials. For instance, carbon bearing materials, deriving from other industrial sectors, such as biomass, residues from food companies, plastic and rubber wastes, represent important materials to be used as partial substitute of fossil materials, such as coal and natural gas. In particular, biomasses or residual plastics, usually landfilled, can be reused as "alternative C sources". Biomass can be used in iron- and steelmaking in order to reduce fossil-based CO_2 emissions [32] (higher than 50%, compared to the current integrated route) and the net increase of direct CO_2 emissions is avoided. Biomass can be used as reducing agent in several iron- and steelmaking processes [33]. For instance, in cokemaking, sintering and in carbon composite agglomerate production, biomass, especially charcoal, can be injected and Pulverized Coal Injection (PCI) can be replaced with high carbon content charcoal. Furthermore, due to its significant economic and ecological potential, the substitution of fossil fuels in metallurgical processes, such as in EAF steelmaking, with biochar-agglomerates has been investigated [34]. In addition, the simultaneous conversion and utilisation of carbon dioxide and plastics into fuels/chemicals in high temperature iron and steel processing can be significantly effective [35]. Moreover, replacing 100% of injection carbon and charge carbon in EAF steelmaking with renewable bio-carbon can produce more than 50% reduction of greenhouse gas emission from the EAF steelmaking [36]. For instance, waste plastics and other materials used in steelmaking processes can reduce ~30% of CO_2 emissions compared to the use of fossil carbon sources [37]. CO_2 reformation with the CH_4 from the waste plastics used in high temperature processes generates fuel gases and reducing gases (i.e., hydrogen and carbon monoxide). These innovative uses represent important aspects pushing toward the creation of new local economies.

3. Recent Developments in Slags Reuse

Among by-products resulting in the iron and steel production, slags represent the main by-products (90% by mass). Worldwide, more than 400 million tonnes of iron and steel slags are produced each year. About 24.6 M tonnes of BF slag and 18.4 M tonnes of steelmaking slags are produced every year in Europe [38]. They are mainly used in the building sector, as aggregates and cement components in hydraulic engineering, and for metallurgical use, and about 9% of steel slags are internally stored, while about 14% are landfilled. A better knowledge of their formation, composition and physical properties is fundamental for increasing slags reuse both internally and in different field of applications. In particular, knowing the phase compositions and, consequently, applying ad hoc stabilisation methods make slags suitable for its reuse and/or inert disposal [39]. Recently, some critical aspects on the steel slags use have been highlighted. Such aspects concern their volume instability (due to the free lime exposure to moisture) and their leaching behaviour (due to the content of metals that can cause water or soil pollution) [40]. On the other hand, steel slags can be used for replacing natural sand as aggregate in cement, often in combination with its CO_2 sequestration properties. However, due to their variability composition that can affect the final product, it can be difficult to internally reuse slags. In order to overcome this issue, a general purpose-monitoring tool was developed and exploited for the simulation and the feasibility assessment of the replacement of lime and dolime with LF slag with or without the partial recovery of EAF slag for the production of two steel families [41]. A small increase of 3–4% of the electric energy has been detected, but compensated through the reduction of about 14–16% of non-metallic raw materials.

The mixture of BF slags and Portland cement with other steelmaking by-products, such as Electrostatic Precipitator dusts (ESP), BF sludge and BOF sludge have shown significant properties, including up to 90% immobilisation of hazardous elements. Moreover, with organic additives (e.g., citric acid) added to the mixture, hazardous constituents can be liberated or immobilised [42].

3.1. Metal Recovery

Steel slags mainly contain iron oxides that can be recycled by reduction methods and metallic iron can settle down to the bottom of the reactor. Recently, a re-melting and reducing treatment, in order to recycle iron and modify the chemical composition of the residue, has been carried out. In particular, SiO_2 can promote reduction and separation of iron and slag. With the 10% of SiO_2, melting time of 30 min, coke of 5%, metallisation rate was 87.30% and metal recovery rate was 96.45%. Furthermore, by reaching a slag with good stability, these results can be improved. The reuse of slag can produce other significant advantages, such as the reduction of its dumping, of occupation of land, and of the environmental pollution. In addition, the use the energy of molten slag could reduce the process costs [43]. Moreover, bioleaching can be used for recovering metals, depending on the slag composition. Recently, a test on BOF slag for bacterial leaching and recovery of aluminium (Al), chromium (Cr) and vanadium (V) has been performed [44]. Batch test results have shown a significant bioleaching of Al, Cr and V more from steel slag than in control treatments. In addition, the culture supernatant could be used for an upscaled industrial application in order to recover metals. On the other hand, the removal and recovery percentages of metals from the leachate have been relatively modest, due to the high concentration of competing ions (SO_4^{2-}, PO_4^{3-}) in the culture medium. About this topic, other methods, such as selective precipitation, could improve the performance of the resin. In BOF slag, the mineralogy and element availability, such as chromium (Cr), molybdenum (Mo) and vanadium (V), could provide information about its possible environmental impact. A Sequential Extraction Procedure (SEP), four-fraction-based, combined with X-Ray Diffraction (XRD), of two BOF slags has been applied [45]. In particular, the four fractions are: F1 (water soluble), F2 (acid soluble), F3 (reducible) and F4 (residual). The results have shown that Cr and Mo primarily occurred in F4, rather than immobile elements under natural conditions, strongly bound into/onto Fe minerals. In addition, V has been more mobile with proportional higher findings in the two fractions F2 and F3. By applying the X-ray diffraction, V has resulted bound into Ca minerals (larnite, hatrurite, kirschsteinite and calcite) and to Fe minerals. Nevertheless, the total amount of recovery cannot be considered an indicator of the availability of analysed elements and, in addition, it did not correspond to the leaching of elements from BOF slag.

3.2. Removal of Harmful Elements

By-products can also be used for removal of harmful elements and compounds that can negatively affect the environment. On this subject, a recent investigation concerned the Induction Furnace (IF) steel slag-based application for removing Cr(VI) ions (hexavalent chromium) from aqueous solution in laboratory conditions at room temperature [46]. Results have shown that the alkali activated steel slag can induce the rate of adsorption of Cr(VI) on the surface of the adsorbent and, consequently, Cr(VI) can be effectively removed from aqueous solution. In addition, a recent study concerned a suitable application of the Water-Spray EAF (WS-EAF) slag. In particular, results have shown that WS-EAF slag can be a promising material for removal, by adsorption, Cd (II) and Mn(II) from aqueous solutions and from industrial wastewater [47]. In addition, other iron- and steelmaking by-products, such as BF slag, dust from the bag filters in the coking installation and dust from the liquid sludge from the scrubber, have shown a relevant ability for removing tricholoroethylene (TCE) from the groundwater. In particular, according to its composition and porosity, BF slag has showed the highest catalytic activity to degrade TCE by using hydrogen peroxide [48]. Recently, some industrial by-products, such as steel slag, iron filings and three recycled steel by-products, have been tested for their abilities for phosphate adsorption, showing a higher ability to do so compared to three natural minerals (limestone, zeolite and calcite) [49]. Due to the strong chemical bonds between phosphate and steel by-products, the adsorbed phosphate can be released to the solution. Consequently, they can be considered as alternative and low-cost effective adsorption media for phosphate removal from subsurface drainage.

3.3. Waste Heat Recovery

Another recent important use of BF and steel slags consists in the high potential for the energy consumption reduction in the steel sector, through heat recovery from hot slags. On this purpose, some high value-added applications (e.g., cutting edge surface coating technologies) can be performed. In particular, molten slag, with a temperature range of 1723–1923 K and with a slag enthalpy of ~1.6 GJ per tonne of slag, can be used for waste heat recovery by applying different methods [50]. A Slag Carbon Arrestor Process (SCAP), which uses slag for catalysing conversion of tar and Coke Oven Gas (COG) into hydrogen-rich fuel gas, has been recently introduced for recovering waste heat in steelmaking processes [51]. A multi-stage system for effective recovery of the waste heat from BF slag and its environmental and economic impacts has been proposed. In addition, the solidification of molten slag droplets in centrifugal granulation for heat recovery though an enthalpy based mathematical model has been studied. This resulted in low crystal phase content, the desired parameter for high quality heat recovery. Another technique concerns the use of a gravity bed waste heat boiler, by associating heat recovery efficiency with decreasing slag particle diameter. A further application concerns the use of slag as energy storage material in Thermal Energy Storage (TES) systems, which are widely used in Concentrated Solar Power Plants (CSPs) to collect energy [52]. On this subject, EAF slag has shown microstructural and thermal properties that make it usable in TES systems. Another method [53] involves mechanical destruction of liquid slag film by impingement with solid slag particles that have previously been solidified and sieved in the same device with simultaneous heat recovery in a fluidised bed, which produces dry and dust-free granules. Additionally, investigations were carried out where liquid slag was poured into moulds of slag caster and steel spheres were added. The liquid slag heated the steel spheres. When the slag solidified, the hot spheres were recovered and used to produce hot air.

3.4. Ceramic Tile and Biomedical Applications

A further use of the EAF slag has been recently investigated, in particular its recycling as a green source in ceramic tile production. Because of its chemical composition, EAF slag has been considered also in the biomedical applications, due to the bioactivity and biocompatibility of Fluorapatite-based glass ceramics for some applications (e.g., bone replacement, dental and orthopaedic applications). In addition, the possible optoelectronic applications, due to their chemical and crystallographic structure similar to the apatite structure of the bone, have also been investigated [52]. Further tests have been carried out in China concerning the use hot-poured converter slag in ceramic materials. Fly ash, microsilica and quartz have been mixed and, after heating at 1100–1200 °C, the ceramics have been sintered [54].

4. By-Products Reuse in Cement Production

Over the last few decades, the slags and other by-products reuse in cement production has allowed coping with the increased demand of cement, resulting in reduction of environmental issues, natural resources exploitation and costs. In particular, BF slag presents specific characteristics, achieved after a rapid cooling by water quenching, resulting in a glassy and granular form, that make it an excellent material for producing Portland cement. Among slags, the Ground Granulated Blast furnace Slag (GGBS) presents structural and durable properties. These features make it suitable to cement concrete, resulting in an eco-friendly and economical material when the replacement of cement by GGBS lies between 40% and 45% by weight. In addition, although ultrafine GGBS can improve strength and durability, the addition of new materials to GGBS can increase these properties [55]. Furthermore, the mechanical properties in Self-Compacting Concretes (SCC), replaced by Granulated Blast Furnace Slag (GBFS) in the limestone aggregate from 0% to 60%, have been analysed [56]. Going into detail, it was found that the replacement in mortar the 50% of cement with ground GBFS produced a compressive strength similar to the reference mortar, containing 100% of cement. In addition, the formation of new compounds as well as the formation of stronger bond, due to the paste richer in Si, can improve the mechanical properties of aggregates. On the other hand, due to the reactivity of the slags, replacing sand with the slag can improve compressive strength of the concrete. The resulting reductions in energy and raw material consumptions and in greenhouse gas emissions as well as the preservation of the destruction of natural quarries represent the main advantages of the use of this by-product instead of natural raw materials. However, slag cooled by water quenching can be inefficient for heat recovery and can produce harmful wastes such as H_2S, heavy metals and SO_2. For this reason, new methods have been developed (e.g., dry granulation) for preparing molten slag, providing a material with better properties. Among them, the use of insoluble chemical activators aims at improving the hydraulic activity of slag blended Portland cement, which results in the improvement of the compressive strength, obtained through the application of optimum proportions of chlorine chemical activators and quality improvers. In addition, tests on the temperature effect on the binders of BF slag and metakaolin (MK) have been carried out [52]. Alkaline activated aluminosilicate precursors, also known as geopolymers, have been tested as a cementing material alternative to Portland cement, due to their high mechanical characteristics and durability. BF slag mortar activated with Olive-stone Biomass Ash (OBA) presented a lower zeolite content and average pore diameter compared to commercial industrial reagents and other processes that result in high CO_2 emissions. Recent studies on the improvement of BF slag use in cement production have been carried out in both Europe and China. Due to the increasing amount of industrial by-products, linked to the increasing steel production, an investigation of the production of a novel green cement containing superfine particles with high volume fly ash and BF slag addition has been performed [57]. This novel green cement can present significant properties, such as better mechanical properties with respect to commercial blended cement as well as better hydration properties than Ordinary Portland Cement (OPC).

To sum up, the use of BF slag is mainly devoted to replace cement in concrete, but steel slags are mostly used as a filler material in embankment construction, due to their relatively low hydraulicity and problems with their volumetric expansion. However, recent progresses in the slag quenching process have also allowed the improvement of steel slag properties. In particular, BOF slag, EAF slag and LF slag, currently used in road construction, asphalt concrete, agricultural fertiliser and soil improvement, can be valuable materials for cement clinker production. This potential use can result in environmental and economic advantages. Recent studies have been focused on the challenge aiming at using steel slags as cement replacement and aggregate in cement concrete. To this purpose, due to the low cementitious ability of steel slags in concrete and the requirement of their activation, suitable aging/weathering and treatments have been carried out for improving the hydrolyses of free-CaO and MgO, aiming to mitigate their instability [58]. In addition, the use of steel slag aggregates with wastewater in concrete has been shown to be possible without any significant deterioration of fresh and hardened concrete properties [59]. On this subject, a comparative study, evaluating the mechanical properties of concrete containing EAF oxidising slag, steel slag and GBFS, has been performed [60]. Replacing cement with EAF oxidising slag has produced the delayed of the hydration reaction at early ages, without significant issues in setting time, shrinkage or strength development. Additionally, research was carried out to adapt the chemistry and mineralogy of BOF slags to improve the hydraulic properties, by the reduction of iron oxides [61] and the addition of raw materials as sand and clay followed by heating and a rapid cooling to gain hydraulic active phases.

Another important by-product generated in the iron- and steelmaking processes, the BF flue dust, mainly contains significant quantities of iron oxides and coke fines. Recently, its potential use in replacing the traditional fuel and raw materials in cement production in India has been studied [62]. The application of magnetic separation to reduce the iron content in the flue dust has been evaluated. Although the results have shown that it does not effectively segregate the iron in the flue dust and the energy content does not increase, the cost analysis has shown that flue dust can be effectively used by the cement industry. This can produce advantages for both steel and cement industries.

In the last few years, other by-products coming from different sectors (e.g., steel fibre, asphalt, slag, asbestos, lead, dry sludge, wet sludge, fly ash, bagasse ash, red mud, plastic, glass etc.) have been tested for concrete preparation [63]. These tests have included compressive strength, flexural strength and slump value, which aim at finding out the most suitable by-product to replace natural materials.

5. By-Products Reuse in Road Construction

Industrial by-products can also be used in road construction with good results without compromising the quality and the performance of the road and with the advantage of reduction of their disposal to the landfills. In this regard, the recycling of some industrial by-products to replace conventional natural aggregates can be used for producing hydraulically bound mixtures for road foundations [64,65]. Recent achievements have been shown by combining Foundry Sands (FS), EAF steel slags and bottom ash from Municipal Solid Waste Incineration (MSWI) in five different proportions to be applied in road foundations [66]. Additionally, in Asian countries, such as Vietnam, under local legislation, steel slags have been considered as a solid waste to be processed and landfilled. However, over the last few years, this concept has been revised and now steel slags are considered as a normal or non-deleterious solid waste. This new consideration has paved the way to important studies aiming at steel slag reuse in the construction sector, as a replacement for mineral aggregate, in Hot Mix Asphalt (HMA) [67]. In particular, two HMA mixtures using steel slag have passed the Marshall stability and flow test requirements. In addition, its skid resistance for the surface course has satisfied the national specification for asphalt. Finally, the pavement sections with the surface course of steel slag HMA has showed a significant higher modulus than the conventional one. Only the roughness of the paved surface course has not met the requirement of the specification.

Recent studies on the recycling by-products in road construction have taken also into account other materials, associated with steel by-products. On this subject, aramid fibre, a synthetic fibre chemically produced through the reaction between amine group and carboxylic acid group, is currently applied as a reinforced material to improve the asphalt mixtures performance [68]. The replacement of natural coarse aggregate with EAF steel slag in the asphalt mixture reinforced by aramid fibre has been assessed. The results have shown that asphalt layer thickness and transportation costs have been reduced. In addition, the immersion of steel slag aggregate in water for 6 months can reduce (by 68%) the content of free lime and free magnesia, in order to avoid the expansion volume due to their characteristics. A further combination, including steel slag and bottom ash, has been studied as aggregate in asphalt pavement. After the characterisation of physical, chemical and morphological features, a comparison of the bottom ash and the steel slag with the conventional granite aggregate was performed. The results have highlighted that ash has presented weaker characteristics in terms of strength than the steel slag; on the other hand, steel slag, due to its content of iron oxide, has shown to be much stronger than granite. In addition, lower silica content in steel slag and bottom ash has shown to be potentially more resistant to moisture damage than granite [69]. BF slag has also been taken into account in recent studies as an alternative substitute of natural crushed aggregate. To this purpose, three asphalt concrete types, with different maximum aggregate particle size and one stone mastic asphalt mixture for low noise surface layers, have been selected [70]. It has been found that using BF slag aggregates in asphalt mixtures does not influence the quality or durability of the mixture. However, in some cases, it could even improve its properties. Due to its physical properties and mineralogical and chemical composition, BF slag can be used as granular aggregate in the production of HMA. The effect on the resistance of HMA, after the replacement of the coarse fraction of a natural aggregate with BF slag, has been evaluated by performing specific tests. Although significant enhancement in the properties of the HMA mixture has been detected, when limestone is totally replaced, the adhesive properties of the asphalt-aggregate system will be worse [71].

The improvement of the reuse efficiency of steel slag through the composition adjustment and activation of steel slag has been studied in China. This has resulted in an optimal slag-based composite with improved cementation efficiency, in the suppression of the swelling potential and in strength improvement. Furthermore, the Chinese standard for first-class road/highway has been satisfied [72]. A recent work concerns the use of BOF slag as coarse aggregate as well as the Blast Furnace Dust (BFD) as a fine aggregate for manufacturing asphalt hot mixes for pavements [73]. The physical characteristics and the susceptibility to water and plastic deformation of each type of mixture have been assessed. The use of BOF slag and the feasibility of BFD use as fine aggregate resulted useful for partially replacing conventional aggregates in road paving. Finally, waste foundry sand and BF steel slag have been tested in an Indian study for gradation, specific gravity, morphology, chemical composition and compaction as well as engineering properties, shear strength and permeability [74]. In addition, compaction behaviour of these analysed by-products has resulted to be similar to granular soils and they have been tested for fill applications. Tests on leachate behaviour has allowed assessing the environmental impact of their use.

6. By-Products Use as Liming and Amending Materials

The use of by-products from the steel sector for soil amendment is a consolidate practice in some countries, not only in Europe but also in other worldwide countries. In particular, over the last few decades, the use of steel slags as a liming material to raise the pH in acidic soils and to improve the physical properties of soft soils have been deepened and consolidated [75]. Recently, a study based on the application of BOF slag to alkaline soils, affected by excess sodium, has been carried out through lysimeter trials, by producing significant results in decreasing the exchangeable sodium content of saline sodic soil irrigated with saline water. The observed effects of BOF slag application have been mainly related to the improvement of the yields, thanks to the reduction of the negative effect of sodium [76]. An example of these achieved results is shown in Figure 5. Furthermore, the assessment of the technical and economic viability of a slag treatment plant has been carried out in order to obtain an amendment material to be sold in the fertiliser market [77]. The high porosity and large surface area of steel slags make them successful materials to also be used in marine environment for coral reef repairing [78] and for building artificial reefs. In addition, they can be used for H_2S and metalloids adsorption in marine environments [79].

Figure 5. Improvement of the tomato yields, due to the reduction of the negative effect of sodium after BOF slag application.

However, due to the content of trace amounts of heavy metals, the application of the BOF slag as liming material need to be carefully analysed. Among different metals that can be present in the slags, chromium (Cr), under environmental conditions, exists in two stable oxidation states, +III and +VI. CrIII is an essential nutrient, while CrVI is highly toxic. In soils, soluble CrIII is oxidised to CrVI by manganese (hydr)oxides (MnO_2). Due to its liming properties, BOF slag can reduce the CrVI, while adding synthetic $Mn^{IV}O_2$ promotes oxidation of CrIII. Generally, the oxidation risk of CrIII present in BOF slag to CrVI, promoted by MnO_2 present in the soil, is very low, due the low solubility of CrIII in soil [80]. Recent studies have shown that the Linz-Donawitz (LD) converter slag applied as amending material at a rate of 2.0×10^3 kg/ha in submerged rice cropping systems has significantly ($p < 0.05$) allowed the increase of grain yield by 10.3–15.2% [81].

In addition, CH_4 emissions have been mitigated by 17.8–24.0%, and inorganic as concentrations in grain have decreased by 18.3–19.6%. Furthermore, higher yield (due to the increase of photosynthetic rates) has been achieved, with the increase of nutrients availability to the rice plant. Finally, the observed decrease of CH_4 emissions could be due to the higher Fe availability in the slag amended soil. In this context, it works as an alternate electron acceptor by suppressing CH_4 emissions. The LD slag is also an amendment material in decreasing the arsenic uptake by rice. This effect could be due to the more Fe-plaque formation adsorbing more arsenic and the competitive inhibition of arsenic uptake with higher availability of Si [81]. On the other hand, the effect of iron materials from the casting industry on the arsenic mobility in two soils by a long-term (about 100 days) flooded soil incubation experiment has been evaluated [82]. In particular, immobilising arsenic in soils can be achieved using iron materials, resulting in the reduction of arsenic concentration in rice grains.

7. Reuse of Other By-Products

Other by-products, although produced in lower quantities during the iron- and steelmaking routes, are currently studied in order to be reused in a sustainable way. Among them, mill scale is a by-product formed during the hot rolling process. It can be a potential material used as Bipolar plates (BPP), a component of Proton Exchange Membrane Fuel Cells (PEMFC). As mill scale presents high iron content, it can be a source for current collector in BPP. This can allow providing a contribute in decreasing the overall cost of PEMFC-based fuel cell systems. For this application, the mill scale powder is sieved, mechanically alloyed with the carbon source and pressed under inert gas atmosphere. Through the analysis by XRD, optical microscopy, scanning electron microscopy and micro-hardness measurement, the study of the of powder particles structural changes has been performed [83]. The achieved results have displayed the potential application of the mill scale to BPP. Nevertheless, these results should be confirmed by further investigations and evaluations in the next future.

Concerning solid and gaseous steel by-products, their potential use as raw materials and reducing gases has been studied in order to be used for synthesising rich iron bearing products like iron powder.

For producing this value-added product, a chemical reduction technique is suitable through the use of steel by-products. The optimisation of the process parameters of the applied techniques aims to produce pure iron powders [84]. Furthermore, by applying the Sequential Chemical Extraction (SE), analysis information on the composition of solid steel processing by-products can be provided, resulting in the possibility to their classification process to improve the environmental protection. In particular, SE can provide more refined classification through information for potential reusing and then reducing hazardous materials. On this subject, the distribution of potentially toxic elements such as zinc, lead and copper between sensitive and immobile phases has been obtained [85].

Further by-products that can be efficiently reused include fly ash, coming from burning pulverised coal in electric power generating plants, and GGBS, obtained by quenching molten BF slag to produce a glassy, granular product, then dried and ground into a fine powder. Over the last few years, the use of fly ash bricks has increased, due to their long-term performance and their good mechanical and durability properties. Replacing the 50% with fly ash can improve the strength of concrete blocks, and can allow to achieve the maximum compressive strength and split tensile strength. Furthermore, the addition of 30% of glass powder can increase the compressive and flexural strength [86].

The mix of Steel Furnace Slag (SFS), Coal Wash (CW) and Rubber Crumbs (RC) have been recently tested for achieving an energy-absorbing capping layer with the same or higher properties compared to conventional subballast [87]. The analysis of seven parameters (i.e., gradation, permeability, peak friction angle, breakage index, swell pressure, strain energy density and axial strain under cyclic loading) have been carried out. Results have shown that a mixture with SFS:CW = 7:3 and 10% RC (63% SFS, 27% CW, and 10% RC) represents the best mixture for subballast.

Refractory materials play a primary role in the steel sector, as they are present in the main furnace of the iron- and steelmaking routes, from the BF to the Casting Machine, including furnaces, converters, vessels, nozzles etc. The current internal recycling of spent refractories includes their use as slag formers or conditioners or the partial substitution of raw materials in the mixtures for new refractories. An example of the use of spent refractories as slag conditioner and, in particular, in order to decrease the slag power to corrode the ladle refractories is provided in Figure 6.

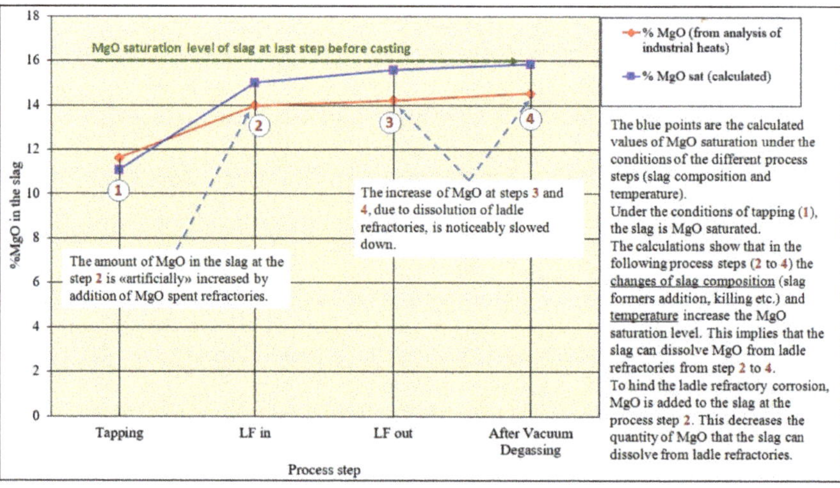

Figure 6. Use of spent refractories for slag conditioning with the aim to decrease the slag power to dissolve ladle refractories (comparison between calculated and experimental values. The experimental values have been obtained from an Italian steel plant).

As far as this last aspect is concerned, the Nippon Steel & Sumitomo Metals Corporation has developed a virtuous model about a method resulting in a recycled material to be used for on-site additions to monolithic refractories or for concretes [88]. Furthermore, the strategic alliances between steel producers, refractory recyclers and refractory manufacturers paved the way to a circular economy approach, also favouring the IS implementation [89]. This also provides the possibility to the external recycling of refractories, such as in the glass and cement sectors [90]. On this subject, the steel sector has been recently committed to the mutual exchange of materials among a steelmaking plant, vendors and other business partners in a network of industrial companies.

8. Conclusions

The performed study concerns the analysis of potential and future applications of iron- and steelmaking by-products, by highlighting both positive and critical aspects as well as the cooperation actions with other sectors, according to the principles of the IS. On this purpose, some significant examples have been provided.

The results discussed in the paper showed that, in order to improve the reuse and recovery rates, it is fundamental to increase the quality of the by-products recovered. New technological solutions need to be developed and implemented, aiming at achieving higher by-product qualities, in order to increase their reuse in an environmental and economic sustainable way. This results in approaching the "zero-waste" goal in the steel sector as well as a viable way for saving natural resources and reducing the environmental impact of the production processes. In addition, replacing natural resources with by-products can also allow saving energy and achieving higher energy efficiency in the production processes. For this reason, the future of the steel industry is closely linked not only with its innovation and the implementation of new technological solutions, but also with the reduction of its negative impacts, making its processes cleaner. On this subject, the steel sector should continue its path to the circular economy, by increasing its by-products reuse in a sustainable way. This includes the reduction of natural resources exploitation, such as CO_2 and pollution reductions, use of alternative raw material with less environmental impact as well as the control of existing impacts.

Author Contributions: Conceptualisation, T.A.B. and V.C.; methodology, T.A.B., V.C., D.A. and R.P.; validation, H.G., S.R., A.M. and U.M; formal analysis, A.M.; investigation, T.A.B. and V.C.; resources, T.A.B. and V.C.; writing—original draft preparation, T.A.B. and V.C.; writing—review and editing, A.M., D.A., S.R. H.G., R.P. and U.M.; supervision, U.M. and R.P.; project administration, V.C.; funding acquisition, V.C. All authors have read and agreed to the published version of the manuscript.

Funding: This research was funded by the European Union through the Research Fund for Coal and Steel (RFCS), Grant Agreement No 839227.

Acknowledgments: The work described in this paper was developed within the project entitled "Dissemination of results of the European projects dealing with reuse and recycling of by-products in the steel sector", REUSteel) (REUSteel GA 839227), which has received funding from the Research Fund for Coal and Steel of the European Union. The sole responsibility of the issues treated in this paper lies with the authors; the Commission is not responsible for any use that may be made of the information contained therein.

Conflicts of Interest: The authors declare no conflict of interest.

Abbreviations

BF	Blast Furnace
BFD	Blast Furnace Dust
BOF	Basic Oxygen Furnace
BPP	Bipolar plates
BTX	benzene, toluene and xylene
CCU	Carbon Capture, Storage and Usage
CDA	Carbon Direct Avoidance
CE	Circular Economy
COG	Coke Oven Gas
CSP	Concentrated Solar Power Plants
CW	Coal Wash
DRI	Direct Reduced Iron
EAF	Electric Arc Furnace
FS	Foundry Sands
GGBS	Ground Granulated Blast furnace Slag
HMA	Hot Mix Asphalt
IF	Induction Furnace
IS	Industrial Symbiosis
LD	Linz-Donawitz
LF	Ladle Furnace
MK	metakaolin
MSWI	Municipal Solid Waste Incineration
OBA	Olive-stone Biomass Ash
OPC	Ordinary Portland Cement
PCI	Pulverized Coal Injection
PEMFC	proton exchange membrane fuel cells
PI	Process Integration
RC	Rubber Crumbs
SCAP	Slag Carbon Arrestor Process
SCC	Self-Compacting Concretes
SE	Sequential Chemical Extraction
SEP	Sequential Extraction Procedure
SFS	Steel Furnace Slag
TCE	Trichloroethylene
TES	Thermal Energy Storage
WS	Water-Spray
XRD	X-Ray Diffraction

References

1. EU Commission. Directive 2008/98/EC of the European Parliament and of the Council of 19 November 2008 on Waste and Repealing Certain Directives (Waste Framework Directive, R1 Formula in Footnote of Attachment II). Available online: http://eur-lex.Europa.Eu/lexuriserv (accessed on 4 March 2020).
2. Worldsteel Association. Steel Industry Co-Products. Available online: https://www.worldsteel.org/en/dam/jcr:1b916a6d-06fd-4e84-b35d-c1d911d18df4/Fact_By-products_2018.pdf (accessed on 20 January 2020).
3. Grillo, F.F.; Coleti, J.; Espinosa, D.C.R.; Oliveira, J.R.D.; Tenório, J.A.S. Zn and Fe recovery from electric arc furnace dusts. *Mater. Trans.* **2014**, *55*, 351–356. [CrossRef]
4. World Steel Association. *Energy Use in the Steel Industry*; World Steel Association: Brussels, Belgium, 2014.
5. Yao, S.; Wu, S.; Song, B.; Kou, M.; Zhou, H.; Gu, K. Multi-objective optimization of cost saving and emission reduction in blast furnace ironmaking process. *Metals* **2018**, *8*, 979. [CrossRef]
6. WorldSteel Association. Sustainable Steel-Indicators 2017 and the Future 2017. Available online: https://www.worldsteel.org/en/dam/jcr:938bf06f-764e-441c-874a-057932e06dba/Sust_Steel_2017_update0408.pdf (accessed on 20 January 2020).

7. Barella, S.; Bondi, E.; Di Cecca, C.; Ciuffini, A.F.; Gruttadauria, A.; Mapelli, C.; Mombelli, D. New perspective in steelmaking activity to increase competitiveness and reduce environmental impact. *La Met. Ital.* **2014**, *16*, 31–40.

8. Matino, I.; Colla, V.; Cirilli, F.; Kleimt, B.; Unamuno Iriondo, I.; Tosato, S.; Baragiola, S.; Klung, J.-S.; Quintero, B.P.; De Miranda, U. Environmental impact evaluation for effective resource management in EAF steelmaking. *Met. Ital.* **2017**, *10*, 48–58.

9. Smol, M. Towards zero waste in steel industry: Polish case study. *J. Steel Struct. Constr.* **2015**, *1*, 2472-0437. [CrossRef]

10. Ning, G.; Zhang, B.; Liu, C.; Li, S.; Ye, Y.; Jiang, M. Large-scale consumption and zero-waste recycling method of red mud in steel making process. *Minerals* **2018**, *8*, 102. [CrossRef]

11. Sarkar, S.; Mazumder, D. Solid waste management in steel industry-challenges and opportunities. *Eng. Technol. Int. J. Soc. Behav. Educ. Econ. Bus. Ind. Eng.* **2015**, *9*, 978–981.

12. Bianco, L.; Porisiensi, S. Trasformazione da lineare a circolare del processo eaf. Esperienza in ferriere nord spa: IL caso della scoria siviera e dei carboni. *La Met. Ital.* **2016**, *108*, 1–26.

13. Ciftci, B. Global steel industry: Outlook challenges and opportunities. In Proceedings of the 5th International Steel Industry & Sector Relations Conference, Istanbul, Turkey, 20 April 2017; Available online: https://www.worldsteel.org/en/dam/jcr:d9e6a3df-ff19-47ff-9e8ff8c136429fc4/International+Steel+Industry+and+Sector+Relations+Conference+Istanbul_170420.pdf (accessed on 4 March 2020).

14. Ansari, N.A. Innovation through Recycling/Minimizing Waste. In Proceedings of the Innovation Forum, Oxford, UK, 7 February 2017.

15. Lombardi, D.R.; Laybourn, P. Redefining industrial symbiosis: Crossing academic–practitioner boundaries. *J. Ind. Ecol.* **2012**, *16*, 28–37. [CrossRef]

16. Rossetti di Valdalbero, D. *The Future of European Steel Innovation and Sustainability in a Competitive World and EU Circular Economy*; European Commission: Bruxelles, Belgium, 2017.

17. Yang, G.C.; Chuang, T.-N.; Huang, C.-W. Achieving zero waste of municipal incinerator fly ash by melting in electric arc furnaces while steelmaking. *Waste Manag.* **2017**, *62*, 160–168. [CrossRef]

18. ArcelorMittal. Steel: Stakeholder Value at Every Stage, Corporate Responsibility 2013. Available online: https://s3-us-west-2.amazonaws.com/ungc-production/attachments/83331/original/Full_online_version_17.5.14.pdf?1400857618 (accessed on 20 January 2020).

19. Matino, I.; Colla, V.; Branca, T.A.; Romaniello, L. Optimization of by-products reuse in the steel industry: Valorization of secondary resources with a particular attention on their pelletization. *Waste Biomass Valorization* **2017**, *8*, 2569–2581. [CrossRef]

20. Karayannis, V. Development of extruded and fired bricks with steel industry byproduct towards circular economy. *J. Build. Eng.* **2016**, *7*, 382–387. [CrossRef]

21. Gutierrez, A.; Miró, L.; Gil, A.; Rodríguez-Aseguinolaza, J.; Barreneche, C.; Calvet, N.; Py, X.; Fernández, A.I.; Grágeda, M.; Ushak, S. Advances in the valorization of waste and by-product materials as thermal energy storage (TES) materials. *Renew. Sustain. Energy Rev.* **2016**, *59*, 763–783. [CrossRef]

22. Matino, I.; Branca, T.A.; Colla, V.; Fornai, B.; Romaniello, L. Assessment of treatment configurations through process simulations to improve basic oxygen furnace slag reuse. *Chem. Eng. Trans.* **2017**, *61*, 529–534.

23. Matino, I.; Branca, T.A.; Fornai, B.; Colla, V.; Romaniello, L. Scenario analyses for by-products reuse in integrated steelmaking plants by combining process modeling, simulation, and optimization techniques. *Steel Res. Int.* **2019**, *90*, 1900150. [CrossRef]

24. Sellitto, M.A.; Murakami, F.K. Industrial symbiosis: A case study involving a steelmaking, a cement manufacturing, and a zinc smelting plant. *Chem. Eng. Trans.* **2018**, *70*, 211–216.

25. Jorge, J. Secondary zinc as part of the supply chain and the rise of EAF dust recycling. In Proceedings of the 19th Zinc & its Markets Seminar, Helsinki, Finland, 6 May 2015.

26. Brunelli, K.; Dabalà, M. Ultrasound effects on zinc recovery from EAF dust by sulfuric acid leaching. *Int. J. Miner. Metall. Mater.* **2015**, *22*, 353–362. [CrossRef]

27. Omran, M.; Fabritius, T.; Heikkinen, E.-P. Selective zinc removal from electric arc furnace (EAF) dust by using microwave heating. *J. Sustain. Metall.* **2019**, *5*, 331–340. [CrossRef]

28. Porzio, G.F.; Colla, V.; Fornai, B.; Vannucci, M.; Larsson, M.; Stripple, H. Process integration analysis and some economic-environmental implications for an innovative environmentally friendly recovery and pre-treatment of steel scrap. *Appl. Energy* **2016**, *161*, 656–672. [CrossRef]

29. Davydenko, A.; Karasev, A.; Glaser, B.; Jönsson, P. Direct reduction of Fe, Ni and Cr from oxides of waste products used in briquettes for slag foaming in EAF. *Materials* **2019**, *12*, 3434. [CrossRef]

30. Varga, T.; Bokányi, L.; Török, T.I. On the aqueous recovery of zinc from dust and slags of the iron and steel production technologies. *Int. J. Metall. Mater. Eng.* **2016**, *2*. [CrossRef]

31. Andersson, A.; Andersson, M.; Mousa, E.; Kullerstedt, A.; Ahmed, H.; Björkman, B.; Sundqvist-Ökvist, L. The potential of recycling the high-zinc fraction of upgraded BF sludge to the desulfurization plant and basic oxygen furnace. *Metals* **2018**, *8*, 1057. [CrossRef]

32. Fick, G.; Mirgaux, O.; Neau, P.; Patisson, F. Using biomass for pig iron production: A technical, environmental and economical assessment. *Waste Biomass Valor.* **2014**, *5*, 43–55. [CrossRef]

33. Suopajärvi, H.; Umeki, K.; Mousa, E.; Hedayati, A.; Romar, H.; Kemppainen, A.; Wang, C.; Phounglamcheik, A.; Tuomikoski, S.; Norberg, N. Use of biomass in integrated steelmaking–status quo, future needs and comparison to other low-Co2 steel production technologies. *Appl. Energy* **2018**, *213*, 384–407. [CrossRef]

34. Kalde, A.; Demus, T.; Echterhof, I.T.; Pfeifer, I.H. Determining the reactivity of biochar-agglomerates to replace fossil coal in electric arc furnace steelmaking. In Proceedings of the 23rd European Biomass Conference and Exhibition, Vienna, Wien, Austria, 1–4 June 2015.

35. Devasahayam, S. Opportunities for simultaneous energy/materials conversion of carbon dioxide and plastics in metallurgical processes. *Sustain. Mater. Technol.* **2019**, *22*, e00119. [CrossRef]

36. Todoschuk, T.; Giroux, L.; Ng, K.W. *Developments of Biocarbon for Canadian Steel Production*; Canadian Carbonization Research Association: Hamilton, ON, Canada, 2016.

37. JISF's. Commitment to a Low Carbon Society. In *Activities of Japanese Steel Industry to Combat Global Warming*; Japan Iron and Steel Federation: Tokyo, Japan, 2018.

38. EUROSLAG—The European Association Representing Metallurgical Slag Producers and Processors. Available online: https://www.euroslag.com/products/statistics/statistics-2016/ (accessed on 20 January 2020).

39. Branca, T.A.; Colla, V.; Valentini, R. A way to reduce environmental impact of ladle furnace slag. *Ironmak. Steelmak.* **2009**, *36*, 597–602. [CrossRef]

40. Fisher, L.V.; Barron, A.R. The recycling and reuse of steelmaking slags—A review. *Resour. Conserv. Recycl.* **2019**, *146*, 244–255. [CrossRef]

41. Matino, I.; Colla, V.; Baragiola, S. Internal slags reuse in an electric steelmaking route and process sustainability: Simulation of different scenarios through the EIRES monitoring tool. *Waste Biomass Valorization* **2018**, *9*, 2481–2491. [CrossRef]

42. Rodgers, K.; McLellan, I.; Cuthbert, S.; Masaguer Torres, V.; Hursthouse, A. The potential of remedial techniques for hazard reduction of steel process by products: Impact on steel processing, waste management, the environment and risk to human health. *Int. J. Environ. Res. Public Health* **2019**, *16*, 2093. [CrossRef]

43. Ma, J.; Zhang, Y.; Hu, T.; Sun, S. Utilization of converter steel slag by remelting and reducing treatment. In *IOP Conference Series: Materials Science and Engineering*; IOP Publishing: Bristol, UK, 2018; p. 022088.

44. Gomes, H.I.; Funari, V.; Mayes, W.M.; Rogerson, M.; Prior, T.J. Recovery of Al, Cr and V from steel slag by bioleaching: Batch and column experiments. *J. Environ. Manag.* **2018**, *222*, 30–36. [CrossRef]

45. Spanka, M.; Mansfeldt, T.; Bialucha, R. Sequential extraction of chromium, molybdenum, and vanadium in basic oxygen furnace slags. *Environ. Sci. Pollut. Res.* **2018**, *25*, 23082–23090. [CrossRef]

46. Baalamurugan, J.; Ganesh Kumar, V.; Govindaraju, K.; Naveen Prasad, B.; Bupesh Raja, V.; Padmapriya, R. Slag-based nanomaterial in the removal of hexavalent chromium. *Int. J. Nanosci.* **2018**, *17*, 1760013. [CrossRef]

47. El-Azim, H.A.; Seleman, M.M.E.-S.; Saad, E.M. Applicability of water-spray electric arc furnace steel slag for removal of cd and mn ions from aqueous solutions and industrial wastewaters. *J. Environ. Chem. Eng.* **2019**, *7*, 102915. [CrossRef]

48. Gonzalez-Olmos, R.; Anfruns, A.; Aguirre, N.V.; Masaguer, V.; Concheso, A.; Montes-Morán, M.A. Use of by-products from integrated steel plants as catalysts for the removal of trichloroethylene from groundwater. *Chemosphere* **2018**, *213*, 164–171. [CrossRef] [PubMed]

49. Sellner, B.M.; Hua, G.; Ahiablame, L.M.; Trooien, T.P.; Hay, C.H.; Kjaersgaard, J. Evaluation of industrial by-products and natural minerals for phosphate adsorption from subsurface drainage. *Environ. Technol.* **2019**, *40*, 756–767. [CrossRef]

50. Sun, Y.; Zhang, Z.; Liu, L.; Wang, X. Heat recovery from high temperature slags: A review of chemical methods. *Energies* **2015**, *8*, 1917–1935. [CrossRef]

51. McDonald, I.; Werner, A. Dry slag granulation with heat recovery. In Proceedings of the AISTech—Iron and Steel Technology Conference Proceedings (Association for Iron and Steel Technology, AISTECH), Indianapolis, IN, USA, 5–8 May 2014; Volume 1, pp. 467–473.

52. Oge, M.; Ozkan, D.; Celik, M.B.; Gok, M.S.; Karaoglanli, A.C. An overview of utilization of blast furnace and steelmaking slag in various applications. *Mater. Today Proc.* **2019**, *11*, 516–525. [CrossRef]

53. Motz, H.; Ehrenberg, A.; Mudersbach, D. Dry solidification with heat recovery of ferrous slag. *Miner. Process. Extr. Metall.* **2015**, *124*, 67–75. [CrossRef]

54. He, M.; Li, B.; Zhou, W.; Chen, H.; Liu, M.; Zou, L. *Preparation and Characteristics of Steel Slag Ceramics from Converter Slag*; Springer: Berlin/Heidelberg, Germany, 2018; pp. 13–20.

55. Saranya, P.; Nagarajan, P.; Shashikala, A. *Eco-friendly Ggbs concrete: A state-of-the-art review. IOP Conference Series: Materials Science and Engineering*; IOP Publishing: Bristol, UK, 2018; p. 012057.

56. Miñano, I.; Benito, F.J.; Valcuende, M.; Rodríguez, C.; Parra, C.J. Improvements in aggregate-paste interface by the hydration of steelmaking waste in concretes and mortars. *Materials* **2019**, *12*, 1147. [CrossRef]

57. Wu, M.; Zhang, Y.; Liu, G.; Wu, Z.; Yang, Y.; Sun, W. Experimental study on the performance of lime-based low carbon cementitious materials. *Constr. Build. Mater.* **2018**, *168*, 780–793. [CrossRef]

58. Jiang, Y.; Ling, T.-C.; Shi, C.; Pan, S.-Y. Characteristics of steel slags and their use in cement and concrete—A review. *Resour. Conserv. Recycl.* **2018**, *136*, 187–197. [CrossRef]

59. Saxena, S.; Tembhurkar, A. Impact of use of steel slag as coarse aggregate and wastewater on fresh and hardened properties of concrete. *Constr. Build. Mater.* **2018**, *165*, 126–137. [CrossRef]

60. Lee, J.-Y.; Choi, J.-S.; Yuan, T.-F.; Yoon, Y.-S.; Mitchell, D. Comparing properties of concrete containing electric ARC furnace slag and granulated blast furnace slag. *Materials* **2019**, *12*, 1371. [CrossRef] [PubMed]

61. Tsakiridis, P.E.; Papadimitriou, G.D.; Tsivilis, S.; Koroneos, C. Utilization of steel slag for portland cement clinker production. *J. Hazard. Mater.* **2008**, *152*, 805–811. [CrossRef] [PubMed]

62. Baidya, R.; Kumar Ghosh, S.; Parlikar, U.V. Blast furnace flue dust co-processing in cement kiln–a pilot study. *Waste Manag. Res.* **2019**, *37*, 261–267. [CrossRef] [PubMed]

63. Babita, S.; Saurabh, U.; Abhishek, G.K.; Manoj, Y.; Pranjal, B.; Ravi, M.K.; Pankaj, K. Review paper on partial replacement of cement and aggregates with various industrial waste material and its effect on concrete properties. In *Recycled Waste Materials*; Springer: Berlin/Heidelberg, Germany, 2019; pp. 111–117.

64. Xiao, Z.; Chen, M.; Wu, S.; Xie, J.; Kong, D.; Qiao, Z.; Niu, C. Moisture susceptibility evaluation of asphalt mixtures containing steel slag powder as filler. *Materials* **2019**, *12*, 3211. [CrossRef]

65. Skaf, M.; Pasquini, E.; Revilla-Cuesta, V.; Ortega-López, V. Performance and durability of porous asphalt mixtures manufactured exclusively with electric steel slags. *Materials* **2019**, *12*, 3306. [CrossRef]

66. Pasetto, M.; Baldo, N. Re-use of industrial wastes in cement bound mixtures for road construction. *Environ. Eng. Manag. J.* **2018**, *17*, 417–426. [CrossRef]

67. Nguyen, H.Q.; Lu, D.X.; Le, S.D. *Investigation of using steel slag in hot mix asphalt for the surface course of flexible pavements. IOP Conference Series: Earth and Environmental Science, 2018*; IOP Publishing: Bristol, UK, 2018; p. 012022.

68. Alnadish, A.; Aman, Y. A study on the economic using of steel slag aggregate in asphalt mixtures reinforced by aramid fiber. *Arpn J. Eng. Appl. Sci.* **2018**, *13*, 276–292.

69. Jattak, Z.A.; Hassan, N.A.; Shukry, N.A.M.; Satar, M.K.I.M.; Warid, M.N.M.; Nor, H.M.; Yunus, N.Z.M. Characterization of industrial by-products as asphalt paving material. In *IOP Conference Series: Earth and Environmental Science, 2019*; IOP Publishing: Bristol, UK, 2019; p. 012012.

70. Vacková, P.; Kotoušová, A.; Valentin, J. Use of recycled aggregate from blast furnace slag in the design of asphalt mixtures. *Waste Forum* **2018**, *1*, 60–72.

71. Rondón-Quintana, H.A.; Ruge-Cárdenas, J.C.; Farias, M.M.D. Behavior of hot-mix asphalt containing blast furnace slag as aggregate: Evaluation by mass and volume substitution. *J. Mater. Civ. Eng.* **2018**, *31*, 04018364. [CrossRef]

72. Wu, J.; Liu, Q.; Deng, Y.; Yu, X.; Feng, Q.; Yan, C. Expansive soil modified by waste steel slag and its application in subbase layer of highways. *Soils Found.* **2019**, *59*, 955–965. [CrossRef]

73. López-Díaz, A.; Ochoa-Díaz, R.; Grimaldo-León, G.E. Use of bof slag and blast furnace dust in asphalt concrete: An alternative for the construction of pavements. *DYNA* **2018**, *85*, 24–30. [CrossRef]

74. Kumar, K.; Krishna, G.; Umashankar, B. Evaluation of waste foundry sand and blast furnace steel slag as geomaterials. In Proceedings of the Geo-Congress 2019: Geoenvironmental Engineering and Sustainability, Philadelphia, PA, USA, 24–27 March 2019.

75. Branca, T.A.; Pistocchi, C.; Colla, V.; Ragaglini, G.; Amato, A.; Tozzini, C.; Mudersbach, D.; Morillon, A.; Rex, M.; Romaniello, L. Investigation of (BOF) converter slag use for agriculture in Europe. *Rev. De Métallurgie Int. J. Metall.* **2014**, *111*, 155–167.

76. Pistocchi, C.; Ragaglini, G.; Colla, V.; Branca, T.A.; Tozzini, C.; Romaniello, L. Exchangeable sodium percentage decrease in saline sodic soil after basic oxygen furnace slag application in a lysimeter trial. *J. Environ. Manag.* **2017**, *203*, 896–906. [CrossRef]

77. Branca, T.A.; Fornai, B.; Colla, V.; Pistocchi, C.; Ragaglini, G. Application of basic oxygen furnace (bofs) in agriculture: A study on the economic viability and effects on the soil. *Environ. Eng. Manag. J. (Eemj)* **2019**, *18*, 1231–1244.

78. Mohammed, T.A.; Aa, H.; Ma, E.E.-A.; Khm, E.-M. Coral rehabilitation using steel slag as a substrate. *Int. J. Env. Prot.* **2012**, *2*, 1–5.

79. Asaoka, S.; Okamura, H.; Morisawa, R.; Murakami, H.; Fukushi, K.; Okajima, T.; Katayama, M.; Inada, Y.; Yogi, C.; Ohta, T. Removal of hydrogen sulfide using carbonated steel slag. *Chem. Eng. J.* **2013**, *228*, 843–849. [CrossRef]

80. Reijonen, I.; Hartikainen, H. Risk assessment of the utilization of basic oxygen furnace slag (BOFS) as soil liming material: Oxidation risk and the chemical bioavailability of chromium species. *Environ. Technol. Innov.* **2018**, *11*, 358–370. [CrossRef]

81. Gwon, H.S.; Khan, M.I.; Alam, M.A.; Das, S.; Kim, P.J. Environmental risk assessment of steel-making slags and the potential use of ld slag in mitigating methane emissions and the grain arsenic level in rice (oryza sativa l). *J. Hazard. Mater.* **2018**, *353*, 236–243. [CrossRef]

82. Suda, A.; Yamaguchi, N.; Taniguchi, H.; Makino, T. Arsenic immobilization in anaerobic soils by the application of by-product iron materials obtained from the casting industry. *Soil Sci. Plant Nutr.* **2018**, *64*, 210–217. [CrossRef]

83. Khaerudini, D.; Prakoso, G.; Insiyanda, D.; Widodo, H.; Destyorini, F.; Indayaningsih, N. Effect of graphite addition into mill scale waste as a potential bipolar plates material of proton exchange membrane fuel cells. In *Journal of Physics: Conference Series, 2018*; IOP Publishing: Bristol, UK, 2018; p. 012050.

84. Sista, K.S.; Dwarapudi, S. Iron powders from steel industry by-products. *Isij Int.* **2018**, *58*, 999–1006. [CrossRef]

85. Rodgers, K.J.; McLellan, I.S.; Cuthbert, S.J.; Hursthouse, A.S. Enhanced characterisation for the management of industrial steel processing by products: Potential of sequential chemical extraction. *Environ. Monit. Assess.* **2019**, *191*, 192. [CrossRef]

86. Sudharsan, N.; Palanisamy, T. A comprehensive study on potential use of waste materials in brick for sustainable development. *Ecol. Environ. Conserv.* **2018**, *24*, S339–S343.

87. Indraratna, B.; Qi, Y.; Heitor, A. Evaluating the properties of mixtures of steel furnace slag, coal wash, and rubber crumbs used as subballast. *J. Mater. Civ. Eng.* **2017**, *30*, 04017251. [CrossRef]

88. Madias, J. A review on recycling of refractories for the iron and steel industry. In Proceedings of the UNITECR 2017—15th Biennial Worldwide Congress, Santiago, Chile, 26–29 September 2017.

89. O'Driscoll, M. Recycling Refractories. Available online: http://imformed.com/wp-content/uploads/2017/07/IMFORMED-Refractory-Recycling-Glass-Int-Mar-2016.pdf (accessed on 20 January 2020).

90. Fasolini, S.; Martino, M. Recovery of Spent Refractories: How to Do It and Using Them as Secondary Raw Materials for Refractory Applications. Available online: http://www.indmin.com/events/download.ashx/document/speaker/8915/a0ID000000ZwxAQMAZ/Presentation (accessed on 20 January 2020).

 *metals*

Review

The Challenge of Digitalization in the Steel Sector

Teresa Annunziata Branca [1], Barbara Fornai [1], Valentina Colla [1,*], Maria Maddalena Murri [2], Eliana Streppa [2] and Antonius Johannes Schröder [3]

[1] Scuola Superiore Sant'Anna, TeCIP Institute, 56124 Pisa, Italy; teresa.branca@santannapisa.it (T.A.B.); barbara.fornai@santannapisa.it (B.F.)

[2] RINA CONSULTING—Centro Sviluppo Materiali S.p.A. (CSM), 00128 Roma, Italy; maria.murri@rina.org (M.M.M.); eliana.streppa@rina.org (E.S.)

[3] Sozialforschungsstelle, Technische Universität Dortmund, D-44339 Dortmund, Germany; antonius.schroeder@tu-dortmund.de

* Correspondence: valentina.colla@santannapisa.it; Tel.: +39-348-071-8937

Received: 27 January 2020; Accepted: 19 February 2020; Published: 21 February 2020

Abstract: Digitalization represents a paramount process started some decades ago, but which received a strong acceleration by Industry 4.0 and now directly impacts all the process and manufacturing sectors. It is expected to allow the European industry to increase its production efficiency and its sustainability. In particular, in the energy-intensive industries, such as the steel industry, digitalization concerns the application of the related technologies to the production processes, focusing on two main often overlapping directions: Advanced tools for the optimization of the production chain and specific technologies for low-carbon and sustainable production. Furthermore, the rapid evolution of the technologies in the steel sector require the continuous update of the skills of the industrial workforce. The present review paper, resulting from a recent study developed inside a Blueprint European project, introduces the context of digitalization and some important definitions in both the European industry and the European iron and steel sector. The current technological transformation is depicted, and the main developments funded by European Research Programs are analyzed. Moreover, the impact of digitalization on the steel industry workforce are considered together with the foreseen economic developments.

Keywords: digitalization; digital technologies; digital transformation; steel industry; digital skills

1. Introduction

Over the last few decades, the European steel sector has undergone relevant transformations. On one hand, this sector has been restructured and consolidated; on the other hand, highly technological production processes and products have been developed. These transformations have considerably impacted the workforce, resulting in numerical reduction and in professional profiles' evolution. In particular, digital transformation and Industry 4.0 contributed to both reducing the need for physical and often cumbersome and repetitive operations and to increasing the demand for highly skilled workforce. In addition, changes in patterns of recruitment as well as in work organization have occurred. In order to be more competitive, the steel sector aims at developing a highly qualified, specialized, and multi-skilled workforce. Nevertheless, due to the skills shortages, recruitment difficulties, and talent management issues, it is important to forecast, identify, and anticipate skill needs.

In order to introduce the current state of this technological transformation, it is important to define the context and to provide some necessary definitions. "Digitization" is defined as *"the action or process of digitizing; the conversion of analogue data (esp. in later use images, video, and text) into digital form"* [1], while the term "Digitalization" refers to the transformation of interactions communications, business functions, and business models into the digital ones. Through the use of "digital technologies" and digitized and natively digital data, digitalization aims at achieving revenue, improving business, replacing/transforming business processes, as well as at creating an environment for digital business. In addition, "digitalization" promotes the integration of digital technologies into areas of a business [2]. On the other hand, "automation" is defined as *"the use or introduction of automatic equipment in a manufacturing or other process or facility"* 1 as well as *"the technology by which a process or procedure is accomplished without human assistance"* [3].

In this context, Industry 4.0 concerns the interoperability, the decentralization of information, the real-time data collection, and the increased flexibility, that represent main aspects of the fourth industrial revolution. This process started with the first industrial revolution, characterized by the mechanization of production performed manually by hand. In this context, steam and water power were used for the mechanization of work [4]. In the second industrial revolution, the introduction of electricity in different processes took place. In particular, in the steel sector, this coincided with the invention of the production area, the improvement of transportation technologies, and the electrification of industrial processes. In addition, the Bessemer process and the open-heart furnace were introduced [5]. The third industrial revolution was characterized by the introduction of Information Technology (IT) and computer technology to automate processes, but still involved human aspects. The use of robots into the processes previously performed by humans and the development of the work based on optimization and the removal of production inefficiencies represent the main aspects [5]. The current industrial revolution, the fourth one, is based on new interconnected technologies in process operations. The term Industry 4.0 was first used in Germany based on initiative of the German Government's High Tech 2020 Strategy [6]. Compared to "Industry 3.0", in Industry 4.0, the machines work autonomously without, or with very limited, human intervention. In particular, the enhancement of automation and connectivity with Cyber Physical Systems (CPS) include smart machines, storage systems, and production facilities capable of autonomously exchanging information, triggering actions, and independently controlling each other. In addition, the Industrial Internet of Things (IIOT) can allow exchange of information provided by sensors that work in real time and transfer data to a local server or a cloud server, where the analysis of the data is performed through the development of predictive models. The final aims can be, for instance, product quality management throughout the entire production chain or early detection and forecasting of anomalies in the processes and prediction of residual lifetime of critical components by means of tools exploiting the data (often Big Data) captured by the sensors (according to the paradigm of Predictive Maintenance).

This review paper derives from an analysis that was performed in the context of a multinational initiative funded by the European Union through the Erasmus Plus framework. This project is devoted to the development of a Blueprint for "New Skills Agenda Steel", is entitled "Industry-driven sustainable European Steel Skills Agenda and Strategy (ESSA)", and has two main aims:

- The proactive identification of skill needs and demands for the construction appropriate training and curricula, including strategizing for the implementation of new vocational education content and pedagogies across the sector, within both companies and education and training institutions;
- The identification, development, and promotion of successful sectoral recruitment and upskilling schemes, and the development of some training tools for efficient management of knowledge fostering talent development and overcoming of recruitment difficulties.

The paper is organized as follows: Section 2 introduces the context of digitalization and provides some important definitions in the European industry and the European iron and steel sector. In Section 3, the main developments achieved within projects that received funding by European Research Programs are analyzed. Section 4 discusses the impact of digitalization on the steel industry workforce and the foreseen future economic developments. Finally, Section 5 provides some concluding remarks.

2. The Framework of the Industrial Digital Transformation

2.1. Digital Transformation in the European Industry

The European industry is deeply committed to integrate the digitalization concept into its production and organization in order to be more competitive in the globalization context. The application of digital technologies allows implementing new processes along the entire value chain, through manufacturing and sales to services. On this subject, digitalization represents a holistic approach covering all areas and functions of a company to exploit digital potentials and analyze each stage of its value chain. For this reason, it is important to underline that digitalization is not a simple transfer from "analogic" to digital data and documents. It is rather the networking between the business processes, the creation of efficient interfaces, and the integrated data exchange and management [7].

The digital transformation is the key aspect of the ongoing industrial revolution [8]. Some new Key Enabling Technologies (KETs) are represented by a. new generation of sensors, Big Data, Machine Learning (ML), Artificial Intelligence (AI), Internet-of-Things (IoT), Internet-of-Services, Mechatronics and Advanced Robotics, Cloud Computing, Cybersecurity, Additive Manufacturing, Digital Twin, and Machine to Machine (M2M) communication. The digital technologies could be ex novo applied to a new plant, or can be adapted to existing plants [9]. In order to achieve industrial production optimization, the application of these technologies aims at developing new skills and new competencies as well as new business models. This will result in the improvement of industrial competitiveness and efficiency, due also to a higher inter-connection and cooperation, sharing resources (e.g., plants, people, and information). On this subject, European industries are committed to addressing some important challenges:

- Ensuring a continuous responsiveness to fulfil the changing future demand and securing the market position;
- Preserving competitiveness with efficient processes, and cost and resources saving;
- Achieving higher product quality;
- Maximizing plant performance, by also minimizing maintenance and low capital lock-up;
- Planning a flexible production by guaranteeing timeliness of delivery [10].

These challenges can be reached through the real-time capability, interoperability, and the horizontal and vertical integration of production systems enabled by Information and Communications Technology (ICT) systems, representing the main features of Industry 4.0 [11]. Furthermore, in order to achieve a flexible production, it is important to also have flexible work, through the self-organization and multi-tasking skills, according to education and lifelong learning initiatives.

Industry 4.0 is based on an intelligent networking of machines, electrical equipment, and modern Information Technology (IT) systems allowing processes optimization and increased productivity of value creation chains [12]. Its strategy is based on intelligent factories exploiting the combination of embedded production system technologies with intelligent production and resulting in a new technological age. In this context, in a recent study, three distinct manufacturing systems in Industry 4.0 have been presented [13]. In the first one, an Intelligent Manufacturing System (IMS), also known as smart manufacturing, the employment of advanced manufacturing technologies, and information enable the optimization of the production process of goods and services. The second manufacturing environment is viewed as IoT-enabled manufacturing, relying on the exploitation of smart manufacturing objects (SMOs); the third major manufacturing is a cloud manufacturing. The intelligent systems involve the development of manufacturing systems that can be organized into a functional network.

Although different possible scenarios have been developed, it has recently estimated that the full shift to Industry 4.0 could need 20 years [14]. This long period can be due to the fact that digital technologies strongly vary with company size. In particular, only 36% of the surveyed European companies with 50–249 employees have implemented industrial robots in their production processes, compared to 74% of European companies with over 1000 employees [15]. However, larger companies have more possibilities to employ some IT/ICT specialists, but Small and Medium-sized Enterprises (SMEs) more intensively use mobile internet and social media (44%) [16].

Although the new digital technologies can make companies more efficient and productive, resulting in opening to new markets, some critical aspects on the future have been recently depicted. According to Pfeiffer [17], Industry 4.0 does not present only positive perspectives, but it can be considered as part of a newly emerging global production regime [18], particularly affecting the low-skilled works and repetitive tasks [19], such as manual operation of specialized machine tools, shot-cycle machine feeding, repetitive packaging tasks, monotonous monitoring tasks, and many warehousing and commissioning functions in logistics [20]. However, even if the digitalization may improve processes, the human experience cannot be replaced. In particular, the mining and metal sector in emerging economies needs to pay gap between highly skilled digital workforce and more traditional workforce [21]. On this subject, evolution in workforce skills are needed, in order to be in line with the industrial requirements. This includes not only attracting and recruiting new talents, but also re-skilling current employees with training programs, which can improve skills in different disciplines, such as science, technology, engineering, and mathematics. In this process, a key aspect is represented by continuous learning, also based on an interdisciplinary perspective. This will produce a future adaptable workforce contributing to increase the competitiveness of the companies as well as their own competitiveness and "appeal" in the work market. Furthermore, re-designing work processes aims at reducing the skill mismatch between jobs and employees and securing their jobs in the future [22].

The implementation of new digital technologies could produce significant improvement in workforce health and safety as well as energy efficiency and CO_2 utilization increase. In addition, innovative and more customized goods and services could be achieved [23]. In the global market and, consequently, in the global competition, the European manufacturing industry will have some market opportunities related to digital technologies. For instance, the 10-year China strategy plan, the *Made in China 2025* [24] strategy, aims at upgrading Chinese industrial base by focusing on 10 key industries. This will offer attractive opportunities for some European businesses in the short and medium term, about critical components, technology, and management skills.

In this context, the European manufacturing industry is leaning towards a greater new production system, increasingly flexible and tailored to the needs of customer. This evolution will preserve its global competitiveness thanks to high-value goods and high-quality products, as well as integration, digitalization, speed, flexibility, quality, efficiency, and security.

2.2. Digitalization in the European Steel Industry

The steel industry is a highly energy intensive industry. However, it is a modern, energy and CO_2 efficient sector, with high value-added production and niche products for the world market, based on an outstanding R&D network [25]. The European steel industry has an annual turnover of EUR 166 billion and it provides the 1.3% of EU GDP. Concerning the workforce, in 2015, it provided 328,000 direct jobs with an even greater number of dependent jobs [26].

The digitalization process in the European steel industry represents a pre-condition for Industry 4.0, as Industry 4.0 is much more than digitalization, but rather a paradigm/philosophy than a technology [27]. The current situation of digitalization in the steel sector is that its workforce is aging. An experienced workforce knows industrial processes very well, but it is less comfortable with digital tools and collaborative work and, sometimes, it is resistant to training and learning activities. For instance, in order to improve the culture of multigenerational digital innovation, the attempt of connection between the workforce under 30 with the older leadership team has been carried out within a Tata Steel site [21]. All over the world, including Europe, innovation, technology, quality, and highly skilled people are the basis for competitiveness [25]. In addition, preserving industrial knowledge and a skilled workforce is an important asset for the iron and steel sector [26]. Nevertheless, also in Europe, the digitalization process lacks uniformity. For instance, in the Czech Republic, the steel industry is still not in line with the other European countries in adopting new technologies and, in particular, in digital transformation, although recent investments have been done for modernizing production and reducing environmental impacts [28].

Due to the complexity of the production processes, the application of new technologies can result in the optimization of the entire steel production. By combining process automation, information technology, and connectivity, the digitalization of the steel production can be beyond a conventional automation of the industrial production. In the future, the intelligent combination of different tools, such as plant and laboratory experiments, physical modelling, and computational modelling will play a significant role in the digitalization process in the steel sector. For instance, recent advances in modeling steel continuous casting have been achieved. As this process involves many interacting phenomena, model verification and validation of model predictions represent the key factors. On this subject, each aspect of models needs to be verified with known solutions and validated with measurements to trust the predictions and improvements arising from modeling studies [29]. The digitalization in the steel production can not only produce quality, flexibility, and productivity, but also ensure the visibility of the real-time operational data and provide insight for a better and faster decision-making along the value chain [30]. Over the last few years, clear implementation of digital technologies has been made in the steel sector. However, the application of a decentralized, unmanned autonomous system, for assembling components, to the continuous process of steel, is difficult and expensive [31]. In addition, the application of digital systems in some production processes, such as Energy management and Water and Wastewater management, aims at achieving continuous improvements in the steel sector regarding quality, costs, energy consumption, and environmental performance. In this context, digital technologies help to adapt and to integrate innovative and emerging techniques to the steel production processes. For instance, novel pollution prevention and control techniques under development and new techniques addressing environmental issues can provide future economic or environmental benefits to this sector [32,33]. In addition, an example on modelling and prediction of the gas consumption to minimize ordered and supplied gas quantity error can be provided. On this subject, a linear regression and the genetic programming approach have been used. The achieved good results can contribute not only to steel production optimization, but they represent a methodological approach, which can be applied to other energy consumption optimizations [34]. Furthermore, as the steel production based on the EAF consumes a large amount of energy, better knowledge of the consumed energy within the EAF operations is fundamental. In a recent study, in order to predict the electric energy consumption during the EAF operation, two models have been considered, one based on linear regression and the second one on genetic programming [35].

The main challenges of the European steel industry on Industry 4.0 are related to legacy equipment, uncertainty about the impact on jobs, and issues of data protection/safety. In addition, results from a recent study [36] showed that the technical barriers are considered less important than the organizational ones. On one hand, internal management represents the driving force for implementing the Industry 4.0 projects, but on the other hand, technology and production are less crucial. Furthermore, the main possible explanations about lack of qualified personnel are the increased use of digital technologies, the lack of suitable educational programs, and the delays in training provision after the introduction of a technological innovation. Nevertheless, one of the main challenges of the European steel industry consists of attracting and retaining qualified personnel [5]. Nevertheless, it is difficult to integrate new technologies and processes among site workers, especially when it comes to older employees. In addition, there is a strong age gap between the workers currently employed and prospective employees creating knowledge transfer issues. Finally, there is a lack of investment in training and education from steelmaking companies and an insufficient amount of in-house training provided by companies.

3. European Research Activities on Digitalization in the Steel Sector

According to [37], the European steel industry faces important challenges due to cost pressure, regulatory requirements, as well as product and service requirements. For this reason, over the last few decades, it has been involved in several policy activities, R&D projects, and patents in the field of digitalization.

Initiatives related to Industry 4.0 also include the Smart Factory Working Group of the ESTEP Platform, founded in 2008 with the former name of "Intelligent Integrated Manufacturing," which published the first edition of a Roadmap for European Steel Manufacturing in 2009 with a vision up to 2020. The ESTEP Working Group covers a broader range of stakeholders and it consists of plant manufacturers and several European Universities and R &D Institutions. In 2018, a workshop on the concept and operational benefits of the Digital Twins in the steel sector was held in Charleroi [38].

3.1. Digitalization & Enabling Technologies

Digital technologies are applied in order to improve the flexibility and the reliability of process and to improve the product quality. In addition, they can be applied for monitoring and assessing the environmental performance of processes, improving control of production and auxiliary processes that have an environmental impact, and providing key performance indicators for resources efficiency [39,40]. Some enabling technologies according to the Use Cases of KETS [41] are:

- Internet of Things (IoT) system: The IoT refers to an inter-networking world where electronic sensors, actuators, or other digital devices are networked and connected with the purpose of collecting and exchanging data [42]. According to [43], the composition of an online monitoring system based on an IoT system architecture can be characterized of four layers: Sensing, network, service resource, and application layers. Such a proposed system has been implemented and demonstrated through a real, continuous steel casting production line and integrated with the TeamCenter platform.

- Big data Analytics and Cloud Computing: In the manufacturing industries, including the steel industry, the conventional database technology can have some difficulties in finishing the capture, storage, management, and analysis of large volumes of structured and unstructured data. Big data analytics is related to algorithms based on historical data identifying quality problems and reducing the product failures. On the steel products, the Big-Data solutions are currently used for quality monitoring and improvement. This technology uses new processing modes in order to obtain significant information from different data types, and, to understand them in-depth, gain insight, and make discoveries for precise decision making. An accurate prediction of surface on steel slab defects can be based on the online collected data from the production line and it is important for adjusting the process online as well as for reducing their occurrence. The main problem is that the samples for normal cases and defects are usually unbalanced. According to [44], a one-class Support Vector Machine (SVM) classifier based on online collected process data and environmental factors for only normal cases was proposed in order to predict the occurrence of defects for steel slabs. ML-based approaches can provide a relevant support in extracting useful information and relevant knowledge from the available data and enabling the development of data-driven models e.g., for a wide variety of applications, such as material properties prediction [45] and product defects detection and identification [46]. Cloud computing gives on-demand computing services with high reliability, scalability, and availability in a distributed environment. Thanks to this technology, everything is treated as a service (i.e., XaaS), e.g., SaaS (Software as a Service), PaaS (Platform as a Service), and IaaS (Infrastructure as a Service) [47].

- Robot-assisted production: This technology is based on the use of humanoid robots in order to perform operations, such as assembly and packaging. Due to an increasing demand for higher quality, faster delivery time, and reduction of cost in the manufacturing industry in the last few decades, automation and robotics have achieved more and more importance. For instance, if in the steelmaking plant, existing technologies are enhanced with robots and automation, an improvement of surface quality of the steel products could be achieved [48,49].

- Production line simulation: In the steel sector, approaches for the simulation optimization solution have been developed. In particular, investigating potential changes to the designs and operations is the aim of the development of decision support systems [50,51]. Novel numerical techniques, such as meshless methods in the simulation systems of the steel sector, have been also exploited. In [52], a rolling simulation system capable of simulating rolling of slabs and blooms, as well as round or square billets, in different symmetric or asymmetric forms in continuous, reversing, or combined rolling has been elaborated. Vertnik et al. [53] developed a meshless Local Radial Basis Function Collocation Method (LRBFCM) for the solution of three-dimensional (3D) turbulent molten steel flow and solidification under the influence of electromagnetic stirring (EMS) and they demonstrate its application to continuous casting process of steel billets.

- Self-Organizing Production: Such technology involves the automatic coordination of machines, leading to the optimization of their utilization and output. The self-organizing production is related to the decentral instead of central solutions. A new combination of resources, equipment, and personnel, based on a close interaction within them with a master computer, is included and an increase of the automation, leading to the real time control of production networks.

- CPS: Is a system where computation, networking, and physical processes are integrated. The physical processes are monitored and controlled by embedded computers and networks, with feedback loops where they affect computations and vice versa [54].

- Smart supply network: Better supply decisions are possible thanks to the monitoring of the entire supply network. Several factors and objectives have to be taken into consideration in a supply chain of a steel industry. The smart supply networks optimize the steelworks production processes from the beginning to the end of products by using models as part of the integrated supply chain.

- Vertical/Horizontal Integration: Horizontal integration concerns the integration between a resource and an information network within the value chain. Vertical integration is related to networked manufacturing systems within the intelligent factories of the future and personalized customer manufacturing [55].

- Predictive Maintenance: It allows repair prior to breakdown thanks to a remote monitoring of equipment. The combination of equipment monitoring together with intelligent decision methods implement the predictive maintenance techniques. In order to support decision-making and to assist steel companies to improve their competitiveness, ML and Data Mining techniques can be used to draw insights from the data and accurately predict results.

- Cyber Security: Such technology should be taken into consideration, especially for the Internet-based services. A procedural model for a Cyber-Security analysis based on reference architecture model Industry 4.0 and a VDI/VDE guideline 2182 are shown for the use case of a Cloud-based monitoring of the production in [56].

- Augmented Work, Maintenance, and Service: The operating guidance, the remote assistance, and the documentation are favored by applying the fourth dimension, which is the use of the augment reality. This is one of the most interesting enabling technology for companies, especially for improving the maintenance services. For instance, remote maintenance based on remote connection can be carried out by a service technician who is virtually connected. This results in travel costs and time saving, and with quick problem solving.

- Self-driving logistics vehicles: Such technology is based on completely automated transportation systems. The use of intelligent software to support intralogistics operations helps companies to improve processes and to make faster them. In the steelworks, the supply and the disposal of raw materials and the transport of intermediate products as well as the removal of finished products and the handling of by-products, for instance bulk material or slag, are very important. The use of an intelligent transport control system can allow one to plan and control the internal transport orders, resulting in an increase of productivity and service levels, cutting costs.

- Digitalization of knowledge management. Due to an increasing competitive market, the steel sector has been committed to facing significant challenges in the digitalization. Although this process has already started, further improvement can be achieved. On this subject, the knowledge and experience of the technical staff represents the basis of this improvements. The main barriers about the usage of this knowledge and experience are represented by their heterogeneous distribution over the individual staff members, human obliviousness, and knowledge erosion by leaving staff members.

3.2. Past and Ongoing Research Activities Funded by the Research Fund for Coal and Steel

In the European steel sector, the most important funding program for the technology development is the Research Fund for Coal and Steel (RFCS) [57]. This funding program concerns aspects related to the innovation in the digitalization of the steel industry as well.

The most active actors among research institutions and steel companies in such projects are: VDEh Betriebsforschungsinstitut BFI, Swerea MEFOS/KIMAB, RINA Consulting-Centro Sviluppo Materiali, Scuola Superiore Sant'Anna, Centre de Recherches Metallurgiques, ArcelorMittal, ThyssenKrupp, as well as Tata Steel and, to some extent, Gerdau and Voestalpine.

The plant manufacturers such as Primetals and SMS Siemag, followed by Danieli (also with its sister company Danieli Automation) are key players to patents and they are seldom involved in those projects.

This study identified 22 RFCS Projects covering aspects of digitalization in the steel industry starting from 2003, considering projects that are ongoing or completed. Figure 1 depicts the identified projects and the enabling technologies that these projects are developing or have developed: In the x-axis, the projects are reported; in the y-axis, the enabling technologies taken into consideration from the project are expressed in percentage.

Among the ongoing projects dealing with the Internet of Things (IoT) systems technology, TrackOpt aims at implementing an automatic ladle tracking system, in order to ensure the tracking of the product from steelmaking via casting to delivery by using a Multi-Objective Optimization (MOO) Framework and big data analytics as well as innovative acoustic sensors.

Quality4.0, NewTech4Steel, Cyberman4.0, PRESED, and Dromosplan are some examples of RFCS Projects related to the Big Data Analytics and Cloud Computing. An adaptive platform is being developed, allowing online analytics of large data streams to realize decisions on product quality and provide tailored information of high reliability in the running Quality 4.0 project. NewTech4Steel is an ongoing project focused on dedicated use cases in the steel industry exploiting all the technological and scientific possibilities offered by the latest technologies concerning data handling and data analysis. Advanced AI and ML-based analytics, also suitable for big data processing, are exploited for process performance monitoring. Cyberman4.0 and CyberPOS deal with CPS. In Cyberman4.0, big-data tools and techniques are applied to merge process and product data in order to forecast quality downgrading, faults, anomalies, and residual life of critical components in order to timely plan suitable and cost-effective maintenance interventions. Cyberman4.0 is also an important example of a Predictive Maintenance approach applied to the rolling area and aims at developing a so-called Integrated Maintenance Model 4.0 (IMM4.0). CyberPOS introduces simulation and verification tools as well as a new IT framework for establishing the feasibility, safety, and benefits of CPPS (Cyber Physical Production System) in the framework of "Steel Industry 4.0 Automation". Moreover, a CPS-based platform for facilities producing long steel products has been developed [58]. PRESED proposed a solution built around Big Data, Feature Extraction, ML, Analytics Server, and Knowledge Management in order to automatically analyze the sensorial time series data.

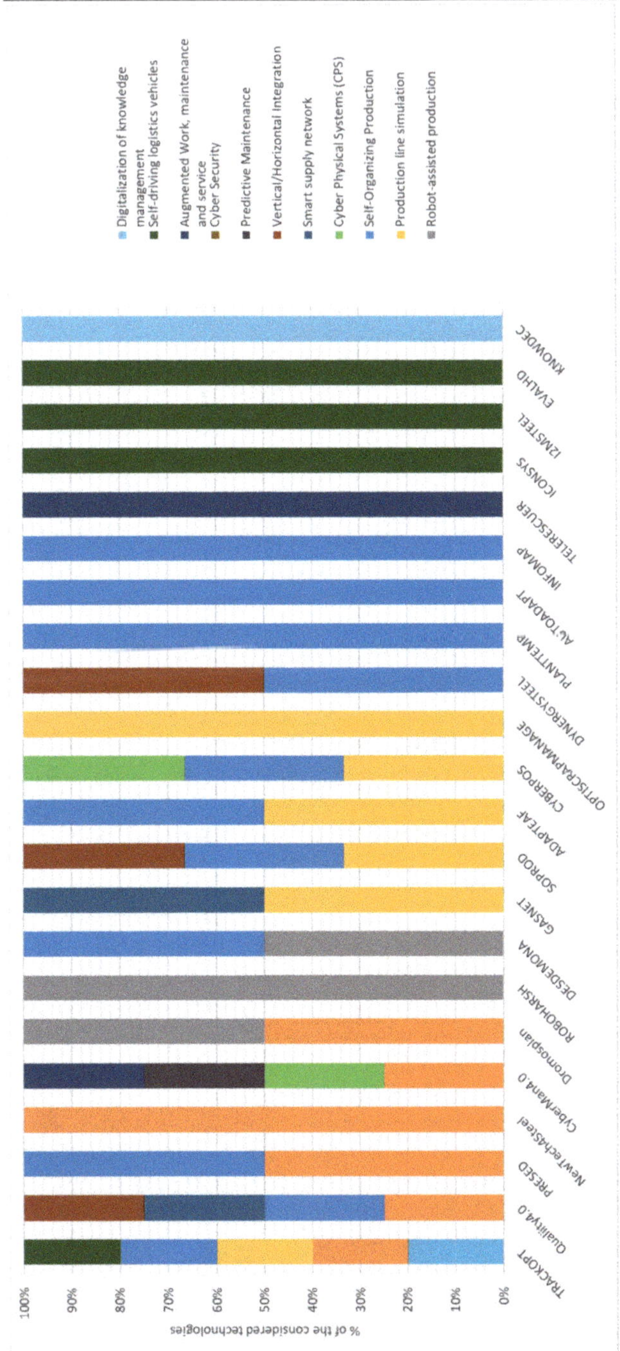

Figure 1. Research Fund for Coal and Steel (RFCS) Projects and the developed enabling technologies.

The projects Dromosplan, RoboHarsh, and Desdemona are related to the robot-assisted production. The ongoing project Dromosplan aims at using Unmanned Aerial Vehicles (UAV) in steel plants in order to replace the human intervention in a number of operations related to the monitoring, maintenance, and safety. New sensor data are being produced in this context in order to prove and evaluate the benefits of UAVs in steelworks [59]. RoboHarsh firstly introduced some concepts of human–robot symbiotic co-operation in the steel industry for the development of a complex maintenance procedure [60,61]. In this project, one of the main results is that the operator role is changing, becoming a supervisor, and, therefore, there is no replacement of the worker but a safer and heavyweight operation reduction. Desdemona is another example of development of procedures for steel defect detection by robotic and automatic systems such as UAVs and ground mobile robots.

In vertical/horizontal integration technology, some examples of RFCS projects are DynergySteel and the abovementioned Quality4.0. In DynergySteel, the simulation, decision support procedures, and control tools have been implemented at several steelmaking plants to improve power management capability and power engagement forecasting [62,63].

Simulation and optimization of the production line is another enabling technology, which handles in several RFCS projects, such as GasNet, SOProd, AdaptEAF, Cyber-POS, as well as OptiScrapManage. GasNet developed a simulation tool of the network of process gas and steam including their generation and flows as well as a multi-level strategy for their optimization. ML-based tools and technologies such as Echo-State and FeedForward Neural Networks, as well as advanced optimization approaches (e.g., Mixed Integer Linear programming) have been used in order to improve energy efficiency and environmental sustainability of the steelmaking processes [64–68]. In SOProd Objected-Oriented Programming (OOP), Python language, LabView, MongoDB, and Optical character recognition are some of the technologies adopted to improve product intelligence and autonomous machine–machine and product–machine communication [69]. In this project, a de-central optimization considering a detailed product and process knowledge facilitates a process self-optimization by using individual product properties and processing information of neighboring processes [58]. Advanced methods for process monitoring and control through multi-criteria approach of performances indicators, together with optimization approaches, have been exploited in OptiScrapManage. Laser scanner and acoustic and hyperspectral sensors are some of the used technologies. By considering the properties of the charged materials, AdaptEAF developed an adaptive online control for the Electric Arc Furnace (EAF), by optimizing the efficiency of the chemical energy input by reducing the total energy consumption and improving the metallic yield.

The project TeleRescuer provides an exemplar application of a special Unmanned Vehicle (UV) within a system for virtual teleportation of rescuers to subterranean areas of coal mines.

A method for the collection, representation, storage, and utilization of the human knowledge to exploit it in computer-based applications has been investigated and implemented by the project "KnowDec". Here, a new approach based on the methodology of knowledge-based decision support system has been developed. The operators of the quality department can capture the experiences concerning the approval of slabs and the collected experiences are stored in the knowledge base, useful for decision support and advices in similar cases.

Most of the abovementioned projects, such as Cyber-POS, TrackOpt, Quality4.0, DynergySteel, AdaptEAF, SOProd, Desdemona, PRESED, InfoMap, PlantTemp, and AutoAdapt deal with the self-organizing production technologies. PlantTemp develops an operator advisory system covering the electric arc furnace and casting processes, meeting the target casting temperature, by saving energy and material consumption. AutoAdapt proposes an expandable system, which aims to apply self-learning methods for adapting such automations to new products and plants. Genetic Algorithms (GA), polynomial models, iterative learning control methods, and feed-forward control are some of the used technologies. In InfoMap, a tool for objective interpretation of maps from different devices along the process route, generating concise data suitable for use within automatic control/advisory systems is developed. Here, Convolutional Neural Networks (CNNs) are applied for flatness defects detection and classification [70] (see also Appendix A).

IConSys implemented an Intelligent Control Station, in order to support decision making in rolling and finishing while the I2MSteel project developed a factory and company-wide automation and information technology for an intelligent and integrated manufacturing steel [71]. EvalHD investigated some aspects related to the implementation of Industry4.0 [72,73].

Figure 2 provides a summary of the abovementioned analysis by showing the number of RFCS Projects for each of the highlighted enabling technology. The identified enabling technologies are inserted in the x-axis and for each technology the number of RFCS projects, which takes into consideration such technologies, is reported in the y-axis.

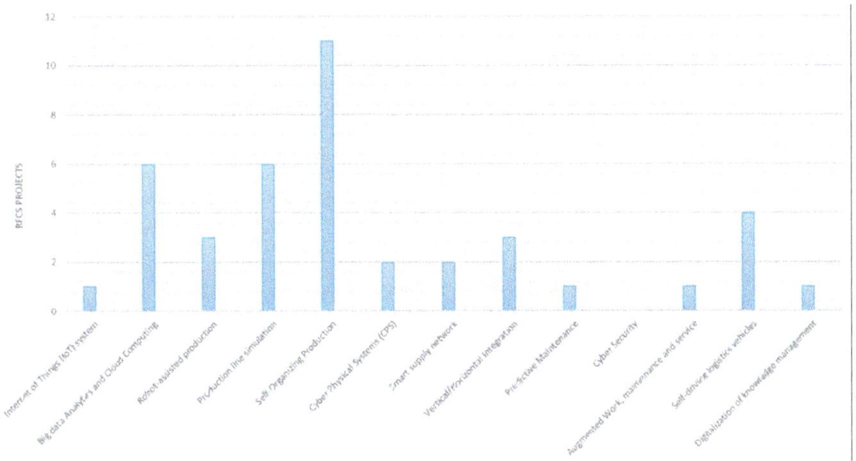

Figure 2. Number of RFCS projects by enabling technologies.

3.3. Other European Funding Programs for Digitalization and Low Carbon Technologies for the Steel Sector

Cost pressure, regulatory, and product/service requirements are some of the European steel industry challenges, which the steel industry has to face. The 7th Framework Program (FP7) (2007–2013) and its successor Horizon 2020 (2014–2020) [74] have been included in addition to the RFCS program by the European Union research and innovation funding program. Most of these projects started between 2014 and 2017. Nevertheless, digitalizing the steel industry started before calling these activities Industry 4.0 [75]. On this subject, some projects, starting in the early 1990s and covering some aspect of digitalization of the steel industry, have been identified, for instance BRICK, OREXPRESS, and TAM. All these projects were funded by EUREKA, which is a pan European network for market-oriented, industrial R&D [76].

As far as the FP7 Projects (2007–2013) are concerned, an example is AREUS, which treated integrated technologies for robotic production systems and robotic manufacturing processes optimization environment. WaterWatt and FACTS4WORKERS are some examples of H2020 projects where digitalization is applied in order to remove market barriers for energy efficient solutions and improve the efficiency for managers and workers within Worker-Centric Workplaces for Smart Factories.

The acronym SPIRE stand for "Sustainable Process Industry through Resource and Energy Efficiency" [77] and refers to a Public Private Partnership (PPP) targeting, within the Horizon 2020 program, the European process industries. DISIRE, CoPro, FUDIPO, MORSE, RECOBA, and COCOP are some SPIRE projects facing digital solutions and with specific demonstration in the steel sector.

Concerning other activities, a project on industry 4.0 has been developed at Dillinger. It is a real-time forecasting project for an "adaptive" Basic Oxygen Furnace (BOF), which "learns" and fine-tunes some settings based on the collected process data [78]. Another project has been led by SSAB and has aimed at making available information and instructions relating to any steel item, regardless of where it is produced. Each link of the production chain can use and accumulate information, by creating a basis for both the circular and platform economy [79].

The circular economy concept promotes the reuse, the refurbishment, and the recycling, maximizing the product life and at the same time keeping products and materials at a high level of utility [80] since an important objective defined in EU Masterplan is a competitive and low-carbon European steel industry [81]. Environmental issues such as CO_2 reduction can have several benefits from the KETs application. Advanced process monitoring and increased quality lead to major efficiency. In the field of CO_2 mitigation technologies, the RFCS and H2020 (2014-2020) programs represent the most important instruments for the EU-funded research projects. Low-carbon steel production requires the development of dedicated technologies.

The current pan-European research for the applicable technologies of CO_2 mitigation is focused on three pathways: Carbon Direct Avoidance (CDA), Process Integration (PI), and Carbon Capture, Storage, and Usage (CCU).

According to Figure 3, several EU Projects have been funded in the Process Integration pathway in order to develop technologies for reducing the use of carbon. For instance, ENCOP dealt with the overall energetic optimization of steel plants [33,82,83]. IDEOGAS focused on injection of reducing gas in the Blast Furnace (BF) and top gas recycling. LoCO2Fe developed a low CO_2 iron and steelmaking integrated process route. The identified low-carbon technologies are inserted in the x-axis and for each technology, the number of projects that takes into consideration such technologies is reported in the y-axis.

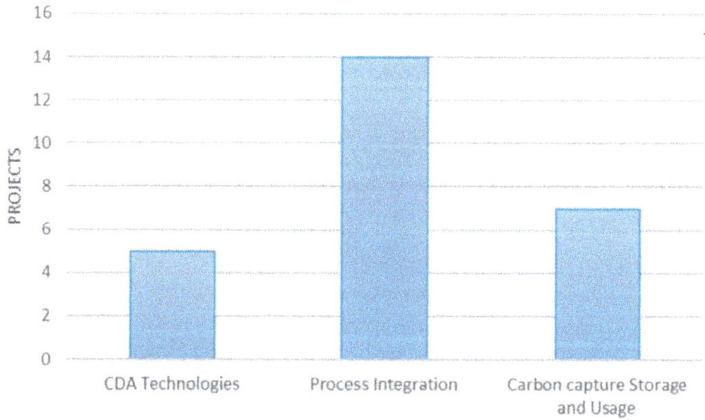

Figure 3. Number of projects related to low-carbon technologies.

CDA technologies mainly consist of iron ore reduction by hydrogen (produced by H_2O electrolysis) and syngas from biomass and Fe reduction by electrolysis. Some examples are HYBRIT, which aims at developing the world's first fossil-free ore-based steel-making technology using hydrogen replacing carbon as reductant, while GrInHy targets the production of a green industrial hydrogen via reversible high-temperature electrolysis designing, manufacturing, and operation of a high-temperature electrolyser.

CCU technologies concern the different methods for carbon capture based on chemical/biological processes of CO_2 conversion and CO_2 capturing by mineral raw materials. Within the CCU technologies, the focus is the conversion of industrial CO_2 into biofuels dealing with the transformation of CO_2 resulting from the iron, steel, cement, and electric power industries into value-added chemicals and plastics. Thanks to the Carbon4PUR project, it has been demonstrated that industrial waste gases such as mixed CO/CO_2 streams can be turned into intermediates for polyurethane plastics useful for rigid foams and coating. It is also possible to recycle carbon into sustainable and advanced bioethanol, as shown in the project Steelanol. FresMe and M4CO2 are other projects dealing with a more efficient CO_2 capture.

4. The Future of Digitalization in the Steel Sector

The steel sector, such as the other European industrial sectors, is committed to understanding the logic of digitalization and, consequently, to implement the digital technologies in its production processes. In the digital transformationm the four levers, resulting from researches carried out on key sectors in German and European economies, which are important for effectively implement the digitalization process, are [84]:

- Digital Data,
- Automation,
- Connectivity,
- Digital Customer Access.

Capturing, processing, and analyzing **digital data** can allow better forecasting of process behavior as well as smarter, easier, and faster decision making. The IoT connects devices equipped with sensors, software, and wireless capabilities, coupled with a growing capacity of data collection and storage. This results in new data sources availability to modern analytical technologies, for pre-processing data faster and in a more detailed way. Concerning the steelmaking processes and products, real-time data allow monitoring both of them. In addition, the use of sensors allows checking a single piece along the production chain: Errors and defects can be easily traced back and eliminated. This can lead to a more efficient production. In addition, data availability and ML, by enabling maintenance work to be anticipated and done before something goes wrong, can produce significant improvements in the equipment maintenance that can be scheduled and remotely checked.

The **automation** concerns the combination of traditional technologies and AI- and ML-based approaches, resulting in systems that work autonomously. In the near future, the automation applications will reduce error rates, increase speed, and cut operating costs. In particular, in the steel sector, the automation of production and consumption will be implemented.

The **connectivity** of separate systems (e.g., interconnecting the entire value chain via mobile or fixed-line high-bandwidth telecom networks) allows overcoming the lack of transparency, resulting in process efficiency improvements. This application in industrial plants is based on the interconnection of production systems facilitated by machine-to-machine (M2M) communications. A better connectivity and data sharing applied to the steelmaking processes aim to reduce some problems linked to remote locations and widespread supply chains. In addition, issues due to market fluctuation and potential hazardous working environments can be overcome.

The **digital customer access** allows the direct access to customers through the mobile internet, providing transparency and new services.

The new reorganization of entire industries and the transformation of business models are enabled through the availability of digital data, automation of production processes, interconnection of value chains, and creation of digital customer interfaces. This can allow steel companies to interact with suppliers and customers in a new and better way.

As far as the implementation of Industry 4.0 in the European steel sector is concerned, some previous and ongoing funded European projects will provide significant results on the digitalization in the near future. Some of them will develop cross-sectorial digital solutions, others are mainly prototype applications and demonstrations. On the other hand, some projects funded by the RFCS can provide further results on the real implementation of the digitalization in the steel sector. On this subject, a work including the publicly funded projects, patent analysis, expert interviews, and a qualitative survey of academics and practitioners related to Industry 4.0 in the steel sector has been performed [36]. Results have shown that transformation of the organizational structure of a company represent the main issues. In addition, Industry 4.0 implementations are required in order to achieve economic benefits to company developments, particularly, improvements in process efficiency and in the development of new business models. Furthermore, Industry 4.0 will improve effectiveness through intelligent support systems for the workforce and the interaction with customers in the organizational domain. In the research approach, the future SPIRE 2050 roadmap, in preparation by the SPIRE Working Group Digital, forecasts an integrated and digital European Process Industry, with new technologies and business models, which will aim to enhance competitiveness and impact for jobs and growth [85]. In the next five years, investments in innovation and digitization will be necessary [37], in order to achieve a level of digitalization to 72% [86]. By providing an example about the future of digitalization in the steel industry, ArcelorMittal group is working in the adoption common platforms and AI algorithms across the whole group and in different business areas [87].

4.1. Digitalization Impact on the Workforce

Significant changes, provided by Industry 4.0 to all aspects of industry structures, also include the workforce dynamics, such as the strategic workforce planning, the right organization structure, developing partnerships, and the technological standardization. The main future directions could be from a human labor-centered production to a fully automated work as well as from monotonous and physical activities to creative ones [88]. Nevertheless, negative changes could also occur, including higher unemployment and widespread workforce de-skilling.

In Germany, in overall sectors, about 23% of the workforce do not have vocational qualifications and in the manufacturing industries, 1.2 million are low-skilled workers. It has been assumed that low-skilled work will be up-skilled, as digitalization will upgrade simple and low-skilled activities and, at the same time, skilled activities will be continuously enhanced [89]. In this context, the industrial low-skilled work will not disappear but rather the level of qualification will steadily rise. It has been foreseen that jobs and skills will be polarized. This thesis consists in the automation of middle-skilled jobs by the use of computers. On the other hand, digitalization increases the productivity of the most skilled jobs, while the low-skilled jobs survive as they cannot be automated. This is because the automated work is concentrated in the middle of the skills distribution [90,91]. In addition, under conditions of digitalization, four development paths for low-skilled work have been identified [90]. The general erosion of low-skilled industrial work and the common idea that simple, routine tasks threatened by the new technologies will probably disappear in the longer term, can be considered only one scenario. In the second path, "upgrading of low-skilled industrial work", a strategy for improving technological product, is paired off with a highly flexible marketing. The third scenario, characterized as "digitalized low-skilled work", shows a high-intensity application of digital technologies and new forms of work (e.g., "crowdsourcing" and "crowdworking") and may also be associated with new forms of low-skilled work. In the fourth scenario, "structurally conservative stabilization of low-skilled work", there is no discernible change in existing employment and organizational structures. The different scenarios show that the potential job losses due to the implementation of the new technologies is controversial. In addition, the consequences for job activities and qualifications are interpreted as the "upgrading" or "polarization" of skills. Nevertheless, significant changes depend on the kind of technology automation and on its implementation process. Consequently, in the medium term, a limited spread of digital technologies is expected [92].

The progressive process of digitalization and automation produce effects and impacts on the employment in the industrial sectors, included the steel industry. In particular, the application of robotics and computerization will increase the creation of new jobs, particularly in IT and data science [14]. For instance, it has been shown that in Germany, only 12% of jobs are endangered by digital automation [93]. In Europe, over 1.5 million net new jobs have been created in the industrial sectors since 2013, with a growth of labor productivity of 2.7% per year on average since 2009, higher than both the US and Korea (0.7% and 2.3%, respectively) [23]. According to the European Centre for the Development of Vocational Training, between 2016 and 2030, over 151 million job openings are expected, with 91% due to the replacement needs (i.e., retirement, migration, movement into other occupations, or workers temporarily leaving the workplace) and the 9% due to new job openings. In the same period, over 1,750,000 jobs will be opened for ICT professionals [94]. Nevertheless, 2.6 million people worked in skill shortage occupations. During 2013, 47,000 vacancies have been estimated, including 25,600 reported as hard-to-fill by employers and around 23,500 as hard-to-fill, because of the lack of skills required [95]. In this regard, companies should develop their future workforce and adopt new business models and organizational structure, in the perspective of Industry 4.0 [96]. While employees need to be re-skilled, according to the requirements of digital economy, new employees need to be educated, according to the requirements of future jobs and skills. For this reason, the achievement of an up-skilled and re-skilled workforce is possible by implementing training programs, based on digital and business topics. This can be done through a life-long learning approach for addressing digital skills, and continuous training activities represent the key aspects for the companies to achieve a successful future [97]. However, companies also face some issues, such as skill mismatches, that refer to a failure of skill supply to meet skill demand, resulting in stopping the economic growth and in limiting the employment and the income opportunities of individuals [98]. The needs of companies, including the steel sector, are mainly focused on horizontal skillsets instead of high specialized profiles, in order to have a workforce flexible and able to move across multiple tasks. The companies need to have stronger horizontal skillsets rather than highly specialized profiles; in particular, workers with transferable skillsets in order to provide a good level of flexibility and coordination across different departments of their companies. In addition, it becomes increasingly important for companies to have employees who are able to move across multiple tasks and intervene in different areas. In addition, due to current job insecurity, transferable cross-functional skills represent a possibility for a greater security for workers [5]. In the process industries, including the steel industry, although 40,000 jobs have been lost in recent years, due to restructuring [99]; digitalization can provide new flexible skills and a workforce able to fast learn new digital technologies. In this context, cognitive sciences play an important role to provide support, combining awareness and knowledge with advanced control algorithms and optimization [100]. In the Industry 4.0, ICT skills are more important than core skills for employees. In particular, employees should not only have hard-skills, but also soft-skills such as collaboration, communication, and autonomy to perform their jobs in hybrid operating systems. In addition, employees should be able to be adaptable to continuous learning in an interdisciplinary perspective. Concerning engineering, the new education requirements are focused on achieving information and knowledge applicable to the business environment, and different disciplines should be able to work together. Through the design of new integrated engineering programs, the gap between universities and the business environment can be overcome. In addition, working in interdisciplinary teams, realizing interdisciplinary tasks, and providing interdisciplinary thinking represent key aspects for the implementation of Industry 4.0 research areas, such as mechatronic engineering, industrial engineering, and computer science [22].

4.2. Digitalization and Economic Impact

The digital economy can offer new opportunities to companies, including the steel sector. It is important to better understand how digitalization is changing the rules of competition, in order to optimize existing business models and to develop new ones. Due to a growth of the third country imports by 16.3% year-on-year, in the final quarter of 2018, a decrease of the domestic deliveries from EU mills to the EU market compared with the same period of 2017 has been revealed [101]. Economic and steel market outlook 2019-2020 European steel is squeezed between rising import pressure and a depressed home market. The main reason for the weakening of the EU economy in 2018, which will at least persist over the first half of 2019, has been the slowing global economic momentum and the related deteriorating contribution from net trade. A digital economy can be successfully achieved through a pan-European coordination based on a harmonized EU-wide approach. On this subject, different actions have to be implemented and, in particular, it is important to outline common standards at European level as well as to share ideas, knowledge, and experiences. A connected economy needs to rely on a strong infrastructure, in order to connect plant and machinery in an extensive and secure way. The digital transformation of the European manufacturing sector should be quickly achieved, in order to increase competitiveness and limit the new competitor actions. Reduction of energy and raw material consumption, lower OPEX, and reduction of losses as well increase of product qualities and productivity are the most important factors related to the innovative technologies in Industry 4.0 [102]. In [103], the recorded scrap information is transferred to EAF for the calculation of the optimized and best melting condition thanks to the detection and recording of volume and weight for each layer of scrap in the bucket. The raw materials, in fact, are a crucial factor and reducing their cost is more effective than acting on the transformation cost. According to [104], the main implementation areas in manufacturing are real-time supply chain optimization, human robot collaboration, smart energy consumption, digital performance management, and predictive maintenance. Especially the predictive maintenance, according to [104], can help not only increasing revenues, by reducing the maintenance costs from 10% to 40% and by reducing the waste from 10% to 20%, but also optimizing planned downtime, limiting unplanned downtime, and an estimation of a reduction of the operating cost by 2% to 10% is also foreseen. Moreover, digital technologies and ML can be useful in the metals industries in order to avoid unplanned shut down time to repair or replace key components, since such breakages are extremely costly. By using predictive maintenance methods, actuators can be replaced before they break [102]. The advanced analytics techniques like AI and ML can automatically help for the quality issues defining the basic causes, optimizing the optimal recipes for new products/grades, and by reducing the rejection rate [105]. The tools exploited in [106] facilitate the production planning by adopting AI and ML and help to improve due date reliability improving the overall economic success of the steelmaking company.

5. Conclusions

Although the steel production is already partly automated, the application of new technologies can further sustain the optimization of its entire production chain. This will allow the steel industry to become smarter in evolving towards Industry 4.0. The implementation of digital technologies, by continuously adjusting and the optimizing the processes online, contributes to improve the flexibility and the reliability of processes, maximizing the yield, and improving the product quality and the maintenance practices. In addition, they also contribute to increase the energy efficiency as well as to monitor and to control the environmental performance of processes in an integrated way.

The analysis reported in this review paper highlights that the challenge of digitalization consists of the integration of all systems and productions units, through three different dimensions: Vertical Integration (Integration of systems across the classic automation levels from the sensor to the ERP system); Horizontal Integration (Integration of systems along the entire production chain); Life-cycle Integration (Integration along the entire lifecycle of a plant from basic engineering to decommissioning) [30] and the Transversal integration (based on the decisions taken during the steel production chain, taking into account technological, economic, and environmental aspects). The digitalization process also requires jobs based on interdisciplinary teams, tasks, and thinking, to provide interdisciplinary skills. These achievements can be possible by integrating new IT, automation, and optimization technologies. Furthermore, Predictive Maintenance techniques can be implemented by equipment monitoring combined with intelligent decision methods. In this context, the application of Data Mining techniques, also based on ML, can allow anticipating maintenance work and scheduling it. In addition, Knowledge Management is a key factor for achieving improvements in the digitalization process.

The future expectations for the steel industry about digitalization include the optimization and the interactions of the individual production units, within the entire production chain (and beyond). This will allow reaching the highest quality, flexibility, and productivity. Furthermore, the following digitalization applications will represent the most important trends in the future:

Adaptive online control, through-process optimization, through-process synchronization of data, zero-defect manufacturing, traceability, and intelligent and integrated manufacturing.

In the coming years, in order to achieve a successful implementation of digitalization, the steel sector has to afford some important challenges, such as the standardization of systems and protocols, work organization and more skilled workers, investments, and research aiming to adopt appropriate frameworks. The implementation of digitalization is expected to generate productivity effects in the industrial sectors and growth in the economy. Concerning the potential consequences of digitalization for industrial workforce, on one hand, new technologies can cause job losses, but on the other hand, higher qualifications can be achieved. Nevertheless, changes will depend on different factors and they are expected to occur in the medium term, leading to some impacts on the industrial workforce. In addition, the steel sector needs to produce within environmental constraints in order to achieve its sustainability. In particular, the pressure of the environmental constraints represents a challenge for the steel sector to implement digital technologies that can help cope with the increasing trend in energy demand and the requirement of adopting low-carbon energy systems. In the coming years, the steel sector should be able to achieve zero waste, zero climate change emissions, and use half its current resources. On this subject, digital technologies can play an important role to enable improvements in sustainability performance, to plan processes in order to better account for demands and opportunities offered by industrial sustainability, and to enable the experimentation with new business models. The transformation of processes for significantly reducing emissions and improving energy efficiency will lead to the circular economy paradigm achievement and, on the other hand, adopting high-performance components, machines, and robots will optimize the materials and energy consumptions.

However, digital transformation and the full implementation of new digital solutions will only be successful if non-technological aspects are also considered in the technological development and implementation, such as framework conditions at European, national, and regional level, market and consumers, human resources, skills, and labor market. These aspects are integrated in the new SPIRE Roadmap 2050. Here, human resources and new (digital) skills especially will play a crucial role for unfolding the potential of new solutions within the companies.

Author Contributions: Conceptualization, V.C, T.A.B., and M.M.M.; methodology, V.C. and M.M.M.; validation, V.C. and A.J.S.; formal analysis, T.A.B. and M.M.M.; investigation, T.A.B., B.F., and E.S.; resources, V.C. and A.S.; writing—original draft preparation, T.A.B., B.F., and VC; writing—review and editing, M.M.M. and A.J.S.; visualization, B.F.; supervision, V.C.; project administration, A.S. and V.C.; funding acquisition, A.J.S. and V.C. All authors have read and agreed to the published version of the manuscript.

Funding: This research was funded by the European Union through the Erasmus Plus Programme, Grant Agreement No 2018-3019/001-001, Project No. 600886-1-2018-1-DE-EPPKA2-SSA-B.

Acknowledgments: The research described in the present paper was developed within the project entitled "Blueprint "New Skills Agenda Steel": Industry-driven sustainable European Steel Skills Agenda and Strategy (ESSA)" and is based on a preliminary deliverable of this project. The ESSA project is funded by Erasmus Plus Programme of the European Union, Grant Agreement No 2018-3019/001-001, Project No. 600886-1-2018-1-DE-EPPKA2-SSA-B. The sole responsibility of the issues treated in the present paper lies with the authors; the Commission is not responsible for any use that may be made of the information contained therein. The authors wish to acknowledge with thanks the European Union for the opportunity granted that has made possible the development of the present work. The authors also wish to thank all partners of the project for their support and the fruitful discussion that led to successful completion of the present work.

Conflicts of Interest: The authors declare no conflict of interest.

Appendix A

Table A1. List of acronyms identified RFCS Projects covering aspects of digitalization in the steel industry.

RFCS Projects				
AdaptEAF	DroMoSPlan	I2MSteel	Plant Temp	Telerescuer
AUTOADAPT	DYNERGYSteel	IConSys	PRESED	TRACKOPT
CYBERMAN4.0	ENCOP	INFOMAP	ROBOHARSH	
Cyber-POS	EvalHD	NEWTECH4STEEL	QUALITY4.0	
DESDEMONA	GASNET	OptiScrapManage	SoProd	

Table A2. List of acronyms of identified other European Funding Projects covering aspects of digitalization in the steel industry.

EUREKA	H2020	SPIRE	FP7
BRICK	WATERWATT	DISIRE	AREUS
OREXPRESS	FACTS4WORKERS	COPRO	
TAM		FUDIPO	
H2PREDICTOR		MORSE	
		RECOBA	
		COCOP	

Table A3. List of acronyms of identified Projects related to the Low Carbon Future Technologies.

CDA	PI	CCU
HYBRIT	IDEOGAS	Biocon-co2
GrInHy	OSMet S2	CarbonNext
H2FUTURE	HISARNA B, C &D	I3UPGRADE
IERO	ACASOS	FresMe
SALCOS	CO2RED	M4CO2
	REGTGF	Carbon4PUR
	RenewableSteelGases	Steelanol
	SHOCOM	
	Torero	
	GREENEAF	
	GREENEAF2	
	ENCOP	
	IDEOGAS	
	LoCO2Fe	
	STEPWISE	

References

1. Schumacher, A.; Sihn, W.; Erol, S. In Automation, Digitization and Digitalization and Their Implications for Manufacturing Processes. In Proceedings of the International Scientific Conference on Innovation and Sustainability, Bucharest, Romania, 28–29 October 2016; pp. 1–6.
2. Clerck, J. Digitization, Digitalization and Digital Transformation: The Differences. i-SCOOP. 2017. Available online: https://www.i-scoop.eu/digital-transformation/digitization-digitalization-digital-transformation-disruption/ (accessed on 20 February 2020).
3. Groover, M.P. *Automation, Production Systems, and Computer-Integrated Manufacturing*; Prentice Hall Press: Upper Saddle River, NJ, USA, 2007.
4. Maintenance Q&As. Available online: https://www.onupkeep.com/answers/predictive-maintenance/industry-3-0-vs-industry-4-0/ (accessed on 6 December 2019).

5. European Commission. *Blueprint for Sectoral Cooperation on Skills-towards an EU Strategy Addressing the Skills Needs of the Steel Sector: European Vision on Steel-Related Skills of Today and Tomorrow—Study*; European Commission: Bruxelles, Belgium, 2019.

6. European Commission. *Germany: Industry 4.0, Digital Transformation Monitor*; European Commission: Bruxelles, Belgium, 2017.

7. Bogner, E.; Voelklein, T.; Schroedel, O.; Franke, J. Study based analysis on the current digitalization degree in the manufacturing industry in Germany. *Procedia CIRP* **2016**, *57*, 14–19. [CrossRef]

8. European Commission. *Communication from the Commission to the European Parliament, the European Council, the Council, the European Economic and Social Committee, the Committee of the Regions and the European Investment Bank Investing in a Smart, Innovative and Sustainable Industry a Renewed EU Industrial Policy Strategy, com/2017/0479 Final*; European Commission: Bruxelles, Belgium, 2017.

9. Beltrametti, L.; Guarnacci, N.; Intini, N.; La Forgia, C. *La Fabbrica Connessa. La Manifattura Italiana (Attra) Verso Industria 4.0*; goWare & Edizioni Guerini e Associati: Milano, Italy, 2017.

10. SMS Group #Magazine. Available online: https://www.sms-group.com/sms-group-magazine/overview/digitalization-in-the-steel-industry/ (accessed on 6 December 2019).

11. Ibarra, D.; Ganzarain, J.; Igartua, J.I. Business model innovation through industry 4.0: A review. *Procedia Manuf.* **2018**, *22*, 4–10. [CrossRef]

12. Stahl Institute VDEh. *Annual Report of Steel Institute Vdeh 2016—Summary of Main Topics of Technical-Scientific Joint Cooperation and Work*; Stahl Institute VDEh: Dusseldorf, Germany, 2016.

13. Zhong, R.Y.; Xu, X.; Klotz, E.; Newman, S.T. Intelligent manufacturing in the context of industry 4.0: A review. *Engineering* **2017**, *3*, 616–630. [CrossRef]

14. Lorenz, M.; Rüßmann, M.; Strack, R.; Lueth, K.L.; Bolle, M. *Man and Machine in Industry 4.0: How will Technology Transform the Industrial Workforce through 2025*; The Boston Consulting Group: Boston, MA, USA, 2015; Volume 2.

15. European Commission. Available online: https://ec.europa.eu/digital-single-market/news/fresh-look-use-robots-shows-positive-effect-automation (accessed on 6 December 2019).

16. European Commission. Digital Single Market. Available online: https://ec.europa.eu/digital-single-market/en/download-scoreboard-reports (accessed on 6 December 2019).

17. Briken, K.; Chillas, S.; Krzywdzinski, M.; Marks, A.; Pfeiffer, S. Industrie 4.0 in the making–discourse patterns and the rise of digital despotism. In *The New Digital Workplace: How New Technologies Revolutionise Work*; Downes, R., Ed.; Red Globe Press: London, UK, 2017.

18. Pfeiffer, S. The vision of "industrie 4.0" in the making—A case of future told, tamed, and traded. *Nanoethics* **2017**, *11*, 107–121. [CrossRef] [PubMed]

19. Botthof, A.; Hartmann, E.A. *Zukunft der Arbeit in Industrie 4.0*; Springer: Berlin, Germany, 2015.

20. Abel, J.; Hirsch-Kreinsen, H.; Ittermann, P. *Einfacharbeit in der Industrie: Strukturen, Verbreitung und Perspektiven*; Edition Sigma: Berlin, Germany, 2014.

21. World Economic Forum. *Digital Transformation Initiative—Mining and Metals Industry*; World Economic Forum: Cologny, Switzerland, 2017.

22. Ustundag, A.; Cevikcan, E. *Industry 4.0: Managing the Digital Transformation*; Springer: Berlin, Germany, 2017.

23. European Commission. *Re-Finding Industry-Report from the High-Level Strategy Group on Industrial Technologies*; European Commission: Bruxelles, Belgium, 2018.

24. Liu, S.X. Innovation design: Made in china 2025. *Des. Manag. Rev.* **2016**, *27*, 52–58.

25. European Commission. *A Blueprint for Sectoral Cooperation on Skills (Wave II)*; European Commission: Bruxelles, Belgium, 2018.

26. European Commission. *Steel: Preserving Sustainable Jobs and Growth in Europe. COM(2016)(155)*; European Commission: Bruxelles, Belgium, 2016.

27. Peters, H. How Could Industry 4.0 Transform the Steel Industry, Steel Times International. In Proceedings of the Future Steel Forum, Warsaw, Poland, 14–15 June 2017.

28. E15CZ. Available online: https://www.e15.cz/byznys/technologie-a-media/ceskym-firmam-muze-ujet-inovacni-vlak-ctvrtina-z-nich-nema-zadnou-koncepci-digitalizace-1351646 (accessed on 6 December 2019).

29. Thomas, B.G. Review on modeling and simulation of continuous casting. *Steel Res. Int.* **2018**, *89*, 1700312. [CrossRef]

30. Herzog, K.; Winter, G.; Kurka, G.; Ankermann, K.; Binder, R.; Ringhofer, M.; Maierhofer, A.; Flick, A. The digitalization of steel production. *BHM Berg Hüttenmännische Mon.* **2017**, *162*, 504–513. [CrossRef]
31. Je-ho, C. The fourth industrial revolution: The winds of change are blowing in the steel industry. *Asian Steel Watch* **2016**, *2*, 6–15.
32. Roudier, S.; Sancho, L.D.; Remus, R.; Aguado-Monsonet, M. *Best Available Techniques (bat) Reference Document for Iron and Steel Production: Industrial Emissions Directive 2010/75/eu: Integrated Pollution Prevention and Control*; Joint Research Centre (Seville site): Seville, Spain, 2013.
33. Porzio, G.F.; Fornai, B.; Amato, A.; Matarese, N.; Vannucci, M.; Chiappelli, L.; Colla, V. Reducing the energy consumption and CO_2 emissions of energy intensive industries through decision support systems—An example of application to the steel industry. *Appl. Energy* **2013**, *112*, 818–833. [CrossRef]
34. Kovačič, M.; Šarler, B. Genetic programming prediction of the natural gas consumption in a steel plant. *Energy* **2014**, *66*, 273–284. [CrossRef]
35. Kovačič, M.; Stopar, K.; Vertnik, R.; Šarler, B. Comprehensive electric arc furnace electric energy consumption modeling: A pilot study. *Energies* **2019**, *12*, 2142. [CrossRef]
36. Neef, C.; Hirzel, S.; Arens, M. *Industry 4.0 in the European Iron and Steel Industry: Towards an Overview of Implementations and Perspectives*; Fraunhofer Institute for Systems and Innovation Research ISI: Karlsruhe, Germany, 2018.
37. Naujok, N.; Stamm, H. Industry 4.0 in Steel: Status, Strategy, Roadmap and Capabilities. In Proceedings of the Future Steel Forum, Warsaw, Poland, 6–7 June 2018; Available online: https://futuresteelforum.com/content-images/speakers/Dr-Nils-Naujok-Holger-Stamm-Industry-4.0-in-steel.pdf (accessed on 20 December 2019).
38. Digital Twin Technology in the Steel Industry. Available online: https://www.estep.eu/assets/Final-Programme-Digital-Twin-WS-21-22-November.pdf (accessed on 20 December 2019).
39. Strategic Research Agenda (SRA). Available online: https://www.estep.eu/assets/SRA-Update-2017Final.pdf (accessed on 21 December 2019).
40. Peters, K.; Malfa, E.; Colla, V. The European steel technology platform's strategic research agenda: A further step for the steel as backbone of EU resource and energy intense industry sustainability. *Metall. Ital.* **2019**, *111*, 5–17.
41. Man and Machine in Industry 4.0: How Will Technology Transform the Industrial Workforce Through 2015? Available online: http://englishbulletin.adapt.it/wp-content/uploads/2015/10/BCG_Man_and_Machine_in_Industry_4_0_Sep_2015_tcm80-197250.pdf (accessed on 21 December 2019).
42. Xia, F.; Yang, L.; Wang, L.; Vinel, A. Internet of things. *Int. J. Commun. Syst.* **2012**, *25*, 1101–1102. [CrossRef]
43. Zhang, F.; Liu, M.; Zhou, Z.; Shen, W. An IoT-based online monitoring system for continuous steel casting. *IEEE Internet Things J.* **2016**, *3*, 1355–1363. [CrossRef]
44. Hsu, C.Y.; Kang, L.W.; Weng, M.F. Big Data Analytics: Prediction of Surface Defects on Steel Slabs Based on One Class Support Vector Machine. In Proceedings of ASME 2016 Conference on Information Storage and Processing Systems, Santa Clara, CA, USA, 20–21 June 2016.
45. Fragassa, C.; Babic, M.; Perez Bergmann, C.P.; Minak, G. Predicting the Tensile Behaviour of Cast Alloys by a Pattern Recognition Analysis on Experimental Data. *Metals* **2019**, *9*, 557. [CrossRef]
46. Tian, S.; Xu, K. An Algorithm for Surface Defect Identification of Steel Plates Based on Genetic Algorithm and Extreme Learning Machine. *Metals* **2017**, *7*, 311. [CrossRef]
47. Xun, X. From cloud computing to cloud manufacturing. *Robot. Comput. Integr. Manuf.* **2012**, *28*, 75–86. [CrossRef]
48. Demetlika, P.; Ferrari, R.; Galasso, L.M.; Romano, F. Robotic system for a "zero-operator" continuous casting floor. In Proceedings of AISTech 2014—Iron and Steel Technology Conference, Indianapolis, IN, USA, 5–8 May 2014.
49. Egger, M.W.; Priesner, A.; Lehner, J.; Nogratnig, H.; Lechner, H.; Wimmer, G. Successful revamping of sublance manipulators fort the LD converters at Voestalpine Stahl Gmbh. In Proceedings of AISTech 2014-Iron and Steel Technology Conference, Indianapolis, IN, USA, 5–8 May 2014.
50. Rauch, Ł.; Bzowski, K.; Kuziak, R.; Uranga, P.; Gutierrez, I.; Isasti, N.; Jacolot, R.; Kitowski, J.; Pietrzyk, M. Computer-Integrated Platform for Automatic, Flexible, and Optimal Multivariable Design of a Hot Strip Rolling Technology Using Advanced Multiphase Steels. *Metals* **2019**, *9*, 737. [CrossRef]

51. Yang, J.; Zhang, J.; Guan, M.; Hong, Y.; Gao, S.; Guo, W.; Liu, Q. Fine Description of Multi-Process Operation Behavior in Steelmaking-Continuous Casting Process by a Simulation Model with Crane Non-Collision Constraint. *Metals* **2019**, *9*, 1078. [CrossRef]

52. Hanoglu, U.; Šarler, B. Hot rolling simulation system for steel based on advanced meshless solution. *Metals* **2019**, *9*, 788. [CrossRef]

53. Vertnik, R.; Mramor, K.; Šarler, B. Solution of three-dimensional temperature and turbulent velocity field in continuously cast steel billets with electromagnetic stirring by a meshless method. *Eng. Anal. Bound. Elem.* **2019**, *104*, 347–363. [CrossRef]

54. Cyber-Physical Systems. Available online: http://cyberphysicalsystems.org (accessed on 21 December 2019).

55. Zhou, K.; Liu, T.; Lifeng, Z. Industry 4.0: Towards Future Industrial. In Proceedings of 12th International Conference on Fuzzy Systems and Knowledge Discovery (FSKD), Zhangjiajie, China, 15–17 August 2015.

56. Flatt, H.; Schriegel, S.; Jasperneite, J.; Trsek, H.; Adamczyk, H. Analysis of the Cyber-Security of industry 4.0 technologies based on RAMI 4.0 and identification of requirements. In Proceedings of IEEE 21st International Conference on Emerging Technologies and Factory Automation (ETFA), Berlin, Germany, 6–9 September 2016.

57. Arens, M.; Neef, C.; Beckert, B.; Hirzel, S. Perspectives for digitising energy-intensive industries—Findings from the European iron and steel industry. In Proceedings of eceee Industrial Summer Study; ECEEE: Stockholm, Sweden, 2018; pp. 259–268.

58. Iannino, V.; Colla, V.; Denker, J.; Göttsche, M. A CPS-based simulation platform for long production factories. *Metals* **2019**, *9*, 1025. [CrossRef]

59. DroMoSplan, Workers Safety Improvement and Significant Reduction of Maintenance Costs by Monitoring and Inspecting Steel Plants with a New Type of Autonomous Flying Drones. Available online: http://www.dromosplan.eu/homepage (accessed on 21 December 2019).

60. Colla, V.; Schroeder, A.; Buzzelli, A.; Abbà, D.; Faes, L.; Romaniello, L. Introduction of Symbiotic Human-robot Cooperation in the Steel Sector: An Example of Social Innovation. *Matériaux Tech.* **2017**, *105*, 505. [CrossRef]

61. Colla, V.; Matino, R.; Faes, A.; Schivalocchi, M.; Romaniello, L. A robot performs the maintenance of the ladle sliding gate. *Stahl Eisen* **2019**, *9*, 44–47.

62. Marchiori, F.; Belloni, A.; Benini, M.; Cateni, S.; Colla, V.; Ebel, A.; Pietrosanti, C. Integrated Dynamic Energy Management for Steel Production. *Energy Procedia* **2017**, *105*, 2772–2777. [CrossRef]

63. Marchiori, F.; Benini, M.; Cateni, S.; Colla, V.; Vignali, A.; Ebel, A.; Neuer, M.; Piedimonte, L. Agent-based approach for energy demand-side management. *Stahl Eisen* **2018**, *138*, 25–29.

64. Colla, V.; Matino, I.; Dettori, S.; Petrucciani, A.; Zaccara, A.; Weber, V.; Romaniello, L. Assessing the efficiency of the off-gas network management in integrated steelworks. *Matériaux Tech.* **2019**, *107*, 502. [CrossRef]

65. Matino, I.; Dettori, S.; Colla, V.; Weber, V.; Salame, S. Forecasting blast furnace gas production and demand through echo state neural network-based models: Pave the way to off-gas optimized management. *Appl. Energy* **2019**, *253*, 113578. [CrossRef]

66. Matino, I.; Dettori, S.; Colla, V.; Weber, W.; Salame, S. Two innovative modelling approaches in order to forecast consumption of blast furnace gas by hot blast stoves. *Energy Procedia* **2019**, *158*, 4043–4048. [CrossRef]

67. Dettori, S.; Matino, I.; Colla, V.; Weber, V.; Salame, S. Neural Network-based modeling methodologies for energy transformation equipment in integrated steelworks processes. *Energy Procedia* **2018**, *158*, 4061–4066. [CrossRef]

68. Colla, V.; Matino, I.; Dettori, S.; Cateni, S.; Matino, R. Reservoir computing approaches applied to energy management in industry. *Commun. Comput. Inf. Sci.* **2019**, *1000*, 66–79.

69. Iannino, V.; Vannocci, M.; Vannucci, M.; Colla, V.; Neuer, M. A multi-agent approach for the self-optimization of steel production. *Int. J. Simul. Syst. Sci. Technol.* **2018**, *19*, 20.1–20.7. [CrossRef]

70. Vannocci, M.; Ritacco, A.; Castellano, A.; Galli, F.; Vannucci, M.; Iannino, V.; Colla, V. Flatness Defect Detection and Classification in Hot Rolled Steel Strips Using Convolutional Neural Networks. *Lect. Notes Comput. Sci.* **2019**, *11507*, 220–234. [CrossRef]

71. Colla, V.; Nastasi, G.; Maddaloni, A.; Holzknecht, N.; Heckenthaler, T.; Hartmann, G. Intelligent control station for improved quality management in flat steel production. *IFAC Pap. OnLine* **2016**, *49*, 226–231. [CrossRef]

72. Brandenburger, J.; Colla, V.; Nastasi, G.; Ferro, F.; Schirm, C.; Melcher, J. Big Data Solution for Quality Monitoring and Improvement on Flat Steel Production. *IFAC Pap. OnLine* **2016**, *49*, 55–60. [CrossRef]

73. Smart Steel: RFCS—Support Steelmaking and Use in the 21th Century. Available online: https://aceroplatea. es/docs/RFCS_SmartSteel2016.pdf (accessed on 21 December 2019).
74. Research and Innovation Funding 2014–2020. Available online: https://ec.europa.eu/research/fp7/index_en. cfm (accessed on 21 December 2019).
75. Hecht, M. Industrie 4.0 der Dillinger Weg. *Stahl Eisen* **2017**, *137*(4), 61–70.
76. Eureka Projects. Available online: http://www.eurekanetwork.org/eureka-projects (accessed on 21 December 2019).
77. SPIRE. Available online: https://www.spire2030.eu/ (accessed on 21 December 2019).
78. Datengetriebenes Prognosemodell für den BOF-Konverter—Anwendung von Data Mining in der AG der Dillinger Hüttenwerke. Available online: https://idw-online.de/en/attachmentdata36931.pdf (accessed on 21 December 2019).
79. SmartSteel Gets Support from Vinnova—The Journey toward the "Internet of Materials" Continues. Available online: https://www.ssab.com/company/newsroom/media-archive/2017/05/05/07/01/smartsteel-gets-support-from-vinnova---the-journey-toward-the-internet-of-materials-continues (accessed on 21 December 2019).
80. Foundation, E.M. *Towards the Circular Economy*; Cowes Ellen MacArthur Foundation: Cowes, UK, 2013.
81. EUROFER. *DISCUSSION PAPER: Towards an EU Masterplan for a Low Carbon-Competitive European Steel Value Chain*; EUROFER: Brussels, Belgium, 2018.
82. Maddaloni, A.; Porzio, G.F.; Nastasi, G.; Colla, V.; Branca, T.A. Multi-objective optimization applied to retrofit analysis: A case study for the iron and steel industry. *Appl. Therm. Eng.* **2015**, *91*, 638–646. [CrossRef]
83. Porzio, G.F.; Colla, V.; Matarese, N.; Nastasi, G.; Branca, T.A.; Amato, A.; Fornai, B.; Vannucci, M.; Bergamasco, M. Process integration in energy and carbon intensive industries: An example of exploitation of optimization techniques and decision support. *Appl. Therm. Eng. J.* **2014**, *70*, 1148–1155. [CrossRef]
84. The Digital Transformation of Industry. Available online: www.rolandberger.com/publications/publication_pdf/roland_berger_digital_transformation_of_industry_20150315.Pdf (accessed on 21 December 2019).
85. SPIRE 2050 Vision. Available online: https://www.spire2030.eu/sites/default/files/users/user85/Vision_Document_V5_Pages_Online_0.pdf (accessed on 9 December 2019).
86. Strategy&—PwC. Available online: https://www.strategyand.pwc.com/gx/en/insights/industry4-0.html (accessed on 9 December 2019).
87. ArcelorMittal. *Integrated Annual Review 2018*; ArcelorMittal: Luxembourg, 2019.
88. Pfeiffer, S. Robots, industry 4.0 and humans, or why assembly work is more than routine work. *Societies* **2016**, *6*, 16. [CrossRef]
89. Guerrieri, P.; Evangelista, R.; Meliciani, V. The economic impact of digital technologies in Europe. *Econ. Innov. New Technol.* **2014**, *23*(8), 802–824.
90. Hirsch-Kreinsen, H. The Future of Low-Skilled Industrial Work. 2016. Available online: https://www.researchgate.net/publication/321635390_THE_FUTURE_OF_LOW-SKILLED_INDUSTRIAL_WORK (accessed on 20 February 2020).
91. European Commission. *Report of the High-Level Expert Group on the Impact of the Digital Transformation on EU Labor Markets*; B-1049; European Commission: Brussels, Belgium, 2019.
92. Hirsch-Kreinsen, H. Digitization of industrial work: Development paths and prospects. *J. Labour Mark. Res.* **2016**, *49*, 1–14. [CrossRef]
93. Bonin, H.; Gregory, T.; Zierahn, U. *Übertragung der Studie von Frey/Osborne (2013) auf Deutschland*; ZEW Kurzexpertise: Mannheim, Germany, 2015.
94. Skills Panorama, Inspiring Choices on Skills and Jobs in Europe. Available online: https://skillspanorama. cedefop.europa.eu/en/analytical_highlights/skills-forecast-key-eu-trends-2030 (accessed on 9 December 2019).
95. European Commission. *Smart Steel: RFCS: Support Steelmaking and Use in the 21th Century*; B-1049; European Commission: Brussels, Belgium, 2016.
96. Karacay, G. Talent development for industry 4.0. In *Industry 4.0: Managing the Digital Transformation*; Springer: Berlin/Heidelberg, Germany, 2018; pp. 123–136.
97. European Commission. *Digital Skills and Jobs Conference—Digital Opportunities for Europe*; European Commission: Brussels, Belgium, 2017.
98. Gambin, L.; Hogarth, T.; Murphy, L.; Spreadbury, K.; Warhurst, C.; Winterbotham, M. *Research to Understand the Extent, Nature and Impact of Skills Mismatches in the Economy*; BIS: London, UK, 2016.

99. European Commission. *Communication from the Commission to the Parliament, the Council, the European Economic and Social Committee and the Committee of Regions Action Plan for a Competitive and Sustainable Steel Industry in Europe/* com/2013/0407 final */*; European Commission: Brussels, Belgium, 2013.

100. DEI WG 2. *Strengthening Leadership in Digital Technologies and in Digital Industrial Platforms—Digitization in the Process Industries through the Spire PPP*; European Commission: Brussels, Belgium, 2016.

101. EUROFER. *Economic and Steel Market Outlook 2019–2020 European Steel Squeezed between Rising Import Pressure and a Depressed Home Market*; EUROFER: Brussels, Belgium, 2019.

102. Herzog, K.; Günther, W.; Kurka, G.; Ankermann, K.; Binder, R.; Ringhofer, M. Primetals Technologies 2018. *The Digital Transformation of Steel Production.* Available online: http://seaisi.org/file/12-2%20The%20Digital%20Transformation%20of%20Steel%20Production.pdf (accessed on 20 February 2020).

103. Danieli Automation Research Center. Application of digital twinning in the melting shop. Experiences. In Proceedings of ESTEP Workshop on Digital Twinning Techniques, Charleroi, Belgium, 21–22 November 2018.

104. McKinsey. *Industry 4.0 after the Initial Hype: Where Manufacturers are Finding Value and How They Can Best Capture It.* McKinsey 2016. Available online: https://www.mckinsey.com/~{}/media/mckinsey/business%20functions/mckinsey%20digital/our%20insights/getting%20the%20most%20out%20of%20industry%204%200/mckinsey_industry_40_2016.ashx (accessed on 20 February 2020).

105. McKinsey. *Unlocking the Digital Opportunities in Metals, Metals and Mining Practice.* McKinsey: 2018. Available online: https://www.mckinsey.com/~{}/media/McKinsey/Industries/Metals%20and%20Mining/Our%20Insights/Unlocking%20the%20digital%20opportunity%20in%20metals/Unlocking-the-digital-opportunity-in-metals_Jan-2018.ashx (accessed on 20 February 2020).

106. Klein, A.; Ptaszyk, K.; Runde, W.; Ohm, T.; Bleskov, I.; Passon, M. Application of advanced artificial intelligence in the manufacturing execution system for metals industry. In Proceedings of the METEC and 4th ESTAD (European Steel Technology and Applications Days) 2019, Dusseldorf, Germany, 24–28 June 2019; VDEh: Dusseldorf, Germany, 2019.

Article

A General Vision for Reduction of Energy Consumption and CO_2 Emissions from the Steel Industry

Lauri Holappa

Department of Chemical and Metallurgical Engineering, Aalto University School of Chemical Engineering, 02150 Espoo, Finland; lauri.holappa@aalto.fi; Tel.: +358-50-560-83-77

Received: 16 July 2020; Accepted: 16 August 2020; Published: 19 August 2020

Abstract: The 2018 IPCC (The Intergovernmental Panel on Climate Change's) report defined the goal to limit global warming to 1.5 °C by 2050. This will require "rapid and far-reaching transitions in land, energy, industry, buildings, transport, and cities". The challenge falls on all sectors, especially energy production and industry. In this regard, the recent progress and future challenges of greenhouse gas emissions and energy supply are first briefly introduced. Then, the current situation of the steel industry is presented. Steel production is predicted to grow by 25–30% by 2050. The dominant iron-making route, blast furnace (BF), especially, is an energy-intensive process based on fossil fuel consumption; the steel sector is thus responsible for about 7% of all anthropogenic CO_2 emissions. In order to take up the 2050 challenge, emissions should see significant cuts. Correspondingly, specific emissions (t CO_2/t steel) should be radically decreased. Several large research programs in big steelmaking countries and the EU have been carried out over the last 10–15 years or are ongoing. All plausible measures to decrease CO_2 emissions were explored here based on the published literature. The essential results are discussed and concluded. The specific emissions of "world steel" are currently at 1.8 t CO_2/t steel. Improved energy efficiency by modernizing plants and adopting best available technologies in all process stages could decrease the emissions by 15–20%. Further reductions towards 1.0 t CO_2/t steel level are achievable via novel technologies like top gas recycling in BF, oxygen BF, and maximal replacement of coke by biomass. These processes are, however, waiting for substantive industrialization. Generally, substituting hydrogen for carbon in reductants and fuels like natural gas and coke gas can decrease CO_2 emissions remarkably. The same holds for direct reduction processes (DR), which have spread recently, exceeding 100 Mt annual capacity. More radical cut is possible via CO_2 capture and storage (CCS). The technology is well-known in the oil industry; and potential applications in other sectors, including the steel industry, are being explored. While this might be a real solution in propitious circumstances, it is hardly universally applicable in the long run. More auspicious is the concept that aims at utilizing captured carbon in the production of chemicals, food, or fuels e.g., methanol (CCU, CCUS). The basic idea is smart, but in the early phase of its application, the high energy-consumption and costs are disincentives. The potential of hydrogen as a fuel and reductant is well-known, but it has a supporting role in iron metallurgy. In the current fight against climate warming, H_2 has come into the "limelight" as a reductant, fuel, and energy storage. The hydrogen economy concept contains both production, storage, distribution, and uses. In ironmaking, several research programs have been launched for hydrogen production and reduction of iron oxides. Another global trend is the transfer from fossil fuel to electricity. "Green" electricity generation and hydrogen will be firmly linked together. The electrification of steel production is emphasized upon in this paper as the recycled scrap is estimated to grow from the 30% level to 50% by 2050. Finally, in this review, all means to reduce specific CO_2 emissions have been summarized. By thorough modernization of production facilities and energy systems and by adopting new pioneering methods, "world steel" could reach the level of 0.4–0.5 t CO_2/t steel and thus reduce two-thirds of current annual emissions.

Keywords: climate warming; carbon footprint; energy saving; emissions mitigation; electricity generation; hydrogen in steelmaking; steel vision

1. Global Challenge of Climate Warming and Its Rationale

Climate change is indisputable. The Intergovernmental Panel on Climate Change's (IPCC) Fifth Assessment Report concluded that, "It is extremely likely that human influence has been the dominant cause of the observed warming since the mid-20th century" [1]. The main culprit is anthropogenic greenhouse gas (GHG) emissions, which have doubled since 1970 due to the rapid population growth, expanded industrialization, and increase in standard of living. The observed growth can be seen in Figure 1 [2]. The scale in the figure is in Gt CO_2 equivalent per year. Carbon dioxide is the most important GHG; its emissions are currently at 37 Gt/year. CO_2 content in the atmosphere increased from 300 ppm in 1950 to the current 410 ppm [3]. Additionally, there are other greenhouse gases that are more potent, albeit in lesser amounts. Methane CH_4 is the most significant, followed by NO_x, and F-bearing gases. The total equivalent GHG emissions are estimated at about 52 Gt/year and 56 Gt/year when land-use, land-use change, and forestry (LULUCF) are taken into consideration.

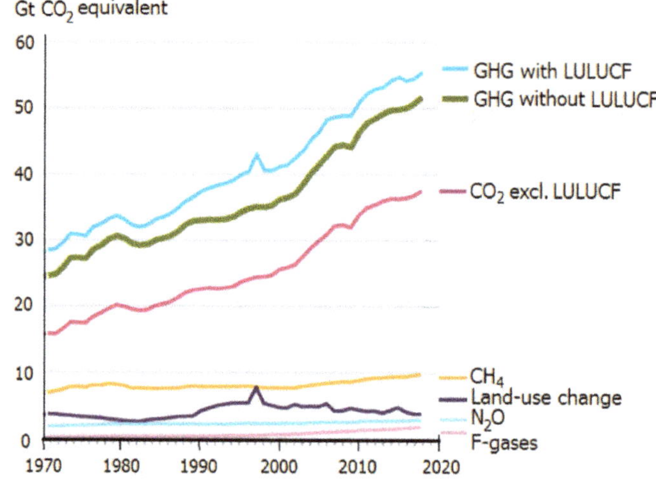

Figure 1. Global greenhouse gas emissions, 1970–2018. Modified from [2].

The recent course of GHG emissions indicate rapid climate warming, 3–5 °C by 2100 (Figure 2). The United Nations' Intergovernmental Panel on Climate Change (IPCC) has stated that CO_2 concentration must be stabilized at 450 ppm to have a fair chance at avoiding global warming above 2 °C, which was set as a limit at the COP 21/CMP 11 Conference in Paris, December 2015 [1]. Later, this target was brought down to 1.5 °C at the COP 24 meeting in 2018 [4]. In Figure 2, feasible future scenarios are shown, together with the last 50 years' history of equivalent CO_2 emissions. Current policies or nationally determined contributions (NDC) are not effective, and more radical actions are needed. The 2 °C pathway means a roughly 50% cut in emissions by 2050, and the 1.5 °C target indicates an 80% cut, respectively.

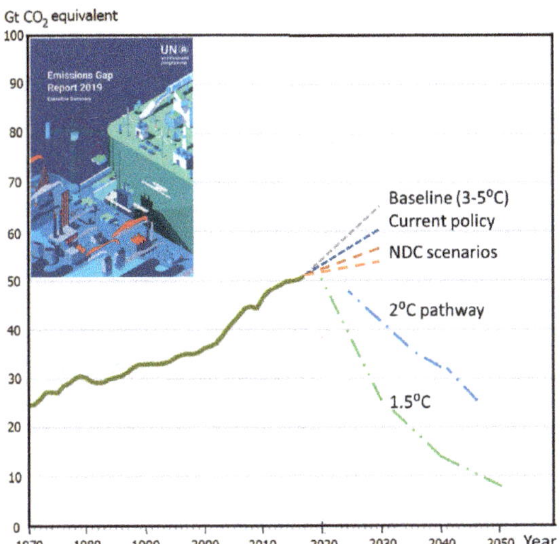

Figure 2. History of global greenhouse gas (GHG) emissions and different scenarios till 2050 [3].

The key question is: how can we stop the growth of CO_2 emissions and drastically lower the curve in the 2020s? The primary fault lies in fossil fuel being a major source of energy. Although renewable energy has gained publicity since the 1990s, its role is still minor—about 14% of all energy supply, whereas fossil energy represents 81% (Figure 3). The remaining 5% is nuclear power. For electricity generation, the corresponding percentage contribution is 26/63/11 for renewables, fossil, and nuclear energy, respectively [4,5].

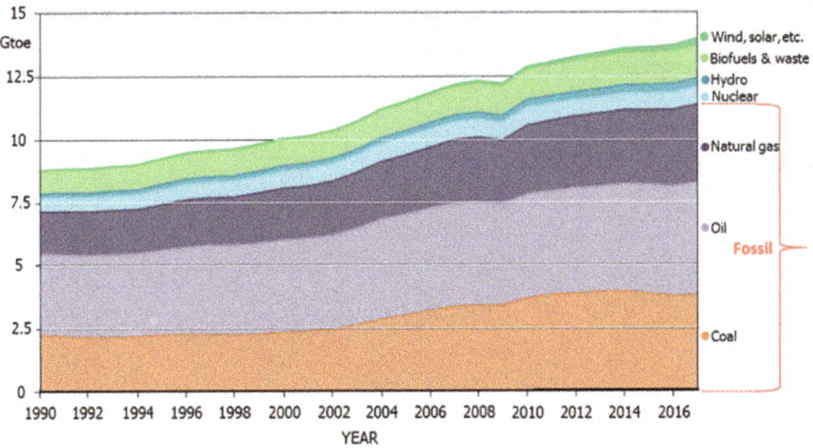

Figure 3. Total primary energy supply by source, globally, 1990–2017 [5].

The global energy production exceeded 14 Gtoe (gigatons oil equivalent) in 2018 [5]. The share of fossil energy was 11.3 Gtoe, which is in accordance with anthropogenic carbon emissions, including fuel combustion and cement production, 37 Gt CO_2/year [2].

2. Progress of the Steel Industry and its Role in Energy Consumption and CO_2 Emissions

The overall progress of world steel production over the last 150 years is shown in Figure 4a [6]. In the early 19th century, the world annual steel production was only a few million tons. After the breakthrough of new technologies, converter processes, and open hearths, production increased and exceeded 30 Mt in 1900. In 1927, steel production reached 100 Mt and 200 Mt in 1951. The next 30 years after the II World War was a period of "new industrial revolution" with innovative novel processes. Extensive investments were made in the steel industry, with Japan, Soviet Union, United States, and South Korea in the vanguard. The annual steel production reached 700 Mt in the 1970s (record 749 Mt in 1979). The growth then stagnated due to economic crises and political changes until the turn of the millennium, when it reached 850 Mt in 2000. This was the overture to the "boom" with China in the forefront. Since then, the world production has doubled and the record so far is 1,869 Mt, attained in 2019 [6]. China´s share is over 50%. Today, China´s domestic steel demand has reached an "established level" and eventual growth is directed towards export. Meanwhile, India has strongly increased steel production and has risen to second place with 111 Mt/2019. It is plausible that in the near future, the consumption in developing countries will grow. Earlier scenarios predicted continuous growth up to 3000 Mt/year in 2050. Owing to the recession period, the stabilization in China, and the newest goals of "stop climate change", the current scenarios are more conservative and an estimate of 2500 Mt in 2050 can be considered realistic [7,8]. Future scenarios until 2050 are sketched accordingly in Figure 4b.

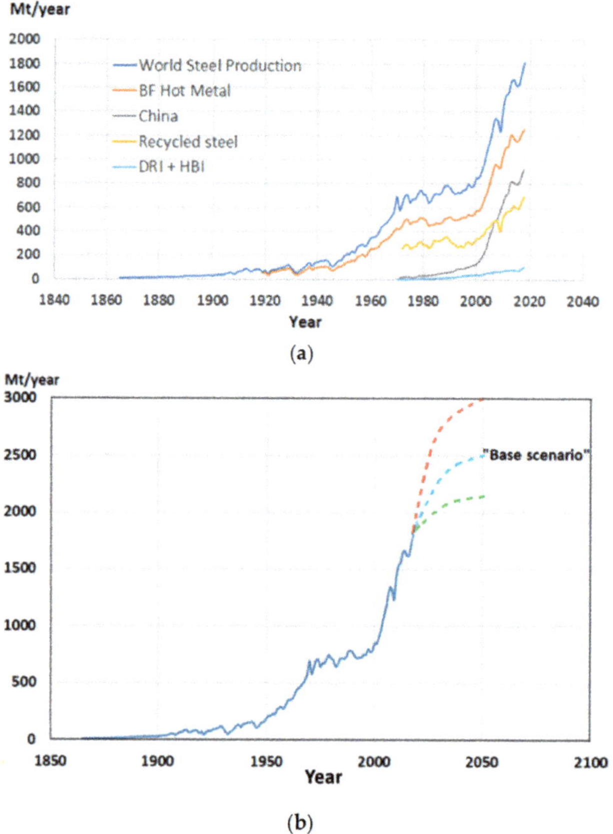

Figure 4. (**a**) World production of steel, BF hot metal, and DRI from 1860 to 2018, including recent steel production in China and estimated recycled steel. (**b**) History of steel production and future scenarios.

Figure 4a also shows the amounts of main raw material of steel: blast furnace hot metal (BFHM), recycled steel (RS), and direct reduced iron (DRI + HBI). Of these, BFHM is mainly charged into converters to make steel, whereas RS and DRI go into electric furnaces. In 2018, 72% of crude steel came from converter processes based on BF hot metal. Electric furnaces produced 27.6%, utilizing recycled steel scrap as the main iron source, with a minor share of direct reduced iron. The balance 0.4% was produced in open hearths, which is currently a declining technique [6].

Ironmaking is an extremely energy-intensive process, utilizing coal as the main primary energy source. The steel production was responsible for 7–9% of direct emissions from the global use of fossil fuel [9]. The specific emission was estimated at 1.85 t CO_2/t steel, corresponding to 3.3 Gt CO_2 in a year at the production rate of 1.8 Gt/a. The predicted growth of steel demand/production by the year 2050 was discussed afore as ending in 2.5 Gt/a (Figure 4b). The "current policy" would lead to annual emission of 4.5–5 Gt CO_2, which would be a disaster, as it is insufficient to stop the growth; therefore, we must radically cut emissions. By tracking the "2 °C pathway", the emissions could be halved and the "1.5 °C pathway" could mean reduction by 80% [3]. In proportion to the steel industry, total emissions should be reduced to 1.5–0.75 Gt CO_2/a, corresponding to specific CO_2 emissions of 0.6 – 0.3 t CO_2/t steel. The fall from 1.85 t CO_2/t steel is, thus, dramatic. A pertinent question follows: By what means can we attain this level by 2050?

The author of this paper examined this problem in 2011 by analyzing production practices in different countries; comparing them with BAT (Best Available Technology) values; estimating emissions from different energy sources (including electricity); and studying the potential of energy-saving actions, and low-carbon and carbon-free innovative technologies [10]. The present contribution, although based on previous studies [10,11], is an updated, generalized version, taking into account the extensive recent developments and numerous works of literature.

3. Review of Means to Cut CO_2 Emissions from the Steel Industry

We have identified several key factors that make it possible to reduce CO_2 emissions from steel production. Some of them are incremental improvements in current processes, whereas the others can be regarded as radical breakthrough technologies for iron/steel making or energy supply/usage. Both incremental and radical improvements are useful and necessary to be implemented due to the huge scale and inertia for change as well as the urgent schedule of having only few decades to achieve the set goal.

3.1. Improving Energy Efficiency

The first key factor is to improve the energy efficiency of current processes—the fastest way to stop the growth of emissions at a moderate cost. A previous comprehensive study by IEA/OECD analyzed steel industries in different countries and showed an energy-saving potential of 4.1 GJ/t steel (corresponding to 20% reduction from the current world average) [12]. The saving potential varied from 1.4 to 8.7 GJ/t in different countries, with the largest savings slated for China, Ukraine, Russia, India, and Brazil. China has made big efforts and decreased its specific energy consumption by 15% from 2006 to 2017 [13]. The total energy consumption or intensity in BF-BOF steel production varied in different countries from 19 to 26 GJ/t crude steel in a recent, rather inclusive benchmarking [14]. Despite great advancements by the world steel industry, energy consumption can still be reduced by 10–15% on average to meet the BAT values by applying best available technologies [15,16]. Even bigger deduction of CO_2 emissions is possible by transfer to low-carbon energy sources.

Any comparison of energy efficiency between different countries or even steel plants is not fair if the boundary conditions (raw materials, energy sources, processes, products etc.) are different. For such comparisons and evaluation of any kind of process changes, it has proved illustrative to set the different process routes (ore-based and scrap-based steel production) at the ends of the *x*-axis and to examine the specific energy consumption (GJ/t steel) against the recycling ratio, defined as percentage of scrap from total Fe input [10,11]. Correspondingly, the specific CO_2 emissions (t CO_2/t steel) can be presented as in Figure 5. The % REC means percentage of recycled steel i.e., scrap. The case % REC = 0 corresponds to 100% ore-based BF–BOF (or in the case of smelting reduction SRF–BOF) route steel production. On the other hand, % REC = 100 means 100% recycled steel (scrap) based production i.e., the EAF route. The direct reduction process (DRI production) cannot be put in a diagram on its own, but the melting stage in EAF, normally with some scrap, must be included. Then, the position on the *x*-axis depends on the scrap/DRI ratio used in the EAF. This kind of representation is apt, e.g., for comparison of the state of the steel industry in different countries as well as evaluation of new development steps. In Figure 5, the present level is outlined. The full BAT line was drawn based on rather conservative data by Worrell et al. [16]. The BAT Line Range was outlined based on different CO_2 emissions from electricity generation, the low line referring to low emission electricity (hydro/nuclear power) and the high line permitting coal/oil/gas as primary energy. In this scale, the current position of "world steel" W is at 1.8 t CO_2/t steel vs. 35% REC [9]. The % REC value takes account of the usage of scrap in BOFs. Hence % REC is notably higher than the percentage of EAF production (see e.g., China). Further, Japan, the European Union, Germany, France, Canada, the United States, and Italy were evaluated in Figure [13,14,17–20]. The drop in overall world average from 1.8 t CO_2/t steel to the BAT level would mean a reduction by 15–20%. This could be achieved by modernizing plants, adopting best available technologies, and closing old-fashioned units in China and other countries—i.e., a certain "low-carbon retrofit".

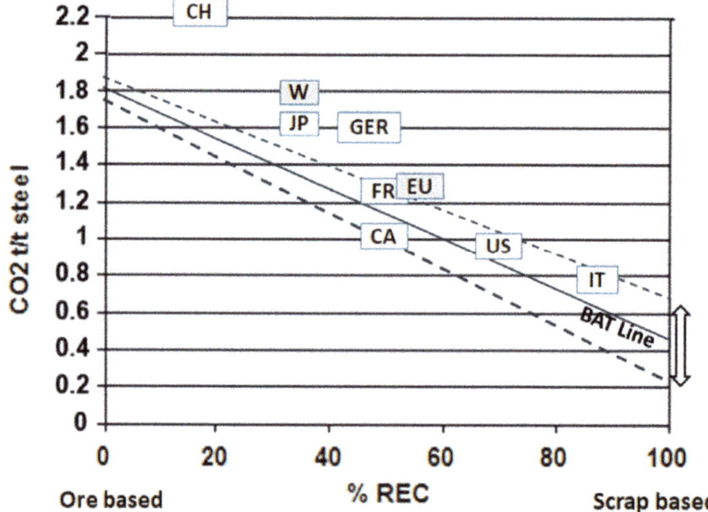

Figure 5. Specific CO_2 emissions from fossil fuels and electricity in iron and steel production as a function of recycled steel ratio (% REC). BAT line and range were approximated based on published data [13,14,16–20]. The country codes in the figure are revealed and values commented in the text.

The positions of world, EU, and different countries in Figure 5 are only an approximate, as both the published CO_2 emissions and estimated % REC values were not necessarily based on equivalent premises. As remarked before, the % REC also rules in the scrap used in converters, whereas DRI is counted as "ore-based" iron raw material. A position in relation to the BAT line relates to the technology level but can also incorporate other factors. For instance, Canada's outstanding position is partly owing to the substantial share of natural gas as the primary energy source. In contrast, Germany's relatively high value is a result of the big role of coal in ironmaking as well as in electricity generation. Generally, in cases with high EAF share, electricity generation emissions have a strong influence, illustrated by the expanding BAT range.

3.2. Potential Means to Mitigate CO_2 Emissions in Ore-Based Production by Improving and Modifying Current Technology

Numerous R&D programs and projects have been carried out over the last 10-15 years or are on-going. The European ULCOS program, COURSE50 in Japan, the POSCO program in Korea, the Australian CO_2 Breakthrough Program (CO2BTP, ISP), and the AISI CO_2 Breakthrough Program in North America as renowned examples [21–30]. In addition, comparable projects have been reported in China, India, and Taiwan [13,17,31–34]. Without going into details of the programs, the main improvements that could have a potential influence on the reduction of CO_2 emissions are summarized below:

- Transfer to coke dry quenching in coke making (CDQ)
- Production of high strength coke to better utilize hydrogen reduction in BF
- Top-pressure Recovery Turbine (TRT) technology in BFs and dry dedusting systems for BFG and BOFG (blast furnace and converter off-gases)
- Incorporation of oxygen enrichment technology in hot stoves
- Integrated optimized usage of off-gases in-plant, for electricity generation and district heating
- Top gas recycling in oxygen blast furnace (TGR-OBF)
- Enhanced utilization of unused steel plant waste heat (off-gases from sintering, coke plant, hot stoves, converters, reheating furnaces, etc.)
- Heat recovery from slags

The applicability of these technologies has been tested at least in pilot scale, and some represent well-established technologies (e.g., CDQ, TRT). By nature, technical improvements result in energy saving, thus indirectly reducing CO_2 emissions. The influences are proportionately minor, but significant on a global scale. Integrated systems with heat recovery, optimized internal recycling, substitution of air with oxygen; and actions that increase the role of hydrogen as a fuel and reductant are powerful.

Top gas recycling, combined with an oxygen blast furnace (TGR-OBF), was a central concept in the European ULCOS program [21–23]. The idea was presented in the early 1970s [35], but proved to cause big changes in gas flow, heat distribution, and reaction zones, compared to air or O_2-enriched conventional (max. 30% O_2) furnaces. Coupling with top gas recycling seemed to solve these problems by increasing the blast gas volume—a part of the top gas is recycled back to the furnace to utilize its CO and H_2 as reducing agents and fuel. In order to enable rational recycling, the CO_2 (and H_2O) in the recycled top gas must be removed (CCS). With a small amount of inert gas N_2, the blast temperature can be decreased, which means savings in hot stoves. The reduction in CO_2 emissions was estimated to be in the range of 10–25% [22] (Figure 6). However, the TGR-OBF-CCS technology is relatively complicated, electricity consumption increases, and less electric power can be generated [27,36–38]. An industrial TGR-OBF installation was put into operation in Anshan, China, in 2012 [39].

In conventional ironmaking, hydrogen is retained in the background in BFs, as the volatile hydrogen-rich matter in the coal-to-coke process is removed in coke oven gas (COG), which is an important heat source and energy storage in an integrated steel plant. Interest in co-injection of COG in BFs has recently increased [24–26]. COG decreases coke consumption and CO_2 emissions in BF, whilst also cutting off a bit of the total energy supply of the integrate, which then must be replaced by another energy source. Another efficient way to decrease coke consumption is by powdered coal injection (PCI) through BF tuyeres. This idea dates back to the early 1980s when it was first put into operation in Japan as a substitute for oil that had increased in price [40]. PCI is now a common practice—up to 30% coke can be replaced; higher amounts are possible, but for smooth operation of the BF process, the properties of coke are extremely critical. Replacement of coke with coal decreases the flame temperature, the effect of which is compensated by oxygen enrichment in the hot blast. Generally, the replacement ratio of typical coke-substitutes (kg coke/kg substitute) is less than 1.0 (typically in the range of 0.7–0.9). At high injection rates, the ratio tends to decrease, which means an increase in total energy consumption and hence a positive effect on CO_2 emissions is lost. Instead of coal powder, other powdered materials can be used e.g., waste plastics, char, sawdust, and other biomasses, providing sufficiently high energy content and low impurity. Waste plastic injection was started in Bremen in 1995 [41] and has become a common practice in Japan as well.

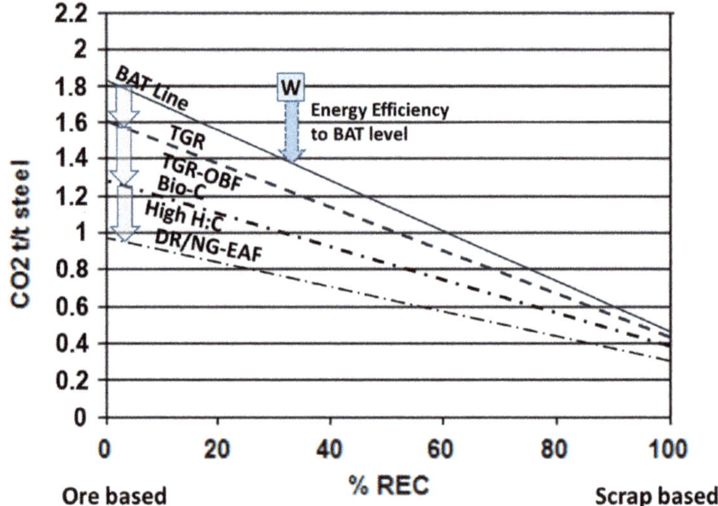

Figure 6. Influence of different means to mitigate CO_2 emissions in ore-based production by improving and modifying current technology. Acronyms are explained in the text. Outlined based on literature data [21–40].

A reawakened trend is to introduce "renewable biomaterials" as a solution to the problem of global CO_2 emissions. Since 2000, several extensive studies on the potential of wood as a substitute for fossil energy in iron and steel production have been conducted e.g., [42–46]. The main interest has been in countries with a high percentage of forests and strong tradition of the wood industry, or countries with high potential for wood plantation (for example, Finland, Sweden, Australia, Brazil). Possible sources of charcoal could be both waste wood from harvesting and industries and large-scale wood plantation, which has been in industrial use in Brazil for pig iron and ferroalloys production since the 1970s [47,48]. Charcoal is made of planted eucalyptus wood by pyrolysis. Pig iron is produced in mini blast furnaces, where 100% charcoal operation is possible. The overall coke/charcoal ratio in Brazilian pig iron is about 80/20. In big BFs, a certain amount, approximately 250 kg coke/t HM, is necessary for smooth operation of the high shaft reactor with complex chemical reactions, counter-current solid-liquid/gas flows, and heat transfer phenomena. Except for substituting charcoal for coal injection and nut coke in BF top charging, it can be used in sintering, as a carbon carrier in EAFs, ladle metallurgy etc. Charcoal is an excellent material owing to its chemical purity and high heat value compared with coal. In addition, the carbon in biomasses ends up in the atmosphere as CO_2, but can be compensated by planting new trees, which then bind CO_2 from the atmosphere and liberate O_2 via photosynthesis. The whole cycle is not emission-free, as both the plantation and usage with preparation processes need energy and produce emissions. In Figure 6, the maximal potential of biomass on CO_2 emissions was estimated as being 40%. In reality, any substantial deployment is limited to countries and areas where the circumstances are favorable for massive, ecologically reasonable biomass production. On a global scale, socio-economic issues, land use, requirements for food production, etc. restrict any extensive 'renaissance' of charcoal blast furnaces. We remember how the early history of ironmaking was based on charcoal and this was a cogent reason why forests disappeared widely in Great Britain and the European continent in the Middle and New Ages by the end of 1800, when the use of coke emerged and created preconditions for the 'steel revolution' [49]. The dominance of coke has lasted for over 200 years but is now ending and novel remedies must be discovered. The 'return of biomass' can be a reasonable supplementary recourse for the steel industry in restricted regions/circumstances, provided that the production is ecological, low-emitting, and economical, wherein carbon-tax is decisive.

3.3. Potential Means to Reduce Emissions by CO_2 Capturing and Storage as well as Utilization

The idea of capturing CO_2 is well-known and has been proven since the 1930s to purify natural gas, hydrogen, and other gas streams on an industrial scale. In 1972, the first commercial-scale operations were performed to inject captured CO_2 into underground storage [50]. CO_2 capture and storage has now become a well-established practice in oil and gas production, led by the U.S. and followed by several oil-producing countries. During the last decades, the interest in the applications of this procedure has extended to energy production, cement, and chemical and metallurgical industries. Different techniques have been investigated and developed to capture CO_2: chemical absorption (amines like MEA), physical absorption (PSA), or membrane or cryogenic distillation-based separation process. Numerous technologies are under development, in pilot scale or close to demo level. The capacity of projects currently operating or under construction is about 40 Mt CO_2 per annum [50]. The IEA's World Energy Outlook 2019 presents a scenario wherein CO_2 capture, utilization, and storage (CCUS) could provide 9% of the cumulative emissions reduction between now and 2050 [51]. The annual capture was calculated as 2.8 Gt CO_2. On the other hand, estimates by the IPCC of the amount of CO_2 that must be stored using CCS are much higher—around 5 to 10 Gt/a by 2050—in order to limit global warming to 1.5 °C [52]. As for the technical realization, the most common course is geological sequestration, generally connected to oil and gas fields, when oil recovery can be notably increased by pumping CO_2 into the empty wells (enhanced oil recovery EOR). This improves the combination economy, but at the same time tends to accelerate the use of fossil energy with a negative impact on climate warming. The storage capacity of the empty wells is limited, and more capacity has been traced in appropriate geological formations. Rough estimates of the disposable "saturation" capacity ranges in hundreds of years. Nonetheless, CCS cannot be a long-term solution to the climate problem, but only a temporary mitigation method. A better sustainable way is mineral sequestration, in which CO_2 is exposed to react with suitable minerals (Mg, Ca-silicates, serpentines etc.) to form stable carbonates [53,54]. When mineralization is performed in situ, the CO_2 must be separated from the off gas and transported to the mine for the mineralization process. At the plant site, mineralization can be performed directly from the off gas. In such cases, it is reasonable to make mineralization e.g., by using steelmaking slags [54–56]. Altogether, the entire process of CCS consumes a remarkable amount of energy, although less-energy consuming methods for CO_2 capture are under development [50].

Principally, a better method might be carbon capture combined with carbon utilization, CCU. One method is the aforementioned EOR use, but more genuine are applications where CO_2 is used as an input to the production of value-added products, resulting in real emissions decrease, as the gas is permanently stored. Traditionally, CO_2 is used in beverage production, metal welding, cooling, and fire suppression. In those applications, captured CO_2 is used to replace CO_2 specially made for that purpose. An example of 'true GHG use' is in horticulture, where the greenhouse atmosphere is enriched with CO_2 to accelerate photosynthesis and growth rate. The captured CO_2 can then displace the gas produced from fossil origin for that purpose. Another example is saline algae cultivation, which could utilize and bind CO_2 and produce raw materials for various bioproducts [57]. Captured CO_2 is already in use for natural gas processing and urea fertilizer production [50]. According to a new 'waste-to-energy' concept, CO_2 should be converted into hydrocarbons and then into liquid or gaseous fuel whenever it displaces a product of fossil fuel origin, chemicals, plastics, and building materials. In different industrial sectors, there are numerous, potential uses for captured CO_2 [58]. The estimated amounts of possible reuse of CO_2 in different applications vary from tens to hundreds of millions, eventually billions of tons per year. Although these are big quantities, CCU will not be a true solution to CO_2 emissions but only a complementary, short-term way. Further research in the field to develop competitive processes and products, as also emphasized in the reports by the Global CCS Institute and IEA [50,52], is important.

The review above concerns the general status of CCS, CCU, and their combination CCUS. With regard to iron and steel production, CO_2 capture has been examined widely and tested but not yet widely deployed. Different techniques have been scrutinized to capture CO_2 from blast furnace gas by applying chemical absorption or physical absorption/adsorption [21,22,25,31,37–39,59–62]. In addition, CO_2 capture from the "end of pipe" off-gas of the whole plant has been examined. In general, these capture techniques require a lot of energy. Therefore, an essential objective is to reduce energy consumption in CCUS. Potential means are, for example, to utilize sensible heat of slags by applying a high-efficiency heat exchanger. For that purpose, process gases coming from the coke plant, blast furnace (BF), and basic oxygen furnace (BOF) containing CO_2, CO, CH_4, and H_2 can be recovered [27,60]. A potential 'waste-to-energy' technology is gas fermentation, which uses works arising gases (WAGs) to produce ethanol. LanzaTech in Illinois U.S. has developed a microbial bioreactor system capable of direct gas fermentation to produce ethanol from carbon-containing gases like integrated iron and steel plant off-gases [63,64]. A demonstration scale plant has been in operation in Shougang China since 2018, and a bigger demonstration plant with an estimated capacity of 64,000 tons ethanol/year is under construction at Arcelor Mittal Ghent, Belgium [64]. This STEELANOL project is supported by the EC Horizon2020 program [65]. Comparable polygeneration concepts involving biomass used in a steel plant and utilization of off gases to methanol production, district heat, and electricity have been proposed e.g., in [66,67].

3.4. Hydrogen Economy—Definitive Solution Toward Carbon Neutral Society?

The actions discussed above to improve or modify the current BF-based ironmaking technology can reduce CO_2 emissions maximally by 50%, i.e., to the specific emission level of 1000 kg CO_2/1ton steel. In the global overall context, this is not enough; a more radical leap is vital. The wide-ranging use of hydrogen as a substitute for coal/coke is a potential solution, both in ironmaking and in energy use, transportation, heating etc. [68–72]. As hydrogen is not a natural resource like coal, it must be first produced. The main technologies to produce hydrogen are based on steam reforming of natural gas or oil, which produce approximately 95% of the current global H_2 production ($\approx$70 Mt/y) [70]. This hydrogen is, by no means, carbon-free and is called "grey hydrogen". In addition to the reformed hydrogen, roughly 50 Mt/y H_2 is used in mixed gases e.g., for direct reduction of iron (DRI) as well as in BFs [69]. In order to produce low-carbon hydrogen through natural gas/oil reforming or coal gasification, CCS must be adopted ("blue H_2"). These technologies are well-established on an industrial scale; future development should include indispensable infrastructure for transmission, storage, and distribution, which will greatly facilitate the emerging hydrogen economy. Instead of fossil-derived hydrogen, the so-called bio-hydrogen is in the spotlight. Various agricultural and industrial residues and municipal wastes are possible raw materials for hydrogen production. Both thermochemical methods and biochemical processes (based on algae, fermentation) have been investigated [71,72]. Additionally, hydrogen production via water electrolysis at room temperature or high temperature (solid oxide electrolyte cell, SOEC) [69] is a well-known principle with different technologies. Combination of SOEC with SOFC (solid oxide fuel cell) might be a true promoter of the development by involving means of energy production, storage, and usage. Hydrogen produced via these technologies as well as bio-hydrogen can be called "green", providing the electric energy is fossil-free.

Along the same lines, the impact of hydrogen on CO_2 emissions from iron and steel production depends greatly on the "purity" of hydrogen, the treatment of the accompanying CO_2 (CCS, CCU), and the carbon footprint of the electricity (grid), which is used in all the stages—from raw materials and energy to final steel. The lowest line in Figure 6 starting from 1000 kg CO_2/t steel describes the current best technology using DR + scrap and average electricity. By moving to "pure" H_2 and "green" electricity, the emissions decrease, both on the primary "ore-based" side and on the secondary "recycled" steel side, as will be discussed later in Paragraph 4. The first industrial scale trials with hydrogen reduction were in the 1950s when the H-iron process was introduced [73]. The idea was to produce fine Fe powder in a fluidized bed at temperatures below 500 °C. Several processes were developed based on the fluidized bed principle and the exclusion of the agglomeration process of iron ore fines as the driving force, but with scant success. For high productivity and metallization, reduction should be made at higher temperatures, which, however, resulted in "sticking" when metallic iron was formed. This is a problem well known since the 1950s; a current publication summarized its phenomena and mechanisms [74]. Among the numerous fluidized bed attempts, the Circored process is still in an active state. It was developed for reformed natural gas but was proved for pure hydrogen as well [75]. Another method for reduction of fine iron ore concentrates is Flash Ironmaking Technology (FIT), developed at the University of Utah by Prof. H-Y. Sohn and his group [30,76,77]. This innovative alternate ironmaking process utilizes a flash reactor, well-known in non-ferrous smelting for Cu and Ni sulfides. The FIT reactor can be operated with different reductant gases, hydrogen or natural gas, and possibly bio/coal gas or a combination. There is an article on the subject in this issue [30]. Overall, just as conventional ironmaking is based on a shaft furnace reactor (BF), the direct reduction takes place in shaft furnaces using natural gas (Midrex, Energiron). This also concerns hydrogen reduction. The European research program ULCOS began in 2004 and had several projects on the application of hydrogen in iron and steel production [21,78–80]. The development work has continued in several projects aiming at pilot and industrial scale installations. The H_2FUTURE project in Austria will prove the PEM electrolyzer for hydrogen production [81], the GrInHy2.0 project in Salzgitter Germany strives to utilize steam from industrial waste heat in SOEC for hydrogen production [82], and the HYBRIT project aims at the development of a fully fossil-free steel production chain [83,84]. HYBRIT (Hydrogen Breakthrough Ironmaking Technology) is a Nordic endeavor with backing from three companies, LKAB (iron ore mining and pelletizing), SSAB (steel manufacturer), and Vattenfall (power utility). The project was launched in 2016 and will continue with pilot reduction trials and hydrogen production and storage until 2025; it will proceed to demonstration plants and industrial scale transformation in 2025-40 and fossil-free production as the final goal in 2045 [85]. The influence of hydrogen deployment on CO_2 emissions are discussed later with respect to the summary part, Paragraph 4.

3.5. Clean Electricity—The Major Energy Form in the Future

As noted before, different ways to generate electricity cause very different CO_2 emissions. Consequently, the influence on total emissions can be substantial, depending on the share of electricity in iron and steel production. The impact is determined in EAF steelmaking, noteworthy in reduction processes, and will be critical in transition to hydrogen reduction, including H_2 production (steam reforming, electrolysis, etc.). In a power station with coal as the primary energy source, specific CO_2 emissions can be over 1000 g CO_2/kWh (mean value 820 g CO_2/kWh), whereas in a modern gas power station, the corresponding figure is below 500 g CO_2/kWh (Table 1). Direct emissions from biomass combustion are relatively high, but a modern dedicated biomass-fired CHP system has much lower emissions. All types of fuel-fired power stations can be reformed to low-emitting units ($\leq$200 g CO_2/kWh) by CO_2 capturing (CCS). However truly low, almost zero emissions, can be attained only via non-fossil/renewable technologies, solar, wind, nuclear, and hydro power stations, where the emissions are in the range of 10–50 g CO_2/kWh (Table 1) [86,87].

Another way of tracking the general global progress is to examine the CO_2 emissions of national electrical grids i.e., emissions per generated kWh in different countries. China is the biggest steel producer and used to have quite high emissions from coal power stations; however, with modernization, the figure fell to 620 g CO_2/kWh in 2017 [88–90]. India, the second-biggest steel producer had 723 g CO_2/kWh, followed by Japan (492 g CO_2/kWh), U.S.A (420 g CO_2/kWh), and the EU (282 g CO_2/kWh). The world overall electricity generation emitted 484 g CO_2/kWh on average. The recent progress has been mildly positive i.e., decreasing, although on the global level, the reduction has been only 10% in 30 years, whereas the demand has more than doubled [89]. Hence, a real radical decline must be reached for. By adopting technologies that utilize non-fossil or renewable primary energy, grid emissions can go an order of magnitude lower. Nordic countries are examples of "green electricity"; Finland, Norway and Sweden reported 83, 19, and 9 g CO_2/kWh for their national grids in 2017 [88]. As a final comment to the reader: the values in the selected studies are comparable with each other, but not necessarily with other studies/references. There are several reasons for this: the calculation methods and system boundaries can vary; sometimes, only "direct" emissions are counted, instead of total "life-cycle" emissions, etc.

Table 1. CO_2 emissions (g/kWh) from electricity generation by using different fuels/technologies in power stations. The values are estimated life-cycle emissions or ranges. IPCC Technical Summary Report, 2014 [87].

Primary Energy—Fossil		Bio	New Technology with CCS	
Coal	Natural gas	Biomass	Coal + CCS	NG + CCS
820	490	740; 230 [1]	160–220	170
Primary Energy	**Renewable/Non-fossil**			
Geothermal	Hydro	Nuclear	Solar	Wind
38	24	12	48	12

[1] Cofiring vs. dedicated.

3.6. Increasing Recycling—A Key Factor in Overall Reduction of CO_2 Emissions

Reduce, reuse, remanufacture, and recycle is a current trend and necessity, which has seen a breakthrough during the last few decennia. Less waste and the reduced need to utilize primary, virgin resources are the main goals. Recycling concerns both short- and long-lived materials. In history, recycling of steel was extremely valued throughout the Iron Age before industrialization. Because steel was a rare and expensive material, village blacksmiths used to store disused steel articles to remanufacture second-hand products, in archaic words, "beat swords into ploughshares". When steel became a mass product, its price fell and remanufacturing almost disappeared. However, collection of scrap and delivery to steel plants has been duly organized for long in industrialized countries. The recycling rate is moderate, and nowadays vigorously increasing e.g., by recovering rebar steel from concrete of demolished buildings. A major part of purchase scrap is used in electric arc furnaces, and a smaller share in converters as a coolant, typically 15–25% of the iron charge, which includes a significant share of the internal plant scrap as reverts from different process stages. Until the turn of the millennium, the global share of EAF was smoothly growing and reached 34% in 2001, but then the rapid growth in China, based on the BF + BOF route, brought the ratio down. Today, EAF´s share is not full 28%, whereas BOF has 72% [6]. The EAF share varies in different countries. It was 100% in 44 countries (in total 44 Mt in 2018), 69% in the United States, 41% in the EU, and 10% in China) [6]. The recent trend of scrap in steel making is shown in Figure 7 [91]. Purchased scrap means external "obsolete" scrap. Another column is internal "own" scrap, which can be counted as *Scrap use—Purchased scrap*. As the exact data of scrap use is missing, the total scrap amount was calculated based on the balance of Fe metallics, including BF hot metal, DRI+HBI, and scrap; and world steel production.

The principle of "circular economy" has recently gained ground. Intensified scrap usage is a self-evident goal. Recent scenarios assume significant growth in the availability and use of steel scrap. Along with the improved consciousness on recycling, another reason is that the sudden growth in steel production and its use in the early 2000 will inevitably reflect in scrap availability a few decades later. As for scrap dynamics, the "age of scrap" or lifetime from production and usage to recycling varies from a few years to decades, or in some constructions, even centuries—as a rule, resulting in 30–40 years [92–94]. Consequently, the amount of scrap should strongly increase from 2020 to 2050 [6,95,96]. This points towards a substantial growth in EAF steel making, whereas BOF production would stay approximately at its present level on a global scale. However, regionally e.g., in China, the availability of scrap will increase and the EAF route will grow remarkably from the current 10% by partial replacement to the BF + BOF route. A general increasing demand will incite recycling and raise collection rates. A scenario for scrap use in steel making is presented in Figure 8 [91,96]. The notation 'Scrap' refers to the usual recycling practice and 'Scrap+' to boosted recycling rate, including strong parallel actions in China as well. In this scenario, the available scrap is estimated to be 1400 Mt in 2050, which is in fair agreement with the estimate of the World Steel Association—1300 Mt [97]. Both these estimates support the forecast that the scrap ratio will rise to the 50% level by 2050. This is extremely important in relation to the issue of CO_2 emissions.

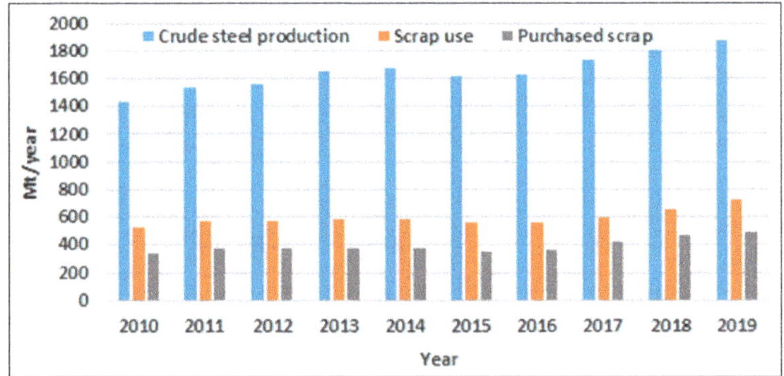

Figure 7. Scrap use for steel production and amount of purchased scrap as well as the total world steel production 2010–2019 [6,91].

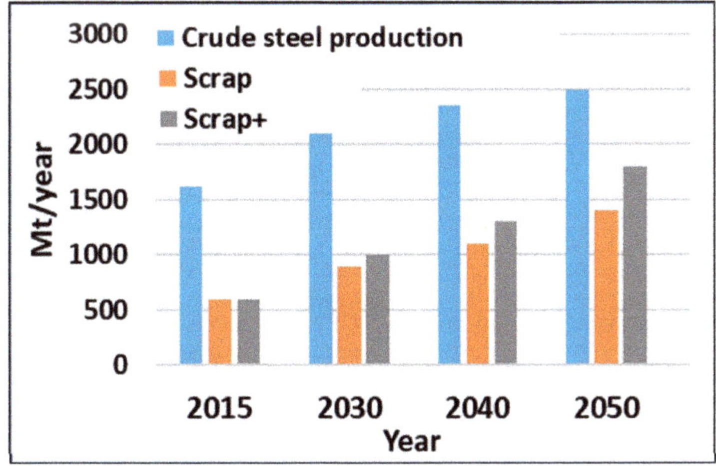

Figure 8. Scenario for scrap use in steel production from 2015 to 2050 related to world steel production. Redrawn based on scenarios in the literature [95–98].

4. Summary of the Means to Cut CO_2 Emissions from the Steel Industry—A General Vision

Earlier in Figure 6, different means to decrease specific CO_2 emissions from conventional ironmaking were examined. It was concluded that by installing and putting into operation the most advanced technologies for improved energy efficiency and heat recovery, along with decreasing the carbon/hydrogen ratio in fuels and reductants, the emissions can be roughly halved to the level of 1000 kg CO_2/t steel. This value is the new "baseline" in Figure 9. In the preceding chapters, several other "key factors" were discussed and their influences are now illustrated. Carbon capture and storage (CCS) and utilization (CCU) are potential ways to decrease CO_2 emissions from steel plants.

Their deployment would be most reasonable and economic in oxy-fuel processes with CO_2-rich off-gases. The same concerns biofuels and reductants. All these means are potential and rational components in the fight against climate change, but they cannot be a final comprehensive solution. In the long term, CCU might become a "viable weapon". Locally, in restricted regions and favorable conditions, the impact can be eminent in the short run. The predictable increase in scrap ratio in steelmaking, on the world level from ≈30% to 50% in 2050, is a very significant change. As such, it will reduce specific emissions in the world scale and cause a marked transmission to EAF production. The emissions from electricity generation are thus emphasized upon. The arrow on the right axis shows the range from fossil electricity (F-E) to world average today (W) and down to renewable electricity (RW-E). Rapid progress in hydrogen economy; production; and use in industries, energy, transportation, and other fields will also give the steel industry a strong push. The breakthrough of hydrogen and non-fossil renewable electricity generation are the key factors, by means of which the universal targets of CO_2 emissions and climate change can be honestly achieved.

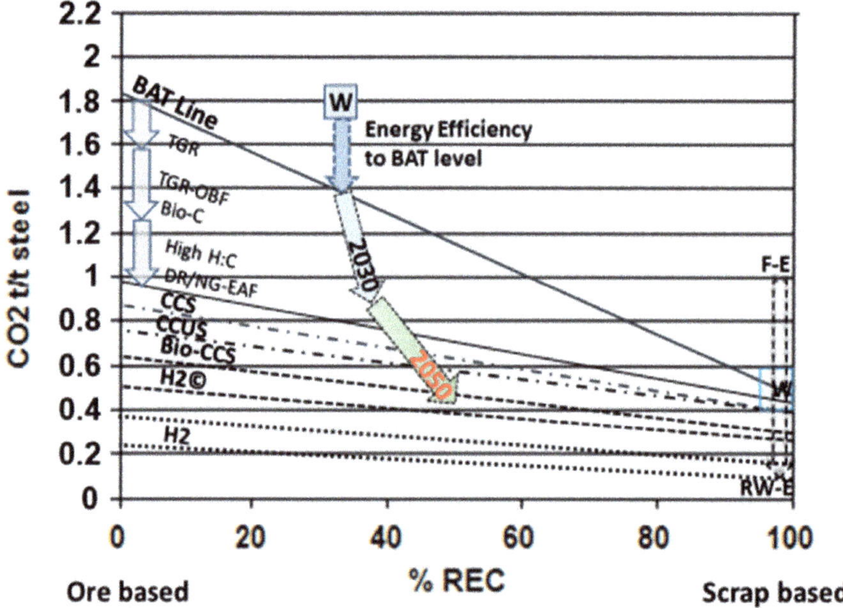

Figure 9. Summary of the potential means to mitigate specific CO_2 emissions in iron/steelmaking. Different lines show the assessed levels achievable by adopting the labelled actions. The evaluation is based on the literature [17,85–91,95–99] and the author´s own assessments. Three arrows starting from the current world position W → 2030 → 2050 show a plausible vision/pathway for the world steel industries. More comments in the text.

The first steps in that direction were taken in ironmaking (HDRI); the efforts toward "green" electricity will have a strong reducing influence on specific CO_2 emissions on both sides of the diagram (Figure 9). The third arrow in the figure ends at the level of 0.4 t CO_2/t steel. By multiplying with the scenario production of 2.5 Gt steel in 2050, the emissions from the steel industries ≈1 Gt CO_2/year are obtained as the outcome, which is roughly 1/3 of the present emissions and thus falls between the climate warming scenarios for 2 °C and 1.5 °C in Figure 2. However, how realistic is this vision or scenario for the steel industry? As stated, this reasoning concerns the whole world steel production i.e., the average specific emission weighted by produced tons. It includes both ore-based producers and scrap-based producers and combinations. Certain proactive companies/plants will be better than the scenario, whereas some plants will fail in attaining the target. The role of big steel producers (China,

India, Russia) is especially very determining. Though every mitigated CO_2 ton is important, the present BAT line is not adequate, and the target must be put on a much lower level. A strong commitment to the '2050 strategy' by enterprises, countries, and organizations like the World Steel Association, the United Nations, and the European Union is the basic necessity and the key to success [9,100].

5. Conclusions

In this review, the position of the steel industry in the global fight against climate warming is discussed. This industrial sector is one of the biggest GHG emitters today and thus has a tremendous task in changing the course toward a low-emitting and carbon neutral industry responding to the challenges defined by the IPCC in 2018. The most essential observations are summarized below:

(1)   The conservative forecast for steel demand/production in 2050 is 2.5 Gt/y. At the same time, the emissions should be cut down by 65–70%, which means that specific emissions, t CO_2/t steel, should be reduced by approximately 75%. This will keep the steel industry in line with the IPCC´s target. This realization demands resolution and a strong commitment to wide-ranging and properly directed investments in reconstruction of the global steel industry. Practical implementation should be performed by steel enterprises and supervised by organizations like the World Steel Association, the United Nations, and the European Union.

(2)   In the short term, the most effective pathway is to improve energy efficiency and mitigate CO_2 emissions by installing best available technologies in existing integrated steel plants (including sintering/pelletizing, coke making, BFs, BOFs, CC, rolling mills, etc.) as well as in EAF plants. Reducing C/H ratio in fuels and reductants, via efficient heat recovery and integrated management of energy flows as well as by adopting CCS from off gases, can decrease specific emissions by 40–50%.

(3)   More radical change is possible only via radical transmission from carbon-based metallurgy (coal, coke, oil, natural gas) to carbon-lean/carbon-free iron making. That can be realized via a combination of hydrogen reduction and melting with low-Carbon/C-free electricity. This provides major reconstruction of steel plants, replacement of blast furnaces with direct reduction furnaces, and production and storage for hydrogen. Pioneering attempts in this direction are under way, but for a wide quantum leap, a bigger wide-ranging transition to hydrogen economy throughout the society is necessary.

(4)   One positive trend that supports the steel industry in its endeavor to attain its targets is the growing share of recycled steel—from 30% to 50% level in steel production. This means a smaller need (share) for ore-based primary iron, higher share of EAF production, and strongly increasing demand for carbon-neutral electricity.

(5)   In order for the steel industry to make these implementations on a broad front and on schedule, certain policy tools are also necessary. Financial support to develop and deploy new technologies should be used as positive incentives. On the other hand, carbon pricing that favors low- and zero-carbon solutions and penalizes CO_2 emissions is indispensable. The system should be transparent and based on both national conventions and wide international treaties.

Funding: My warm thanks to K.H. Renlund Foundation for encouragement to continue my research activities including the role as Guest Editor of this Special Issue.

Acknowledgments: I would like to thank Aalto University, Department of Chemical and Metallurgical Engineering, and Ari Jokilaakso for the opportunity to continue my emeritus career and for the technical support provided. My special gratitude goes out to all my colleagues who joined this effort and for their contributions to this Special Issue. This has been a memorable journey indeed.

Conflicts of Interest: The author declares no conflict of interest.

References

1. IPCC. AR5 SYR Summary for Policymakers. 2014. Available online: https://archive.ipcc.ch/pdf/assessment-report/ar5/syr/AR5_SYR_FINAL_SPM.pdf (accessed on 23 January 2020).
2. Olivier, J.G.J.; Peters, J.A.H.W. *Trends in Global CO_2 and Total Greenhouse Gas Emissions: 2019 Report*; PBL Netherlands Environmental Assessment Agency: The Hague, The Netherlands, 2019; p. 70.
3. UNEP. Emissions Gap Report 2018. United Nations Environment Programme, Nairobi. Available online: https://www.unenvironment.org/resources/emissions-gap-report-2018 (accessed on 6 February 2020).
4. Key World Energy Statistics. Energy and Climate Change. Available online: https://www.iea.org/statistics/kwes/ (accessed on 23 January 2020).
5. Enerdata. Global Energy Statistical Yearbook 2019. Available online: https://yearbook.enerdata.net/total-energy/world-energy-production.html (accessed on 5 March 2020).
6. World Steel Association. World Steel in Figures 2020. Available online: https://www.worldsteel.org/en/dam/jcr:f7982217-cfde-4fdc-8ba0-795ed807f513/World%2520Steel%2520in%2520Figures%25202020i.pdf (accessed on 5 July 2020).
7. Morfeldt, J.; Nijs, W.; Silveira, S. The impact of climate targets on future steel production -an analysis based on a global energy system model. *J. Clean. Prod.* **2015**, *103*, 469–482. [CrossRef]
8. World Steel Association. 2016. Available online: http://www.worldsteel.org/statistics/statistics-archive.html (accessed on 20 April 2017).
9. World Steel Association. *Steel's Contribution to A Low Carbon Future and Climate Resilient Societies—World Steel Position Paper*; World Steel Association: Brussels, Belgium, 2020; ISBN 978-2-930069-83-8. Available online: https://www.worldsteel.org/en/dam/jcr:7ec64bc1-c51c-439b-84b8-94496686b8c6/Position_paper_climate_2020_vfinal.pdf (accessed on 5 March 2020).
10. Holappa, L. Toward Low Carbon Metallurgy in Iron and Steel Making. In Proceedings of the Guthrie Honorary Symposium, Montreal, QC, Canada, 6–9 June 2011; pp. 248–254.
11. Holappa, L. Energy efficiency and sustainability in steel production. In *An EPD Symposium in Honor of Professor Ramana, G. Reddy, Proceedings of the TMS Symposium on Applications of Process Engineering Principles in Materials Processing, Energy and Environmental Technologies, San Diego, CA, USA, 26 February–2 March 2017*; Wang, S., Free, M.L., Alam, S., Zhang, M., Taylor, P.R., Eds.; Springer: Cham, Switzerland, 2017; pp. 401–410.
12. IEA/OECD. Energy Technology Transitions for Industry, Strategies for the Next Industrial Revolution. Available online: http://www.oecd-ilibrary.org/energy/energy-technology-transitions-for-industry_9789264068612-en (accessed on 20 April 2017).
13. Zhang, J.; Liu, Z.; Li, K.; Wang, G.; Jiao, K.; Yang, T. Current status and prospects of Chinese steel industry. In Proceedings of the Current Status and Prospects of Chinese Steel Industry. Scanmet V Conference, Luleå, Sweden, 12–15 June 2016; p. 15.
14. Hasanbeigi, A.; Springer, C. *How Clean is the U.S. Steel Industry? An International Benchmarking of Energy and CO_2 Intensities*; Global Efficiency Intelligence: San Francisco, CA, USA; Available online: https://www.globaleciencyintel.com (accessed on 5 March 2020).
15. The European Commission. Document L:2012:070: TOC. Implementing Decision of 28 February 2012. Available online: http://data.europa.eu/eli/dec_impl/2012/135/oj (accessed on 5 March 2020).
16. Worrell, E.; Price, L.; Neelis, M.; Galitsky, C.; Zhou, N. *World Best Practice Energy Intensity Values for Selected Industrial Sectors*; Ernest Orlando Lawrence Berkeley National Laboratory: Berkeley, CA, USA, 2007; p. 51, LBNL-62806.
17. Zhang, Q.; Xu, J.; Wang, Y.; Hasanbeigi, A.; Zhang, W.; Lu, H.; Arens, M. Comprehensive assessment of energy conservation and CO_2 emissions mitigation in China's iron and steel industry based on dynamic material flows. *Appl. Energy* **2018**, *209*, 251–265. [CrossRef]
18. Bureau of International Recycling. World Steel Recycling in Figures 2011–2015. Available online: https://www.bdsv.org/fileadmin/service/markt_und_branchendaten/weltstatistik_2011_2015.pdf (accessed on 16 March 2020).
19. Eurofer. A Steel Roadmap for A Low Carbon Europe 2050. 2013. Available online: http://www.eurofer.org/News%26Events/PublicationsLinksList/2013-Roadmap.pdf (accessed on 5 March 2020).

20. Wörtler, M.; Schuler, F.; Voigt, N.; Schmidt, T.; Dahlmann, P.; Lüngen, H.-B.; Ghenda, T. Steel's Contribution to a Low Carbon Europe 2050. BCG & VDEh. Available online: https://www.stahl-online.de/wp-content/uploads/2013/09/Schlussbericht-Studie-Low-carbon-Europe-2050_-Mai-20131.pdf (accessed on 16 March 2020).

21. Birat, J.-P. Update on the ULCOS program. In *How will Employment and Labor Markets Develop in the Context of a Transition Towards a Low-Carbon Economy?* ESTEP: Brussels, Belgium, 2014; Available online: http://erc-online.eu/wp-content/uploads/2014/04/2011-00612-E.pdf (accessed on 16 March 2020).

22. van der Stel, J.; Hattink, M.; Zeilstra, C.; Louwerse, G.; Hirsch, A.; Janhsen, U.; Sert, D.; Grant, M.; Delebecque, A.; Diez-Brea, P.; et al. ULCOS top gas recycling blast furnace process (ULCOS TGRBF). In *European Commission EUR 26414*; Publications Office of the European Union: Luxembourg, 2014; p. 47. ISBN 978-92-79-35038-2. [CrossRef]

23. van der Stel, J.; Meijer, K.; Santos, S.; Peeters, T.; Broersen, P. Opportunities for Reducing CO_2 emissions from Steel Industry. In Proceedings of the EMECR 2017 1st International Conference on Energy and Material Efficiency and CO_2 Reduction in the Steel Industry, Kobe, Japan, 11–13 October 2017; pp. 46–49. Available online: http://www.emecr2017.com (accessed on 16 March 2020).

24. Nishioka, K.; Ujisawa, Y.; Tonomura, S.; Ishiwata, N.; Sikstrom, P. Sustainable Aspects of CO_2 Ultimate Reduction in the Steelmaking Process (COURSE50 Project), Part 1: Hydrogen Reduction in the Blast Furnace. *J. Sustain. Met.* **2016**, *2*, 200–208. [CrossRef]

25. Onoda, M.; Matsuzaki, Y.; Chowdhury, F.A.; Yamada, H.; Goto, K.; Tonomura, S. Sustainable Aspects of Ultimate Reduction of CO_2 in the Steelmaking Process (COURSE50 Project), Part 2: CO_2 Capture. *J. Sustain. Met.* **2016**, *2*, 209–215. [CrossRef]

26. Tonomura, S.; Kikuchi, N.; Ishiwata, N.; Tomisaki, S.; Tomita, Y. Concept and Current State of CO_2 Ultimate Reduction in the Steelmaking Process (COURSE50) Aimed at Sustainability in the Japanese Steel Industry. *J. Sustain. Met.* **2016**, *2*, 191–199. [CrossRef]

27. Ariyama, T.; Takahashi, K.; Kawashiri, Y.; Nouchi, T. Diversification of the Ironmaking Process Toward the Long-Term Global Goal for Carbon Dioxide Mitigation. *J. Sustain. Met.* **2019**, *5*, 276–294. [CrossRef]

28. Kim, J.; Ahn, Y.; Roh, T. Low-Carbon Management of POSCO in Circular Economy: Current Status and Limitations. In *Towards a Circular Economy: Corporate Management and Policy Pathways. ERIA Research Project Report 2014-44*; Anbumozhi, V., Kim, J., Eds.; ERIA: Jakarta, Indonesia, 2016; pp. 185–199. Available online: https://www.eria.org/RPR_FY2014_No.44_Chapter_11.pdf (accessed on 16 March 2020).

29. Jahanshahi, S.; Mathieson, J.G.; Reimink, H. Low Emission Steelmaking. *J. Sustain. Met.* **2016**, *2*, 185–190. [CrossRef]

30. Sohn, H.Y. Energy Consumption and CO_2 Emissions in Ironmaking and Development of a Novel Flash Technology. *Metals* **2019**, *10*, 54. [CrossRef]

31. Zhao, J.; Zuo, H.-B.; Wang, Y.; Wang, J.; Xue, Q. Review of green and low-carbon ironmaking technology. *Ironmak. Steelmak.* **2019**, *47*, 296–306. [CrossRef]

32. He, K.; Wang, L.; Li, X. Review of the Energy Consumption and Production Structure of China's Steel Industry: Current Situation and Future Development. *Metals* **2020**, *10*, 302. [CrossRef]

33. Liao, W.-C.; Chen, J.-S.; Liu, C.-H.; Chen, J.-J.; Ko, F.-K. The Inspection of CO_2 Emission Targets of Industry Sector in Taiwan. Available online: https://www.irena.org/assets/IEW/EventDocs/Session_3_Parallel_E_Liao_Wei_Chen_paper.pdf (accessed on 16 March 2020).

34. Krishnan, S.S.; Vunnam, V.; Sunder, P.S.; Sunil, J.V.; Ramakrishnan, A.M. *A Study of Energy Efficiency in the Indian Iron and Steel Industry*; Center for Study of Science, Technology and Policy: Bangalore, India, 2013; p. 132. ISBN 978-81-903613-4-7. Available online: http://www.indiaenvironmentportal.org.in/files/file/A%20Study%20of%20Energy%20Efficiency%20in%20the%20Indian%20IS%20Industry.pdf (accessed on 24 June 2020).

35. Wenzel, W.; Gudenau, H.W.; Fukushima, T. Blast Furnace Operating Methods. US Patent 3884677, 20 November 1973.

36. Zhang, W.; Zhang, J.; Xue, Z.; Zou, Z.; Qi, Y. Unsteady Analyses of the Top Gas Recycling Oxygen Blast Furnace. *ISIJ Int.* **2016**, *56*, 1358–1367. [CrossRef]

37. Arasto, A.; Tsupari, E.; Kärki, J.; Lilja, J.; Sihvonen, M. Oxygen blast furnace with CO_2 capture and storage at an integrated steel mill—Part I: Technical concept analysis. *Int. J. Greenh. Gas Control.* **2014**, *30*, 140–147. [CrossRef]

38. Tsupari, E.; Kärki, J.; Arasto, A.; Lilja, J.; Kinnunen, K.; Sihvonen, M. Oxygen blast furnace with CO_2 capture and storage at an integrated steel mill—Part II: Economic feasibility in comparison with conventional blast furnace highlighting sensitivities. *Int. J. Greenh. Gas Control.* **2015**, *32*, 189–196. [CrossRef]

39. Fu, J.-X.; Tang, G.-H.; Zhao, R.-J.; Hwang, W.-S. Carbon Reduction Programs and Key Technologies in global Steel Industry. *J. Iron Steel Res. Int.* **2014**, *21*, 275–281. [CrossRef]

40. Wakuri, S.; Mochizuki, S.; Baba, M.; Misawa, J.; Anan, K. Operation of Pulverized Coal Injection into Large Blast Furnace with High Top Pressure. *Trans. Iron Steel Inst. Jpn.* **1984**, *24*, 622–630. [CrossRef]

41. Janz, J.; Weiss, W. Injection of waste plastics into the blast furnace of Stahlwerke Bremen. *Revue Métallurgie* **1996**, *93*, 1219–1226. [CrossRef]

42. Norgate, T.; Haque, N.; Somerville, M.; Jahanshahi, S. Biomass as a Source of Renewable Carbon for Iron and Steelmaking. *ISIJ Int.* **2012**, *52*, 1472–1481. [CrossRef]

43. Mathieson, J.G.; Rogers, H.; Somerville, M.; Jahanshahi, S. Reducing Net CO_2 Emissions Using Charcoal as a Blast Furnace Tuyere Injectant. *ISIJ Int.* **2012**, *52*, 1489–1496. [CrossRef]

44. Mousa, E.A.; Wang, C.; Riesbeck, J.; Larsson, M. Biomass applications in iron and steel industry: An overview of challenges and opportunities. *Renew. Sustain. Energy Rev.* **2016**, *65*, 1247–1266. [CrossRef]

45. Suopajärvi, H.; Kemppainen, A.; Haapakangas, J.; Fabritius, T. Extensive review of the opportunities to use biomass-based fuels in iron and steelmaking processes. *J. Clean. Prod.* **2017**, *148*, 709–734. [CrossRef]

46. IEA IETS. Process Integration in the Iron and Steel Industry—IEA IETS Annex XIV Technical Report. Swerea MEFOS Report No: 1406; p. 41. Available online: https://iea-industry.org/app/uploads/mef14064-iets-annex-xiv-technical-report_2014.pdf (accessed on 5 March 2020).

47. Lima Lopes, N.; de Pádua Nacif, A.; de Cássia Oliveira Carneiro, A.; de Assis, J.B.; Costa Oliveira, A. Brazilian Green Steel. Viçosa, Minas Gerais March. 2017. Available online: http://ciflorestas.com.br/arquivos/d_e_e_1099159498.pdf (accessed on 29 May 2020).

48. De Castro, J.A. A Comprehensive Modeling as a tool for Developing New Mini Blast Furnace Technologies Based on Biomass and Hydrogen Operation. Invited Lecture at Nippon Steel, 11th December 2019 on "New trends on the development of green pig iron". 50 Slides. 2019. Available online: https://www.researchgate.net/publication/340564870_Nippon_Steel11th_December_2019_Castro_New_trends_on_the_development_of_green_pig_iron (accessed on 1 June 2020).

49. Williams, M. *Deforesting the Earth: From Prehistory to Global Crisis*; The University of Chicago Press: Chicago, IL, USA, 2003; p. 689. ISBN 0-226-89926-8.

50. Global CCS Institute. The Global Status of CCS: 2019. 2019. Available online: https://www.globalccsinstitute.com/wp-content/uploads/2019/12/GCC_GLOBAL_STATUS_REPORT_2019.pdf (accessed on 1 June 2020).

51. Nuttall, W.J.; Bakenne, A.T. Carbon Capture, Utilisation and Storage. In *Fossil Fuel Hydrogen*, 1st ed.; Nuttall, W.J., Bakenne, A.T., Eds.; Springer: Cham, Switzerland, 2020; pp. 53–67.

52. IEAGHG. Iron and Steel CCS Study (Techno-Economics Integrated Steel Mill). 2013. Available online: https://ieaghg.org/publications/technical-reports/reports-list/9-technical-reports/1001-2013-04-iron-and-steel-ccs-study-techno-economics-integrated-steel-mill (accessed on 5 March 2020).

53. Gerdemann, S.J.; O'Connor, W.K.; Dahlin, D.C.; Penner, L.R.; Rush, H. Ex Situ Aqueous Mineral Carbonation. *Environ. Sci. Technol.* **2007**, *41*, 2587–2593. [CrossRef] [PubMed]

54. Zevenhoven, R.; Fagerlund, J.; Songok, J.K. Review: CO_2 mineral sequestration: Developments toward large-scale application. *Greenh. Gas Sci Technol.* **2011**, *1*, 48–57. [CrossRef]

55. Zevenhoven, R. Metals Production, CO_2 Mineralization and LCA. *Metals* **2020**, *10*, 342. [CrossRef]

56. Said, A.; Laukkanen, T.; Järvinen, M. Pilot-scale experimental work on carbon dioxide sequestration using steelmaking slag. *Appl. Energy* **2016**, *177*, 602–611. [CrossRef]

57. Singh, J.; Dhar, D.W. Overview of Carbon Capture Technology: Microalgal Biorefinery Concept and State-of-the-Art. *Front. Mar. Sci.* **2019**, *6*. [CrossRef]

58. Otto, A.; Grube, T.; Schiebahn, S.; Stolten, D. Closing the loop: Captured CO_2 as a feedstock in the chemical industry. *Energy Environ. Sci.* **2015**, *8*, 3283–3297. [CrossRef]

59. Araki, K. CO_2 Ultimate Reduction in Steelmaking Process (COURSE50 Project). In Proceedings of the EMECR 2017, 1st Int. Conf. on Energy and Material Efficiency and CO_2 Reduction in the Steel Industry, Kobe, Japan, 11–13 October 2017; Paper 11A-KL1. pp. 32–35.

60. Japan Iron and Steel Federation. JISF Long-Term Vision for Climate Change Mitigation—A Challenge Towards Zero-Carbon Steel. 2019. Available online: https://www.jisf.or.jp/en/activity/climate/documents/JISFLong-termvisionforclimatechangemitigation.pdf (accessed on 6 March 2020).
61. JFE Group CSR REPORT. Protecting the Global Environment. Climate Change Mitigation. 2019. Available online: https://www.jfe-holdings.co.jp/en/csr/pdf/2019/2019_09_04.pdf (accessed on 1 June 2020).
62. He, Q.; Shi, M.; Liang, F.; Xu, L.; Ji, L.; Yan, S. Renewable absorbents for CO_2 capture: From biomass to nature. *Greenh. Gases Sci. Technol.* **2019**, *9*, 637–651. [CrossRef]
63. IEA. May 2019 Report: Transforming Industry through CCUS. 2019. Available online: https://webstore.iea.org/download/direct/2778?fileName=Transforming_Industry_through_CCUS.pdf (accessed on 23 June 2020).
64. Handler, R.M.; Shonnard, D.R.; Griffing, E.M.; Lai, A.; Palou-Rivera, I. Life Cycle Assessments of Ethanol Production via Gas Fermentation: Anticipated Greenhouse Gas Emissions for Cellulosic and Waste Gas Feedstocks. *Ind. Eng. Chem. Res.* **2015**, *55*, 3253–3261. [CrossRef]
65. STEELANOL Project 2015-21. Production of Sustainable, Advanced Bio-ethANOL through an Innovative Gas-Fermentation Process Using Exhaust Gases Emitted in the STEEL Industry. On-Going Project under H2020-EU.3.3.3.1. Available online: http://www.steelanol.eu/en (accessed on 24 June 2020).
66. Zhang, Y.; Ni, W.; Li, Z. Research on superclean polygeneration energy system of iron and steel industry. In *Power and Energy Systems, Proceeding of the 7th IASTED International Conference on Power and Energy Systems*; Acta Press: Anaheim, CA, USA, 2004; pp. 152–157.
67. Ghanbari, H.; Saxén, H.; Grossmann, I.E. Optimal design and operation of a steel plant integrated with a polygeneration system. *AIChE J.* **2013**, *59*, 3659–3670. [CrossRef]
68. IEA. The Future of Hydrogen—Report prepared by the IEA for the G20, Japan Seizing Today's Opportunities. 2020. Available online: https://webstore.iea.org/download/direct/2803 (accessed on 24 June 2020).
69. IRENA. Hydrogen: A Renewable Energy Perspective. 2019. Intern. Renewable Energy Agency, Abu Dhabi. Available online: https://www.irena.org/-/media/Files/IRENA/Agency/Publication/2019/Sep/IRENA_Hydrogen_2019.pdf (accessed on 24 June 2020).
70. FCH. Hydrogen Roadmap Europe—A Sustainable Pathway for the European. Available online: https://www.fch.europa.eu/sites/default/files/Hydrogen%20Roadmap%20Europe_Report.pdf (accessed on 24 June 2020).
71. Kapdan, I.K.; Kargi, F. Bio-hydrogen production from waste materials. *Enzym. Microb. Technol.* **2006**, *38*, 569–582. [CrossRef]
72. Dou, B.; Zhang, H.; Song, Y.; Zhao, L.; Jiang, B.; He, M.; Ruan, C.; Chen, H.; Xu, Y. Hydrogen production from the thermochemical conversion of biomass: Issues and challenges. *Sustain. Energy Fuels* **2019**, *3*, 314–342. [CrossRef]
73. Squires, A.M.; Johnson, C.A. The H-iron process. *JOM* **1957**, *9*, 586–590. [CrossRef]
74. Guo, L.; Bao, Q.; Gao, J.; Zhu, Q.; Guo, Z. A Review on Prevention of Sticking during Fluidized Bed Reduction of Fine Iron Ore. *ISIJ Int.* **2020**, *60*, 1–17. [CrossRef]
75. Nuber, D.; Eichberger, H.; Rollinger, B. Circored fine ore direct reduction—The future of modern electric steelmaking. *Stahl Eisen* **2006**, *126*, 47–51.
76. Sohn, H.Y. From Nonferrous Flash Smelting to Flash Ironmaking: Development of an ironmaking technology with greatly reduced CO_2 emissions and energy consumption. In *Treatise on Process Metallurgy*; Elsevier: Oxford, UK; Waltham, MA, USA, 2014; Volume 3 Industrial Processes Part B, Section 4.5.2.2; pp. 1596–1691.
77. Cavaliere, P. *Clean Ironmaking and Steelmaking Processes*; Springer Nature: Cham, Switzerland, 2019; p. 624. Available online: https://link.springer.com/book/10.1007%2F978-3-030-21209-4 (accessed on 5 March 2020).
78. Quader, M.A.; Ahmed, S.; Dawal, S.; Nukman, Y. Present needs, recent progress and future trends of energy-efficient Ultra-Low Carbon Dioxide (CO_2) Steelmaking (ULCOS) program. *Renew. Sustain. Energy Rev.* **2016**, *55*, 537–549. Available online: https://www.sciencedirect.com/science/article/pii/S1364032115011806 (accessed on 1 June 2020). [CrossRef]
79. Spreitzer, D.; Schenk, J.L. Reduction of Iron Oxides with Hydrogen—A Review. *Steel Res. Int.* **2019**, *90*, 17. [CrossRef]
80. Patisson, F.; Mirgaux, O. Hydrogen Ironmaking: How it Works. *Metals* **2020**, *10*, 922. [CrossRef]

81. H2FUTURE. Verbund Solutions GmbH. Available online: https://www.h2future-project.eu (accessed on 24 June 2020).

82. GrInHy. Green Industrial Hydrogen. Salzgitter Mannesmann Forschung GmbH. Available online: https://www.green-industrial-hydrogen.com (accessed on 28 April 2020).

83. Åhman, M.; Olsson, O.; Vogl, V.; Nyqvist, B.; Maltais, A.; Nilsson, L.J.; Hallding, K.; Skånberg, K.; Nilsson, M. Hydrogen Steelmaking for a Low-Carbon Economy. A Joint LU-SEI Working Paper for the HYBRIT Project. EESS Report No 109, SEI Working Paper WP 2018-07. 2018. Available online: https://www.sei.org/publications/hydrogen-steelmaking/ (accessed on 24 June 2020). Hydrogen Steelmaking for a Low-Carbon Economy.

84. Vogl, V.; Åhman, M.; Nilsson, L.J. Assessment of hydrogen direct reduction for fossil-free steelmaking. *J. Clean. Prod.* **2018**, *203*, 736–745. [CrossRef]

85. Pei, M.; Petäjäniemi, M.; Regnell, A.; Wijk, O. Toward a Fossil Free Future with HYBRIT: Development of Iron and Steelmaking Technology in Sweden and Finland. *Metals* **2020**, *10*, 972. [CrossRef]

86. WNA. Comparison of Lifecycle Greenhouse Gas Emissions of Various Electricity Generation Sources. World Nuclear Association (WNA). 2011. Available online: http://www.world-nuclear.org/uploadedFiles/org/WNA/Publications/Working_Group_Reports/comparison_of_lifecycle.pdf (accessed on 1 July 2020).

87. Schlömer, S.; Bruckner, T.; Fulton, L.; Hertwich, E.; McKinnon, A.; Perczyk, D.; Roy, J.; Schaeffer, R.; Sims, R.; Smith, P.; et al. Annex III: Technology-specific cost and performance parameters. In *Climate Change 2014: Mitigation of Climate Change. Contribution of Working Group III to the Fifth Assessment Report of the Intergovernmental Panel on Climate Change*; Cambridge University Press: Cambridge, UK; New York, NY, USA, 2014.

88. EEA. CO_2-Emission Intensity from Electricity Generation. 2018. Available online: https://www.eea.europa.eu/data-and-maps/daviz/sds/co2-emission-intensity-from-electricity-generation-2/@@view (accessed on 1 June 2020).

89. IEA. Tracking the Decoupling of Electricity Demand and Associated CO_2 Emissions, 1990–2017. 2019. Available online: https://www.iea.org/commentaries/tracking-the-decoupling-of-electricity-demand-and-associated-co2-emissions (accessed on 1 July 2020).

90. Otto, A.; Robinius, M.; Grube, T.; Schiebahn, S.; Praktiknjo, A.; Stolten, D. Power-to-Steel: Reducing CO_2 through the Integration of Renewable Energy and Hydrogen into the German Steel Industry. *Energies* **2017**, *10*, 451. [CrossRef]

91. Bureau of International Recycling. World Steel Recycling in Figures 2011–2018. 2019. Available online: https://www.bdsv.org/fileadmin/user_upload/World-Steel-Recycling-in-Figures-2014-2018.pdf (accessed on 28 April 2020).

92. Pauliuk, S.; Milford, R.L.; Müller, D.B.; Allwood, J.M. The Steel Scrap Age. *Environ. Sci. Technol.* **2013**, *47*, 3448–3454. [CrossRef] [PubMed]

93. Geyer, R.; Davis, J.; Ley, J.; He, J.; Clift, R.; Kwan, A.; Sansom, M.; Jackson, T. Time-dependent material flow analysis of iron and steel in the UK. *Resour. Conserv. Recycl.* **2007**, *51*, 101–117. [CrossRef]

94. Gauffin, A.; Andersson, N.Å.I.; Storm, P.; Tilliander, A.; Jönsson, P.G. Use of volume correlation model to calculate lifetime of end-of-life steel. *Ironmak. Steelmak.* **2014**, *42*, 88–96. [CrossRef]

95. Haslehner, R.; Stelter, B.; Osio, N. Steel as a Model for a Sustainable Metal Industry in 2050. 2015. Available online: https://www.bcg.com/publications/2015/metals-mining-sustainability-steel-as-a-model-for-a-sustainable-metal-industry-in-2050.aspx (accessed on 5 March 2020).

96. World Economic Forum. 2015 Mining & Metals in a Sustainable World 2050, REF 250815. Available online: http://www3.weforum.org/docs/WEF_MM_Sustainable_World_2050_report_2015.pdf (accessed on 1 July 2020).

97. Rammer, B.; Millner, R.; Boehm, C. Comparing the CO_2 Emissions of Different Steelmaking Routes. *BHM Berg Hüttenmänn. Monatshefte* **2017**, *162*, 7–13. [CrossRef]

98. Bhaskar, A.; Assadi, M.; Somehsaraei, H.N. Decarbonization of the Iron and Steel Industry with Direct Reduction of Iron Ore with Green Hydrogen. *Energies* **2020**, *13*, 758. [CrossRef]

99. World Steel Association. Global steel industry: Outlook, challenges and opportunities. 5th International Steel Industry & Sector Relations Conference; April 20th, 2017 Istanbul. Available online: https://www.worldsteel.org/en/dam/jcr:d9e6a3df-ff19-47ff-9e8ff8c136429fc4/International+Steel+Industry+and+Sector+Relations+Conference+Istanbul170420.pdf (accessed on 14 August 2020).
100. Elkerbout, M.; Egenhofer, C. Tools to Boost Investment in Low-Carbon Technologies—Five Possible Ways to Create Low-Carbon Markets in the EU. CEPS Policy Insights No 2018/11. Available online: http://www.cepsech.eu/system/tdf/PI2018_11_ME_CE_Tools%20to%20boost%20investment%20in%20low%20carbon%20technologies_0.pdf?file=1&type=node&id=612&force= (accessed on 14 August 2020).

MDPI

St. Alban-Anlage 66

4052 Basel

Switzerland

Tel. +41 61 683 77 34

Fax +41 61 302 89 18

www.mdpi.com

Metals Editorial Office

E-mail: metals@mdpi.com

www.mdpi.com/journal/metals